Foundations of Linear Algebra

Kluwer Texts in the Mathematical Sciences

VOLUME 11

A Graduate-Level Book Series

The titles published in this series are listed at the end of this volume.

Foundations of
Linear Algebra

by

Jonathan S. Golan

Department of Mathematics and Computer Science,
University of Haifa,
Haifa, Israel

KLUWER ACADEMIC PUBLISHERS
DORDRECHT / BOSTON / LONDON

A C.I.P. Catalogue record for this book is available from the Library of Congress.

ISBN 978-90-481-4592-8

Published by Kluwer Academic Publishers,
P.O. Box 17, 3300 AA Dordrecht, The Netherlands.

Kluwer Academic Publishers incorporates
the publishing programmes of
D. Reidel, Martinus Nijhoff, Dr W. Junk and MTP Press.

Sold and distributed in the U.S.A. and Canada
by Kluwer Academic Publishers,
101 Philip Drive, Norwell, MA 02061, U.S.A.

In all other countries, sold and distributed
by Kluwer Academic Publishers Group,
P.O. Box 322, 3300 AH Dordrecht, The Netherlands.

Printed on acid-free paper

This is a translation of the original Hebrew work
Yesodot HaAlgebra HaLinearit,
Dekel Publications, Tel Aviv © 1992
Translation by the author

TABLE OF CONTENTS

PREFACE

This book is an extensively revised version of my textbook *"Yesodot HaAlgebra HaLiniarit"* (The Foundations of Linear Algebra) used at many universities in Israel. It is designed for a comprehensive one-year course in linear algebra (112 lecture hours) for mathematics majors. Therefore, I assume that the student already has a certain amount of mathematical background — including set theory, mathematical induction, basic analytic geometry, and elementary calculus — as well as a modicum of mathematical sophistication.

My intention is to provide not only a solid basis in the abstract theory of linear algebra, but also to provide examples of the application of this theory to other branches of mathematics and computer science. Thus, for example, the introduction of finite fields is dictated by the needs of students studying algebraic coding theory as an immediate followup to their linear algebra studies.

Many of the students studying linear algebra either are familiar with the care and feeding of computers before they begin their studies or are simultaneously enrolled in an introductory computer science course. Therefore, consideration of the more computational aspects of linear algebra – such as the solution of systems of linear equations and the calculation of eigenvalues – is delayed until all students are assumed able to write computer programs for this purpose. Beginning with Chapter VII, there is an implicit assumption that the student has access to a personal computer and knows how to use it. While such access is not necessary for understanding the material in the text, it is important for gaining insight into the theory presented here. This assumption of computer access also allows us to concentrate on the mathematical aspects of linear algebra and not get bogged down on techniques for hand computation which so often take up much of an introductory linear algebra curriculum. However, various topics specificially relating to the computational aspects of linear algebra are included as needed. Access to symbolic algebra software, such as MATHEMATICA or MAPLEV, is not assumed, but, needless to say, that too would be of great advantage in enhancing the student's appreciation of the mathematics involved.

The problems are for the most part those I have collected or thought up over the past twenty years. They range from the straightforward to the challenging and, deliberately, I have not indicated which is which by an asterisk or similar means. Many of them have appeared in print before: in journals such as *American Mathematical Monthly*, *College Mathematics Journal*, or *Mathematics Magazine*; in various mathematics competitions; and in other published or circulated problem

collections. Some of them are special cases of general results in the research liter-
ature, others are well-known "old chestnuts", and several were donated to me over
the years by colleagues and even students. Not a few of the problems originate in
files of old examinations collected at universities around the world which I visited.
Since, over the years, I did not keep track of the sources from which these problems
came, all I can do is offer a collective acknowledgement to all those to whom it is
due. Good problem formulators, like the God of the abbot of Citeaux, know their
own.

Special thanks are due to Prof. Gadi Moran and Prof. Moshe Roitman who
looked over the preliminary versions of the Hebrew edition of this book, as well
as to Prof. Jack Anderson, Prof. David Dobbs and Prof. Robyn Serven who saw
the preliminary versions of the English edition, for their important comments and
suggestions. Also, I owe a debt of gratitude to the many students who suggested
changes and improvements in the material over the years.

Haifa, 1994

BASIC NOTATION AND TERMINOLOGY

Sets will be delimited by braces, { }, and between them we will either enumerate the elements of the set or give a rule for determining whether an element is in the set or not, as in $\{x \mid p(x)\}$, which is to be read as "the set of all x such that $p(x)$". If a is an element of a set A we write $a \in A$; otherwise we write $a \notin A$.

A finite or countable set the elements of which are enumerated in a specific order is a *list*. Often, especially when we are dealing with finite sets, we will be interested in them as lists rather than just as sets. If we were being extremely formal, we would have to devise separate notation to differentiate lists from sets. Instead, in order not to burden our notation unnecessarily, we will make the following convention: when the elements of a finite or countably-infinite set are specifically enumerated, they will be assumed to be ordered as written from left to right. Thus $\{1, 2, 3, 4\}$ and $\{2, 4, 3, 1\}$ denote the same set of objects but the implied order on the set is different in each case. Therefore, as lists, they are distinct.

If A and B are sets then their *union* $A \cup B$ is the set of all elements which are either in A or in B and their *intersection* $A \cap B$ is the set of all elements belonging to both of them. More generally, if $\{A_i \mid i \in \Omega\}$ is a family of sets indexed by some index set Ω then $\bigcup_{i \in \Omega} A_i$ is the set of all elements belonging to at least one of the A_i and $\bigcap_{i \in \Omega} A_i$ is the set of all elements belonging to all of the A_i. If A and B are sets then the *set difference* $A \setminus B$ is the set of all elements of A not belonging to B. The empty set is denoted by $\varnothing$.

A *function* f from a nonempty set A to a nonempty set B is a rule which assigns to each element of the set A a unique element $f(a)$ of the set B. The set A is called the *domain* of f and the set B is called the *range* of f. To denote that f is such a function we write $f: A \rightarrow B$. To denote that an element $b \in B$ is assigned to an element $a \in A$ by f we write $f: a \mapsto b$. (Note the different shape of the arrow!) This notation is particularly helpful in case the action of f is given by a formula. Thus, the function f from the set of integers to itself defined by $f: a \mapsto a^2$ assigns to each integer its square. The set of all functions from a nonempty set A to a nonempty set B will be denoted by B^A.

If $f \in B^A$ and if A' is a nonempty subset of A then the *restriction* of f to A' is the function $f': A' \rightarrow B$ defined by $f'(a') = f(a')$ for all $a' \in A'$.

Functions f and g in B^A are *equal* if and only if $f(a) = g(a)$ for all $a \in A$. In this case we write $f = g$. A function $f \in B^A$ is *monic* if and only if it assigns different elements of B to different elements of A, i.e. if and only if $f(a_1) \neq f(a_2)$ whenever $a_1 \neq a_2$ in A. A function $f \in B^A$ is *epic* if and only if it assigns every element of B to some element of A, i.e. if and only if for each $b \in B$ there exists an element

1

$a \in A$ satisfying $f(a) = b$. A function which is both monic and epic is *bijective*. A bijective function from a set A to a set B determines a one-to-one correspondence between the elements of A and the elements of B. If $f: A \rightarrow B$ is bijective then one can define the *inverse function* $f^{-1}: B \rightarrow A$ by setting $f^{-1}(b) = a$ if and only if $f(a) = b$. A bijective function from a set A to itself is a *permutation* of A. There is always one permutation of any nonempty set A, namely the identity function $a \mapsto a$.

If A, B, and C are nonempty sets and if $f \in B^A$ and $g \in C^B$ are functions then the *composite function* gf is the function in C^A defined by $gf: a \mapsto g(f(a))$. Note that if $f: A \rightarrow B$ is a bijective function then $f^{-1}f$ is the identity function on A and ff^{-1} is the identity function on B.

The *cartesian product* $A_1 \times A_2$ of two nonempty sets A_1 and A_2 is the set of all ordered pairs (a_1, a_2), where $a_1 \in A_1$ and $a_2 \in A_2$. More generally, if $A_1, \ldots, A_n$ are nonempty sets then $A_1 \times \cdots \times A_n$ is the set of all ordered n-tuples of the form $(a_1, \ldots, a_n)$, where $a_i \in A_i$ for each $1 \leq i \leq n$. Note that each ordered n-tuple $(a_1, \ldots, a_n)$ with $a_i \in A_i$ for each i uniquely defines a function f from the set $\{1, \ldots, n\}$ to $A = \bigcup_{i=1}^{n} A_i$ by $f: i \mapsto a_i$. Conversely, each function $f: \{1, \ldots, n\} \rightarrow A$ uniquely defines such an ordered n-tuple, namely $(f(1), \ldots, f(n))$. This suggests a method for defining the cartesian product of an arbitrary family of nonempty sets: if $\{A_i \mid i \in \Omega\}$ is an arbitrary family of nonempty sets and if $A = \bigcup_{i \in \Omega} A_i$ then $\prod_{i \in \Omega} A_i$ is defined to be the set of all functions $f: \Omega \rightarrow A$ satisfying the condition that $f(i) \in A_i$ for all $i \in \Omega$. The existence of such functions is guaranteed by an axiom of set theory known as the Axiom of Choice. A certain amount of controversy surrounds this axiom and there are mathematicians who prefer to make as little use of it as possible. However, we will need it throughout this book and will always assume that it holds.

In the above construction we did not assume that the sets A_i are distinct. Indeed, they may very well be all the same. If $A_i = A$ for each $i \in \Omega$ then we write A^Ω instead of $\prod_{i \in \Omega} A_i$. This fits our previous notation, since A^Ω is just the set of all functions from Ω to A. If $\Omega = \{1, \ldots, n\}$ we write A^n instead of A^Ω. Thus, A^n is just the set of all ordered n-tuples $(a_1, \ldots, a_n)$ of elements of A.

Finally we make consistent use of the following standard notation:

$\mathbb{P}$ will denote the set of all positive integers;

$\mathbb{N}$ will denote the set of all nonnegative integers;

$\mathbb{Z}$ will denote the set of all integers;

$\mathbb{Q}$ will denote the set of all rational numbers;

$\mathbb{R}$ will denote the set of all real numbers;

$\mathbb{C}$ will denote the set of all complex numbers.

CHAPTER I

FIELDS AND INTEGRAL DOMAINS

A *binary operation* on a nonempty set A is a function from $A \times A$ to A. A nonempty set F on which we have defined two binary operations, $(a, b) \mapsto a + b$ (called *addition*), and $(a, b) \mapsto ab$ (called *multiplication*), is a *field* if and only if the following conditions are satisfied:

(1) *(associativity of addition and multiplication)* $a + (b + c) = (a + b) + c$ and $a(bc) = (ab)c$ for all $a, b, c \in F$.

(2) *(commutativity of addition and multiplication)* $a + b = b + a$ and $ab = ba$ for all $a, b \in F$.

(3) *(existence of neutral elements with respect to addition and multiplication)* There exist distinct elements of $0_F \neq 1_F$ in F having the property that $a + 0_F = a$ and $a1_F = a$ for all $a \in F$.

(4) *(existence of additive and multiplicative inverses)* For each $a \in F$ there exists an element $-a \in F$ satisfying $a + (-a) = 0_F$ and for each $0 \neq a \in F$ there exists an element $a^{-1} \in F$ satisfying $aa^{-1} = 1_F$.

(5) *(distributivity of multiplication over addition)* $a(b + c) = ab + ac$ for all $a, b, c \in F$.

The abstract theory of fields is due to Heinrich Weber at the end of the 19th century, based on previous work done by Leopold Kronecker and Richard Dedekind.

Let F be a field. For $a, b \in F$ we write $a - b$ instead of $a + (-b)$. If n is a positive integer we write na instead of $\overbrace{a + \cdots + a}^{n \text{ times}}$ and a^n instead of $\overbrace{a \cdot \ldots \cdot a}^{n \text{ times}}$. We will also write $(-n)a$ instead of $n(-a)$ and a^{-n} instead of $(a^{-1})^n$. We also set $0a$ to be 0_F and a^0 to be 1_F.

While we have not assumed so, it is clear that the neutral elements with respect to addition and multiplication are in fact unique. Indeed, if 0_F and $0'_F$ neutral elements of a field F with respect to addition then, by definition, $0_F = 0_F + 0'_F = 0'_F$. Similarly, if 1_F and $1'_F$ are neutral elements of F with respect to multiplication then $1_F = 1_F 1'_F = 1'_F$.

As a consequence of the associativity and commutativity of addition one sees that if $(a_1, \ldots, a_n) \in F^n$ for some positive integer n then the sum $a_1 + \cdots + a_n$ and the product $a_1 \cdot \ldots \cdot a_n$ are unambiguously defined elements of F. We often denote this sum and product by $\sum_{i=1}^n a_i$ and $\prod_{i=1}^n a_i$ respectively. By convention, the sum of an empty set of elements of F is taken to be 0_F and the product of an empty set of elements of F is taken to be 1_F.

EXAMPLE

Certainly one of the most familiar fields is the field $\mathbb{R}$ of *real numbers*, namely those numbers which can be represented in the form of infinite decimal expansions. In school you learned how to add and multiply real numbers, and made use of the above properties. Another familiar field is the field $\mathbb{Q}$ of *rational numbers*, namely those numbers of the form $\frac{m}{n}$, where m and n are integers and $n \neq 0$. The field $\mathbb{Q}$ is a *subfield* of $\mathbb{R}$, namely it is a subset of $\mathbb{R}$ closed under addition and multiplication and containing the additive and multiplicative inverses of each of its members. It is not equal to $\mathbb{R}$ since, as the followers of the Greek mathematician/mystic Pythagoras already knew, the real number $\sqrt{2}$ is not rational. These two structures are, indeed, the inspiration for the abstract notion of a field.

EXAMPLE

Let $\mathbb{C} = \mathbb{R} \times \mathbb{R}$ be the set of all ordered pairs of real numbers and define operations of addition and multiplication on $\mathbb{C}$ by setting $(a, b) + (c, d) = (a + c, b + d)$ and $(a, b) \cdot (c, d) = (ac - bd, ad + bc)$. It is fairly straightforward to check that these operations define on $\mathbb{C}$ the structure of a field. The neutral element of $\mathbb{C}$ with respect to addition is $(0, 0)$ and its neutral element with respect to multiplication is $(1, 0)$. Indeed, you can check that an element $(a, b) \in \mathbb{C}$ has an additive inverse $-(a, b) = (-a, -b)$ and, if it differs from $(0, 0)$, it has a multiplicative inverse

$$(a, b)^{-1} = \left(\frac{a}{a^2 + b^2}, \frac{-b}{a^2 + b^2} \right).$$

The field $\mathbb{C}$ is called the field of *complex numbers*. The subset of $\mathbb{C}$ consisting of all of those elements of the form $(a, 0)$ is a subfield of $\mathbb{C}$ which we identify with the field of real numbers $\mathbb{R}$ by associating to each real number a the complex number $(a, 0)$. It is also standard to denote the complex number $(0, 1)$ by i. This number has the property that $i^2 = (-1, 0)$ and so i is frequently written as $\sqrt{-1}$. Thus an arbitrary complex number $(a, b) = (a, 0) + (0, b)$ is very often written as $a + bi$ or $a + b\sqrt{-1}$. The field $\mathbb{C}$ is of extreme importance in mathematics, though this is not the place to pursue this in detail.

If $z = a + bi \in \mathbb{C}$ we will write $\bar{z} = a - bi$ and $|z| = \sqrt{z\bar{z}} = \sqrt{a^2 + b^2}$. Note that $|z|$ is always a nonnegative real number, which is in fact positive if $z \neq 0_{\mathbb{C}}$. Moreover, if $z, z' \in \mathbb{C}$ then

(1) $$|z + z'| \leq |z| + |z'|,$$
(2) $$|zz'| = |z||z'|$$
(3) $$|z| = |\bar{z}|$$
(4) $$z^{-1} = \bar{z}/|z|^2 \quad \text{whenever} \quad z \neq 0_{\mathbb{C}}$$

EXAMPLE

We certainly know how to define addition and multiplication on the set $\mathbb{Z}$ of all integers, but these operations do not define the structure of a field on $\mathbb{Z}$ since not every nonzero integer has a multiplicative inverse which is again an integer.

EXAMPLE

Let p be a prime integer and let $F = \mathbb{Q} \times \mathbb{Q}$ be the set of all pairs (a, b) of rational numbers. Define addition and multiplication on F by setting $(a, b) + (c, d) = (a + c, b + d)$ and $(a, b) \cdot (c, d) = (ac + bdp, ad + bc)$. These operations define on F the structure of a field, the neutral element with respect to addition is $(0, 0)$ and the neutral element with respect to multiplication is $(1, 0)$. If $(a, b) \in F$ then its additive inverse is $-(a, b) = (-a, -b)$ and if it is not equal to 0_F then its multiplicative inverse is

$$(a, b)^{-1} = \left(\frac{a}{a^2 - pb^2}, \frac{-b}{a^2 - pb^2} \right).$$

The set of all elements of F of the form $(a, 0)$ is a subfield of F which we identify with $\mathbb{Q}$ by associating each rational number a with the element $(a, 0)$ of F. Moreover, the element $(0, 1) \in F$ satisfies the condition $(0, 1)^2 = (p, 0)$ and so it is usual to denote this element by $\sqrt{p}$ and to denote the field F by $\mathbb{Q}(\sqrt{p})$. Again, we note that each element (a, b) of this field can be written as $a + b\sqrt{p}$, where a and b are rational numbers.

EXAMPLE

Let n be an integer greater than 1 and let $F = \{0, 1, \ldots, n - 1\}$. Given a nonnegative integer k, we can divide it by n and obtain a remainder (perhaps equal to 0), which we will denote by $[k]$. Then $[k] \in F$ for all nonnegative integers k and each element of F is of the form k for some (not necessarily unique) nonnegative integer k. Define operations of addition and multiplication on F by setting $[k] + [h] = [k + h]$ and $[k] \cdot [h] = [k \cdot h]$. (Note that the operations *inside* the brackets refer to ordinary addition and multiplication of integers!) If n is a prime integer, then these operations define on F the structure of a field, which we denote by $\mathbb{Z}/(n)$. If n is not prime, then F is not a field.

Finite fields were first studied by the French mathematical genius Evariste Galois. In his honor, such fields are known as *Galois fields*. The theory of Galois fields is beyond the scope of this course, but one should point out that there exist finite fields which are not of the form $\mathbb{Z}/(n)$. Indeed, if n is a prime integer and k is an arbitrary integer there always exists a field having precisely n^k elements.

The last example above points out an interesting property of fields: given a field F, it is quite possible that there exists a positive integer n for which $n1_F = 0_F$. For example, in the field $\mathbb{Z}/(2)$ we have $1_{\mathbb{Z}/(2)} + 1_{\mathbb{Z}/(2)} = 0_{\mathbb{Z}/(2)}$. The least such positive integer, if one in fact exists, is called the *characteristic* of the field. If $n1_F \neq 0_F$ for all positive integers we say that the field F has characteristic 0. The fields $\mathbb{Q}$, $\mathbb{R}$, and $\mathbb{C}$ all have characteristic 0. If the characteristic of a field is positive then it must be prime (though we will not prove this here).

(1.1) PROPOSITION. *If a and b are elements of a field F then:*

(1) *There exists a unique element c of F satisfying $a + c = b$;*

(2) *If, in addition, $a \neq 0_F$ then there exists a unique element d of F satisfying $ad = b$.*

PROOF. (1) Pick $c = b - a$. Then

$$\begin{aligned}
a + c &= a + (b - a) \\
&= a + [b + (-a)] \\
&= a + [(-a) + b] \\
&= [a + (-a)] + b \\
&= 0_F + b \\
&= b,
\end{aligned}$$

proving the existence of an element satisfying the given condition. If c' is any element of F satisfying $a + c' = b$ then

$$c' = 0_F + c' = (-a + a) + c' = -a + (a + c') = -a + b = b - a,$$

establishing uniqueness.

(2) Pick $d = a^{-1}b$. Then $ac = a(a^{-1})b = (aa^{-1})b = 1_F b = b$, proving the existence of an element satisfying the given condition. If d' is any element of F satisfying $ad' = b$ then $d' = 1_F d' = (a^{-1}a)d' = a^{-1}(ad') = a^{-1}b$, establishing uniqueness. $\square$

(1.2) PROPOSITION. *If a, b, c are elements of a field F then:*

(1) $0_F a = 0_F$;

(2) $(-1_F)a = -a$;

(3) $a(-b) = -ab = (-a)b$;

(4) $-(-a) = a$;

(5) $(-a)(-b) = ab$;

(6) $-(a + b) = -a - b$;

(7) $a(b - c) = ab - ac$;

(8) *If $a \neq 0_F$ then $(a^{-1})^{-1} = a$;*

(9) *If $a, b \neq 0_F$ then $(ab)^{-1} = a^{-1}b^{-1}$;*

(10) *If $a + c = b + c$ then $a = b$;*

(11) *If $c \neq 0_F$ and $ac = bc$ then $a = b$;*

(12) *If $ab = 0_F$ then $a = 0_F$ or $b = 0_F$.*

PROOF. (1) Since $0_F a + 0_F a = (0_F + 0_F)a = 0_F a$ we can add $-(0_F a)$ to both sides of the equation and obtain $0_F a = 0_F$.

(2) Since, by (1), we have

$$(-1_F)a + a = (-1_F)a + 1_F a = (-1_F + 1_F)a = 0_F a = 0_F$$

and also $-a + a = 0_F$ it follows from Proposition 1.1(1) that $(-1_F)a = -a$.

(3) From (2) we see that $a(-b) = a[(-1_F)b] = (-1_F)(ab) = -(ab)$ and $(-a)b = [(-1_F)a]b = (-1_F)(ab) = -(ab)$.

(4) Since $a + (-a) = 0_F = -(-a) + (-a)$, this follows from Proposition 1.1(1).

(5) From (3) and (4) we have $(-a)(-b) = a[-(-b)] = ab$.

(6) Since $(a+b)+(-a-b) = a+b+(-a)+(-b) = 0_F$ and $(a+b)+[-(a+b)] = 0_F$ this follows from Proposition 1.1(1).

(7) From (3) we have $a(b - c) = ab + a(-c) = ab + [-(ac)] = ab - ac$.

(8) Since $(a^{-1})^{-1}(a^{-1}) = 1_F = a(a^{-1})$ this follows from Proposition 1.1(2).

(9) Since $(a^{-1}b^{-1})(ab) = a^{-1}ab^{-1}b = 1_F = (ab)^{-1}(ab)$ this follows from Proposition 1.1(2).

(10) This follows immediately from adding $-c$ to both sides of the equation.

(11) This follows immediately from multiplying both sides of the equation by c^{-1}.

(12) If $b \neq 0_F$ then by (1) it follows that, after multiplying both sides of the equation by b^{-1} we obtain $a = 0_F$. $\square$

As we have seen, the set $\mathbb{Z}$ of all integers is not a field. This set does, however, satisfy condition (12) of Proposition 1.2. If we replace the condition that every nonzero element have a multiplicative inverse by this weaker condition, retaining all of the other conditions listed in the definition of a field, we obtain another important algebraic structure known as an *integral domain*. Thus, all fields are integral domains, while $\mathbb{Z}$ is an integral domain which is not a field.

EXAMPLE

Let F be an integral domain and let X be an element not in F. A *polynomial* over F in X is an expression of the form $p(X) = \sum_{i \geq 0} a_i X^i$, where the a_i are elements of F only finitely-many of which are not equal to 0_F. These elements a_i are called the *coefficients* of the polynomial and the element X is called the *indeterminate* of the polynomial. The greatest nonnegative integer i satisfying the condition $a_i \neq 0_F$, if such an element exists, is called the *degree* of the polynomial and is denoted by $deg(p)$. If no such element exists, i.e. if $a_i = 0_F$ for all i, then we say that $p(X)$ is the *zero polynomial* and set its degree to be $-\infty$. If $p(X) = \sum_{i \geq 0} a_i X^i$ is a polynomial having degree $n \geq 0$ we often ignore the coefficients of $p(X)$ of index greater than n and write $p(X)$ as $\sum_{i=0}^{n} a_i X^i$. The coefficient a_n is called the *leading coefficient* of the polynomial. A polynomial having leading coefficient 1_F is *monic*. It is also standard to identify a polynomial $a_0 X^0$ of degree 0 with the element a_0 of F and to identify the zero polynomial with the element 0_F of F.

EXAMPLE (Continued)

The set of all polynomials in X over an integral domain F is denoted by $F[X]$. On this set we can define operations of addition and multiplication as follows: if $p(X) = \sum_{i \geq 0} a_i X^i$ and $q(X) = \sum_{i \geq 0} b_i X^i$ are elements of $F[X]$ then $p(X) + q(X)$ is defined to be the polynomial $\sum_{i \geq 0} c_i X^i$, where $c_i = a_i + b_i$ for each $i \geq 0$. The polynomial $p(X)q(X)$ is defined to be $\sum_{i \geq 0} d_i X^i$, where $d_i = \sum_{h+j=i} a_h b_j$ for all $i \geq 0$. We note that the zero polynomial is the neutral element of $F[X]$ with respect to addition and the neutral element with respect to multiplication is 1_F. It is quite straightforward to check that $F[X]$ satisfies all of the conditions for being an integral domain except perhaps condition (12) of Proposition 1.2. To check this condition, suppose that we are given polynomials $p(X) = \sum_{i \geq 0} a_i X^i$ and $q(X) = \sum_{i \geq 0} b_i X^i$ in $F[X]$, neither of them equal to the zero polynomial, and having degrees k and n respectively. Then the coefficient of X^{k+n} in the polynomial $p(X)q(X)$ equals $a_k b_n$, and this is not 0_F since $a_k \neq 0_F$ and $b_n \neq 0_F$. Therefore $p(X)q(X)$ is not the zero polynomial.

Integral domains are sufficiently "like" fields that any integral domain R can be embedded in a field "of fractions" which is constructed in a manner akin to the manner that the field of rational numbers is constructed from the integral domain of integers. We will not pursue this construction here but the interested student can find it most books on modern algebra.

A well-known property of the integers is the following: if n and k are integers, with $k \neq 0$, then there exist unique integers u and v satisfying $n = ku + v$ and $|v| < |k|$. We now show that something of the same sort holds for polynomials over a field.

(1.3) PROPOSITION. *Suppose F is a field and $p(X), q(X) \in F[X]$, where $q(X)$ is not the zero polynomial. Then there exist unique polynomials $u(X)$ and $v(X)$ in $F[X]$ satisfying $p(X) = q(X)u(X) + v(X)$ and $deg(v) < deg(q)$.*

PROOF. Let $p(X) = \sum_{i \geq 0} a_i X^i$ and $q(X) = \sum_{i \geq 0} b_i X^i$. If $p(X)$ is the zero polynomial or if $deg(p) < deg(q)$, we can choose $u(X)$ to be the zero polynomial and $v(X) = p(X)$, and we are done. Thus we can assume that neither of these conditions hold, i.e. that $n = deg(p) \geq deg(q) = k$. We now proceed by induction of n. If $n = 0$ then k must also equal 0. We can therefore choose $u(X)$ to be the polynomial $a_0 b_0^{-1}$, which is of degree 0, and choose $v(X)$ to be the zero polynomial, and we are again done.

Assume therefore that $n > 0$ and that the proposition has been established for all polynomials of degree less than n. Let $h(X)$ be the polynomial $p(X) - a_n b_k^{-1} X^{n-k} q(X)$. If this is the zero polynomial, then we can choose $u(X)$ to be $a_n b_k^{-1} X^{n-k}$ and choose $v(X)$ to be the zero polynomial, and be finished. Otherwise, we note that $deg(p) > deg(h)$ and so by the induction hypothesis there exist polynomials $v(X)$ and $u'(X)$ in $F[X]$ satisfying $h(X) = q(X)u'(X) + v(X)$ and $deg(h) > deg(v)$. Then

$$p(X) = [a_n b_k^{-1} X^{n-k} + u'(X)]q(X) + v(X)$$

and $deg(p) > deg(v)$ so we can define $u(X) = a_n b_k^{-1} X^{n-k} + u'(X)$ and again are finished.

All that is left to show is uniqueness. Indeed, assume that

$$p(X) = q(X)u_1(X) + v_1(X) = q(X)u_2(X) + v_2(X),$$

where both $v_1(X)$ and $v_2(X)$ have degree less than the degree of $q(X)$. If $u_1(X) = u_2(X)$ then $v_2(X) - v_1(X)$ is the zero polynomial and so $v_1(X) = v_2(X)$, and we have proven uniqueness. Otherwise, since $F[X]$ is an integral domain, we see that $q(X)[u_1(X) - u_2(X)] = v_2(X) - v_1(X)$ is not the zero polynomial. But the degree of the right-hand polynomial is, by assumption, less than the degree of the left-hand polynomial, so we have a contradiction. Thus we must have uniqueness. $\square$

Let us reemphasize that if F is a field then the elements of $F[X]$ are formal expressions and not functions. However, each polynomial $p(X) = \sum_{i \geq 0} a_i X^i$ in $F[X]$ defines a function from F to itself by *evaluation*, i.e. the function $c \mapsto p(c) = \sum_{i \geq 0} a_i c^i$. A function of this form is called the *polynomial function* defined by the polynomial $p(X)$. If $p(c) = 0$ we say that the element c of F is a *zero* of the polynomial $p(X)$. Different polynomials may define the same polynomial function. For example, if $F = \mathbb{Z}/(2)$ then the polynomials X^2 and X^4 in $F[X]$ both define the same polynomial function from F to itself, namely the identity function.

Problems

1. Let F be a field and let $G = F \times F$. Define operations of addition and multiplication on G by setting

$$(a, b) + (c, d) = (a + c, b + d)$$

and

$$(a, b) \cdot (c, d) = (ac, bd).$$

Is G, together with these operations, a field?

2. Let $F = \mathbb{Z}/(11)$ and let $G = F \times F$. Define operations of addition and multiplication on G by setting

$$(a, b) + (c, d) = (a + c, b + d)$$

and

$$(a, b) \cdot (c, d) = (ac + 7bd, ad + bc).$$

Is G, together with these operations, a field?

3. Show that the set of all real numbers of the form $a + b\sqrt{2} + c\sqrt{3} + d\sqrt{6}$, where $a, b, c, d \in \mathbb{Q}$, forms a subfield of $\mathbb{R}$.

4. Show that the field $\mathbb{R}$ has infinitely-many subfields.

5. Let $t = \sqrt[3]{2} \in \mathbb{R}$ and let S be the set of all real numbers of the form $a + bt + ct^2$, where $a, b, c \in \mathbb{Q}$. Is S a subfield of $\mathbb{R}$?

6. Show that every complex number $z \in \mathbb{C}$ satisfies

$$z^4 + 4 = (z - 1 - i)(z - 1 + i)(z + 1 + i)(z + 1 - i).$$

7. Find the set of all complex numbers z satisfying each of the following conditions:

    (i) $z = (2 + 3i)(-4 + i)$;
    (ii) $z = (1 + 2i)^6$;
    (iii) $z = (1 - i)(1 + 2i)^{-1}$;
    (iv) $z = (3 + i)^{-1}$;
    (v) $|z| + z = 2 + i$;
    (vi) $z^4 = -4$;
    (vii) $z^4 = 2 - (\sqrt{12})i$;
    (viii) $z = [\frac{1}{2}(-1 + i\sqrt{3})]^3$;
    (ix) $z^2 = \frac{1}{2}(1 + i\sqrt{3})$.

8. Show that any complex number $z = a + bi$ satisfies $(\sqrt{2})|z| \geq |a| + |b|$.

9. If z_1 and z_2 are complex numbers, show that $|z_1|^2 + |z_2|^2 - z_1\bar{z}_2 - \bar{z}_1 z_2 = |z_1 - z_2|^2$.

10. Is $F = \{a + bi \in \mathbb{C} \mid a, b \in \mathbb{Q}\}$ a subfield of $\mathbb{C}$?

11. Find the following elements of $\mathbb{Q}(\sqrt{7})$:

    (i) $(1 + \sqrt{7})(-2 + 3\sqrt{7})$;
    (ii) $(1 - 2\sqrt{7})(2 - \sqrt{7})^{-1}$;
    (iii) $(2 + 3\sqrt{7})^{-1}$.

12. Find the following elements of $\mathbb{Z}/(11)$:

    (i) $7 + 8$;
    (ii) $7 \cdot 8$;
    (iii) 7^{-1};
    (iv) $7 \cdot 8^{-1}$;
    (v) $7^{-1} + 8^{-1}$.

13. For any prime integer p, find all subfields of the field $\mathbb{Z}/(p)$.

14. Let w be the real number $\frac{1}{2}(1 + \sqrt{5})$ and let S be the set of all real numbers of the form $k + nw$, where $k, n \in \mathbb{Z}$. Show that S, on which the usual addition and multiplication of real numbers is defined, is an integral domain. Is it a field?

15. Let F be an integral domain having only finitely-many elements. Show that F is a field.

16. Show that $\mathbb{Z}/(6)$ is not an integral domain.

17. Show that the set of all real numbers of the form $a + b\sqrt{5}$, where a and b are integers, is an integral domain but not a field.

18. Let R be an integral domain and let $0_R \neq a \in R$. Show that $ab = ac$ in R if and only if $b = c$.

19. Let F be a field and let $f(X), g(X) \in F[X]$. Show that

$$deg(fg) = deg(f) + deg(g).$$

20. Let F be a field and let $f(X), g(X) \in F[X]$. Show that

$$deg(f + g) \leq max\{deg(f), deg(g)\}$$

and give an example in which this inequality is proper.

21. Find polynomials $u(X)$ and $v(X)$ in $\mathbb{Q}[X]$ satisfying

$$X^4 + 3X^3 = (X^2 + X + 1)u(X) + v(X).$$

22. Let $F = \mathbb{Z}/(2)$. Find polynomials $u(X)$ and $v(X)$ in $F[X]$ satisfying

$$X^5 + X^2 = (X^3 + X + 1)u(X) + v(X).$$

23. Let $F = \mathbb{Z}/(7)$. Find a nonzero polynomial $f(X) \in F[X]$ such that the polynomial function defined by f is the 0-function.

CHAPTER II

VECTOR SPACES

Let F be a field. A nonempty set V together with a function $(v, v') \mapsto v + v'$ from $V \times V$ to V (called *vector addition*) and a function $(a, v) \mapsto av$ from $F \times V$ to V (called *scalar multiplication*) is a *vector space* over the field F if and only if the following conditions are satisfied:

(1) (*associativity of vector addition*) $(v_1 + v_2) + v_3 = v_1 + (v_2 + v_3)$ for all $v_1, v_2, v_3 \in V$.

(2) (*commutativity of vector addition*) $v_1 + v_2 = v_2 + v_1$ for all $v_1, v_2 \in V$.

(3) (*existence of an element neutral with respect to vector addition*) There exists an element 0_V of V satisfying $v + 0_V = v$ for all $v \in V$.

(4) (*existence of inverses with respect to vector addition*) For each element $v \in V$ there exists an element of V which we will denote by $-v$ and which satisfies $v + (-v) = 0_V$.

(5) (*distributivity of scalar multiplication over vector addition*) $a(v_1 + v_2) = av_1 + av_2$ for all $a \in F$ and all $v_1, v_2 \in V$.

(6) (*distributivity of scalar multiplication over scalar addition*) $(a_1 + a_2)v = a_1 v + a_2 v$ for all $a_1, a_2 \in F$ and all $v \in V$.

(7) (*associativity of scalar multiplication*) $a_1(a_2 v) = (a_1 a_2)v$ for all $a_1, a_2 \in F$ and all $v \in V$.

(8) (*existence of an element neutral with respect to scalar multiplication*) The element $1_F \in F$ satisfies $1_F v = v$ for all $v \in V$.

The elements of V are called *vectors* and the elements of the field F are called *scalars*. These terms were coined by the British physicist James Clerk Maxwell. The abstract theory of vector spaces was first introduced by the American Josiah Willard Gibbs in 1881. (In 1844, the German teacher Hermann Grassmann had studied essentially similar structures, but his work was not understood at the time and had little influence.)

If v_1 and v_2 are vectors in a vector space V we simplify our notation by writing $v_1 - v_2$ instead of $v_1 + (-v_2)$. Also, as a consequence of the associativity and commutativity of vector addition one sees that if $(v, \ldots, v_n) \in V^n$ for some positive integer n then the sum $v_1 + \cdots + v_n$ is an unambiguously defined element of V. We often denote this sum by $\sum_{i=1}^{n} v_i$. By convention, the sum of an empty set of elements of V is taken to be 0_V. Note that the distributivity of scalar multiplication over vector addition and an easy induction argument yield $a \sum_{i=1}^{n} v_i = \sum_{i=1}^{n} av_i$ for any scalar $a \in F$. Similarly, if $a_1, \ldots a_n$ are scalars and $v \in V$ is a vector then $(\sum_{i=1}^{n} a_i)v = \sum_{i=1}^{n} a_i v$.

EXAMPLE

As an immediate consequence of the definitions, we see that a field F is a vector space over itself, when we take vector addition and scalar multiplication to be the operations of addition and multiplication already defined on the field.

EXAMPLE

Let F be a field and let Ω be a nonempty set. For each $i \in \Omega$, assume that we have a vector space V_i over F, addition in which will be denoted by $+_i$. We can define the structure of a vector space on the set $\prod_{i \in \Omega} V_i$ in the following manner. If f and g are elements of $\prod_{i \in \Omega} V_i$ we define the function $f + g$ to be that function which assigns to each $i \in \Omega$ the element $f(i) +_i g(i)$ of V_i. Similarly, if $c \in F$ we define cf to be that function which assigns to $i \in \Omega$ the element $c(f(i))$ of V_i. It is straightforward, if somewhat tedious, to verify that $\prod_{i \in \Omega} V_i$ is indeed a vector space over F. Thus, the neutral element of this space is the function which assigns to each $i \in \Omega$ the neutral element of V_i and the negative of a function f is the function which assigns to each $i \in \Omega$ the negative of $f(i)$ in V_i. The vector space $\prod_{i \in \Omega} V_i$ is called the *direct product* of the vector spaces V_i over the field F. If $\Omega = \{1, \ldots, n\}$ we will often write $V_1 \times \cdots \times V_n$ instead of $\prod_{i=1}^{n} V_i$. If each of the vector spaces V_i is equal to a given vector space V then we write V^Ω instead of $\prod_{i \in \Omega} V_i$.

The two examples given above, used recursively, allow us to construct more and more complicated examples of vector spaces. In particular, let us look at the second of these constructions in more detail.

EXAMPLE

Let V be a vector space over a field F, and let n be a given positive integer. The elements of V^n are just n-tuples of elements of V. We will enclose them in brackets rather than parentheses in order to emphasize that we are talking about vectors and not just elements of an arbitrary set. Vector addition and scalar multiplication in the vector space V^n are defined componentwise:

$$[v_1, \ldots, v_n] + [y_1, \ldots, y_n] = [v_1 + y_1, \ldots, v_n + y_n]$$

and

$$c[v_1, \ldots, v_n] = [cv_1, \ldots, cv_n].$$

Elements of V^n are known as *row vectors of length n* over the space V. In a similar manner, we can define the vector space of *column vectors of length n* over V, the elements of which are of the form $\begin{bmatrix} v_1 \\ \vdots \\ v_n \end{bmatrix}$. Vector addition and scalar multiplication are again defined componentwise.

An important special case of this construction occurs when we choose $V = F$.

EXAMPLE

Let V be a vector space over a field F. If k and n are positive integers and if Ω is the set of all pairs of integers of the form (i, j), where $1 \leq i \leq k$ and $1 \leq j \leq n$, then the functions f from Ω to V correspond bijectively to arrays of elements of V of the form

$$\begin{bmatrix} v_{11} & \cdots & v_{1n} \\ \vdots & & \vdots \\ v_{k1} & \cdots & v_{kn} \end{bmatrix}.$$

Such arrays are called k *by* n *matrices* over the space V. The set of all k by n matrices $[v_{ij}]$ over V is denoted by $\mathcal{M}_{k \times n}(V)$. The elements v_{ij} of such a matrix are called its *entries*. In the special case $k = n$, the set of entries $\{v_{11}, v_{22}, \ldots, v_{nn}\}$ is called the *principal diagonal* or *main diagonal* of the matrix.

Vector addition and scalar multiplication in this vector space are again defined componentwise:

$$\begin{bmatrix} v_{11} & \cdots & v_{1n} \\ \vdots & & \vdots \\ v_{k1} & \cdots & v_{kn} \end{bmatrix} + \begin{bmatrix} y_{11} & \cdots & y_{1n} \\ \vdots & & \vdots \\ y_{k1} & \cdots & y_{kn} \end{bmatrix} = \begin{bmatrix} v_{11} + y_{11} & \cdots & v_{1n} + y_{1n} \\ \vdots & & \vdots \\ v_{k1} + y_{k1} & \cdots & v_{kn} + y_{kn} \end{bmatrix}$$

and

$$c \begin{bmatrix} v_{11} & \cdots & v_{1n} \\ \vdots & & \vdots \\ v_{k1} & \cdots & v_{kn} \end{bmatrix} = \begin{bmatrix} cv_{11} & \cdots & cv_{1n} \\ \vdots & & \vdots \\ cv_{k1} & \cdots & cv_{kn} \end{bmatrix}.$$

We always denote the neutral element of $\mathcal{M}_{k \times n}(V)$ by $\mathbf{O}$, where the size of this matrix is assumed to be known from the context.

Note that the vector space $\mathcal{M}_{1 \times n}(V)$ is the same as the vector space V^n of row vectors of length n and the vector space $\mathcal{M}_{k \times 1}(V)$ of column vectors of length k defined in the previous example.

EXAMPLE

Let V be a vector space over a field F. In the consideration of a vector space of the form V^Ω, the set Ω itself may very well have an algebraic structure. For example, the set $\mathbb{R}^{\mathbb{R}}$ of all functions from the field $\mathbb{R}$ of real numbers to itself is a vector space over $\mathbb{R}$ with addition and scalar multiplication defined componentwise: if f and g are functions from $\mathbb{R}$ to $\mathbb{R}$ and if $c \in \mathbb{R}$ then $(f + g)(a) = f(a) + g(a)$ and $(cf)(a) = c(f(a))$ for all $a \in \mathbb{R}$. This vector space plays an important role in calculus. Similarly, if $\mathbb{N}$ is the set of all nonnegative integers then $\mathbb{R}^{\mathbb{N}}$ is the vector space of all infinite sequences of real numbers, which too has an important role in calculus.

EXAMPLE

If F is a field then the integral domain $F[X]$ is a vector space over F, with addition defined as before and scalar multiplication defined by setting $c(\sum_{i \geq 0} a_i X^i) = \sum_{i \geq 0}(ca_i)X^i$. This notion of multiplication is compatible with polynomial multiplication defined before, when we consider nonzero scalars as being polynomials of degree 0 and identify the scalar 0_F with the zero polynomial.

EXAMPLE

Let A be a nonempty set. Each subset B of A determines a function χ_B from A to the field $F = \mathbb{Z}/(2)$, called the *characteristic function* on B, defined as follows:

$$\chi_B(a) = \begin{cases} 1 & \text{if } a \in B \\ 0 & \text{if } a \notin B. \end{cases}$$

Conversely, any function ϕ from A to F determines a unique subset $B = \{a \in A \mid \phi(a) = 1\}$ such that $\phi = \chi_B$. Thus we can effectively identify the family of all subsets of A with the vector space F^A over F. Addition in F^A can be described as follows: if B and C are subsets of A then the set

$$B + C = \{a \in B \cup C \mid a \notin B \cap C\}$$

is called the *symmetric difference* of B and C and it is easily verified that $\chi_B + \chi_C = \chi_{B+C}$ in F^A. The neutral element with respect to vector addition is just the function $\chi_\varnothing$, which sends every element of A to 0. The additive inverse of an element of F^A is itself: $\chi_B + \chi_B = \chi_\varnothing$ for any subset B of A. Scalar multiplication is trivial:

$$c\chi_B = \begin{cases} \chi_\varnothing & \text{if } c = 0 \\ \chi_B & \text{if } c = 1. \end{cases}$$

As in the case of fields, the neutral element of a vector space V with respect to addition is unique. Also, a straightforward adaptation of the proofs of Proposition 1.1 and Proposition 1.2 allows us to establish the following results:

(2.1) PROPOSITION. *If v and w are elements of a vector space V over a field F then there exists a unique element y of V satisfying $v + y = w$.*

(2.2) PROPOSITION. *If v and w are elements of a vector space V over a field F and if a is an element of F then:*

(1) $a0_V = 0_V$;
(2) $0_F v = 0_V$;
(3) $(-1_F)v = -v$;
(4) $(-a)v = -(av) = a(-v)$;
(5) $-(-v) = v$;
(6) $av = (-a)(-v)$;

(7) $-(v + w) = -v - w;$

(8) $a(v - w) = av - aw;$

(9) *If* $av = 0_V$ *then* $a = 0_F$ *or* $v = 0_V$.

Let V be a vector space over a field F. A nonempty subset W of V is a *subspace* of V if and only if W is a vector space with respect to the operations of vector addition and scalar multiplication defined in V. Before giving examples of subspaces, let us note that to check whether a given nonempty subset of a vector space V is indeed a subspace, it is not necessary to check all eight conditions in the definition of a vector space.

(2.3) PROPOSITION. *Let* V *be a vector space over a field* F *and let* W *be a nonempty subset of* V. *Then* W *is a subspace of* V *if and only if it is closed under taking vector addition and scalar multiplication.*

PROOF. If W is a subspace of V then it is surely closed under taking vector addition and scalar multiplication. Conversely, suppose that it is so closed. If we choose a particular element $w_1 \in W$ then $0_V = 0_F w_1 \in W$ and it remains true that $0_V + w = w$ for all $w \in W$ since that is what happens in V. Therefore 0_V acts as the neutral element of W with respect to vector addition. Moreover, if $w \in W$ then $-w = (-1_F)w \in W$ and this is the additive inverse of w in W as well as in V. All of the other conditions: associativity and commutativity of vector addition, distributativity of scalar multiplication over vector addition and of scalar multiplication over scalar addition, and the existence of an element neutral with respect to scalar multiplication, hold in W because they hold in V. $\square$

Every vector space V over a field F has two subspaces: V itself and $\{0_V\}$, known respectively as the *improper subspace* and the *trivial subspace*. Any subspace of V other than V itself is a *proper subspace*; any subspace of V other than $\{0_V\}$ is a *nontrivial subspace*. If v is an arbitrary element of a vector space V over a field F then $Fv = \{av \mid a \in F\}$ is a subspace of V, which is nontrivial if and only if $v \neq 0_V$.

EXAMPLE

If $\{V_i \mid i \in \Omega\}$ is a family of vector spaces over a field F then

$$\coprod_{i \in \Omega} V_i = \left\{ f \in \prod_{i \in \Omega} V_i \;\middle|\; f(i) = 0_{V_i} \text{ for all but finitely-many elements } i \in \Omega \right\}$$

is a subspace of $\prod_{i \in \Omega} V_i$, called the *coproduct* of the vector spaces V_i. This subspace is improper if the set Ω is finite; otherwise it is proper. If all of the V_i are equal to a given vector space V then we write $V^{(\Omega)}$ instead of $\coprod_{i \in \Omega} V_i$. This is a proper subspace of V^{Ω} when Ω is infinite.

EXAMPLE

If $\{V_i \mid i \in \Omega\}$ is a family of vector spaces over a field F and if Λ is a proper subset of Ω then $\{f \in \prod_{i \in \Omega} V_i \mid f(i) = 0_{V_i} \text{ for all } i \in \Omega \backslash \Lambda\}$ is a proper subspace of $\prod_{i \in \Omega} V_i$. Thus, in particular, if $k < n$ are positive integers and then we can identify the vector space V^k with the subspace of V^n of the form $\{[v_1, \ldots, v_k, 0_V, \ldots, 0_V] \mid v_i \in V\}$.

EXAMPLE

We have already noted that the set $\mathbb{R}^{\mathbb{R}}$ of all functions from the field of real numbers to itself is a vector space over $\mathbb{R}$. This space has several important subspaces, including the space of all everywhere continuous functions, the space of all everywhere differentiable functions, and the space of all everywhere analytic functions. Similarly, if c is a given real number then $\{f \in \mathbb{R}^{\mathbb{R}} \mid f(c) = 0\}$ is a subspace of $\mathbb{R}^{\mathbb{R}}$, as are the set of all functions continuous at c, the set of all functions differentiable at c, and the set of all functions analytic at c.

(2.4) PROPOSITION. *Let V be a vector space over a field F and let $\{V_i \mid i \in \Omega\}$ be a family of subspaces of V. Then $\bigcap_{i \in \Omega} V_i$ is also a subspace of V.*

PROOF. Set $W = \bigcap_{i \in \Omega} V_i$. If $w, w' \in W$ then $w, w' \in V_i$ for each $i \in \Omega$ and so, since V_i is a subspace, we see that $w + w' \in V_i$ for all $i \in \Omega$. Thus $w + w' \in W$. Similarly, if $a \in F$ then $aw \in V_i$ for each $i \in \Omega$ and so $aw \in W$. Thus W is closed under taking vector sums and scalar products, and so is a subspace of V. $\square$

Unions of subspaces do not behave as nicely. Indeed, we note that if $\{V_i \mid i \in \Omega\}$ is a family of subspaces of a vector space V over a field F then it is not true in general that $\bigcup_{i \in \Omega} V_i$ is a subspace of V.

We now come to a very important method of constructing subspaces of vector spaces. If D is a nonempty subset of a vector space V over a field F then an element v of V is a *linear combination* of elements of D if and only if there exist elements $v_1, \ldots v_n$ of D and scalars $a_1, \ldots, a_n$ in F satisfying $v = a_1 v_1 + \cdots + a_n v_n$. The set of all linear combinations of elements of D will be denoted by FD. This notation is consistent with that which we have already introduced, for if $D = \{v\}$ then $FD = \{av \mid a \in F\}$, which we have already denoted by Fv. For technical reasons, it is convenient to set $F\emptyset = \{0_V\}$.

Let us look at a few specific examples of linear combinations.

EXAMPLE

If $V = \mathbb{R}^3$ and $D = \{[1, 2, -1], [3, -2, 5]\}$ then the linear combinations of elements of D are those vectors in V of the form

$$a_1[1, 2, -1] + a_2[3, -2, 5] = [a_1 + 3a_2, 2a_1 - 2a_2, -a_1 + 5a_2],$$

where $a_1, a_2 \in \mathbb{R}$. Thus, $[-5\frac{1}{2}, 5, -10\frac{1}{2}] = \frac{1}{2}[1, 2, -1] + (-2)[3, -2, 5]$ is a linear combination of elements of D while $[0, 0, 1]$ is not. Indeed, suppose we could write

$$[0, 0, 1] = [a_1 + 3a_2, 2a_1 - 2a_2, -a_1 + 5a_2]$$

for real numbers a_1 and a_2. Then we must have $2a_1 - 2a_2 = 0$ and so $a_1 = a_2$. But then $0 = a_1 + 3a_2 = 4a_2$ implies that $a_1 = a_2 = 0$, which is incompatible with the condition that $-a_1 + 5a_2 = 1$.

EXAMPLE

If F is a field then every element of the vector space $F[X]$ is a linear combination of members of the infinite set $D = \{1, X, X^2, X^3, \ldots\}$.

Note that a vector can often be shown to be a linear combination of elements of a given set of vectors in several ways.

EXAMPLE

If $D = \{[0, 0, 4], [2, 2, 0], [2, 0, 0], [1, 1, 1]\} \subseteq \mathbb{R}^3$ then

$$
\begin{aligned}
[4, 2, 4] &= 1[0, 0, 4] + 1[2, 2, 0] + 1[2, 0, 0] \\
&= (-1)[2, 2, 0] + 1[2, 0, 0] + 4[1, 1, 1] \\
&= \frac{1}{2}[0, 0, 4] + 1[2, 0, 0] + 2[1, 1, 1] \\
&= \frac{3}{2}[0, 0, 4] + 2[2, 2, 0] + 1[2, 0, 0] + (-2)[1, 1, 1].
\end{aligned}
$$

(2.5) PROPOSITION. *If D is a subset of a vector space V over a field F then:*

(1) *FD is a subspace of V containing D;*

(2) *Any subspace of V containing D also contains FD;*

(3) *FD equals the intersection of all subspaces of V containing D.*

PROOF. If $D = \varnothing$ then FD is just the trivial subspace of V, by definition. Hence we can assume that $D \neq \varnothing$. As an immediate consequence of the definition, we note that FD is closed under taking vector sums and scalar products, which proves (1). If W is any subspace of V containing D and if $v_1, \ldots, v_n$ are elements of D, then $a_1 v_1 + \ldots a_n v_n \in W$ for all $a_1, \ldots, a_n \in F$ and so $FD \subseteq W$. This proves

(2). As a consequence of (2), we note that FD is contained in the intersection of all subspaces of V containing D. But FD it itself such a subspace by (2), and so contains this intersection. Therefore we have equality, proving (3). $\square$

If D is a subset of a vector space V over a field F then $FD = D$ precisely when D is itself a subspace of V. The subspace FD of V is called the subspace of V *generated* or *spanned* by D. The set D is called a *generating set* or *spanning set* for this subspace.

EXAMPLE

If F is a field and $V = F^3$ then

$$D_1 = \{[1,0,0],[0,1,0],[0,0,1]\}$$

is a generating set for V. The set

$$D_2 = \{[0,1,1],[1,0,1],[1,1,0]\}$$

is also a generating set for V on the condition that F does not have characteristic 2. If F has characteristic 2 then D_2 generates a proper subspace of V for, in this case, the vector $[1,1,1]$ is not a linear combination of elements of D_2. For all fields F, the set
$$D_3 = \{[1,0,0],[0,1,0],[1,1,0]\}$$
is not a generating set for V since $[0,0,1]$ is not a linear combination of elements of D_3.

Our intuition tells us that the subspace generated by a subset D of a vector space V should not increase if we add to D vectors which are already linear combinations of elements of D. This correct statement is embodied in the following proposition.

(2.6) PROPOSITION. *Let V be a vector space over a field F and let $D \subseteq D' \subseteq FD$ be subsets of V. Then $FD' = FD$.*

PROOF. Since FD is a subspace of V containing D', we know from Proposition 2.5 that $FD' \subseteq FD$. On the other hand, any linear combination of elements of D is also a linear combination of elements of D' since $D \subseteq D'$ and so $FD \subseteq FD'$. Thus we have equality. $\square$

A vector space is *finitely generated* over its field of scalars if and only if it has a finite generating set. Many of the vector spaces we will consider are finitely generated over their field of scalars.

EXAMPLE

The field $\mathbb{R}$ is finitely generated as a vector space over itself, having generating set $\{1\}$; it is not finitely generated as a vector space over $\mathbb{Q}$.

EXAMPLE

If F is a field and n is a positive integer then the vector space F^n is finitely generated over F, since the set $\{[1,0,\ldots,0],[0,1,0,\ldots,0],\ldots,[0,\ldots,0,1]\}$ is a generating set for it. More generally, if V is a vector space over a field F having a generating set $\{v_1,\ldots,v_k\}$ then V^n is also finitely generated over F and it has a generating set consisting of the kn vectors

$$\{[v_1,0,\ldots,0],\ldots,[v_k,0,\ldots,0],\ldots,[0,\ldots,0,v_1],\ldots,[0,\ldots,0,v_k]\}.$$

EXAMPLE

Similarly, if k and n are positive integers and if V is a vector space finitely generated over a field F then $\mathcal{M}_{k\times n}(V)$ is finitely generated over F. In particular, $\mathcal{M}_{k\times n}(F)$ is always finitely-generated over F.

EXAMPLE

If F is a field then the vector space $F[X]$ is not finitely generated over F. If Ω is an infinite set then F^Ω and $F^{(\Omega)}$ are not finitely generated over F.

Let V be a vector space over a field F and let $\{W_i \mid i \in \Omega\}$ be a family of subspaces of V. We define the subset $\sum_{i\in\Omega} W_i$ of V to be the set of all vectors in V which can be written in the form $w_{i_1} + \cdots + w_{i_n}$, where $\{i_1,\ldots,i_n\}$ is a finite subset of Ω and $w_{i_j} \in W_{i_j}$ for all $1 \le j \le n$. In other words,

$$\sum_{i\in\Omega} W_i = F\left(\bigcup_{i\in\Omega} W_i\right).$$

Actually, we note something more: if D_i is a generating set of W_i for each $i \in \Omega$ then $\bigcup_{i\in\Omega} D_i$ is a generating set for $\sum_{i\in\Omega} W_i$, namely

$$\sum_{i\in\Omega} W_i = F\left(\bigcup_{i\in\Omega} D_i\right).$$

As a special case, we note that if W_1 and W_2 are subspaces of a vector space over a field F then $W_1 + W_2$ is the set of all vectors in V of the form $w_1 + w_2$, where $w_1 \in W_1$ and $w_2 \in W_2$. Therefore, if both W_1 and W_2 are finitely generated, so is $W_1 + W_2$.

(2.7) PROPOSITION. *Let V be a vector space over a field F and let $\{W_i \mid i \in \Omega\}$ be a family of subspaces of V. Then:*

(1) *W_h is a subspace of $\sum_{i\in\Omega} W_i$ for all $h \in \Omega$;*
(2) *If Y is a subspace of V satisfying the condition that W_h is a subspace of Y for each $h \in \Omega$ then $\sum_{i\in\Omega} W_i \subseteq Y$.*

PROOF. Part (1) is an immediate consequence of the definition. As for (2), assume that Y satisfies the stated condition. If $\{i_1, \ldots, i_n\} \subseteq \Omega$ and if $w_{i_h} \in W_{i_h}$ for all $1 \le h \le n$ then $w_{i_h} \in Y$ for all h and so $\sum_{h=1}^n w_{i_h} \in Y$. Similarly, if $c \in F$ then $cw_{i_h} \in W_{i_h}$ for all $1 \le h \le n$ so $\sum_{h=1}^n cw_{i_h} \in Y$. Thus $\sum_{i \in \Omega} W_i$ is a subspace of Y. $\square$

(2.8) PROPOSITION. *Let V be a vector space over a field F and let W_1, W_2, and W_3 be subspaces of V. Then:*

(1) $(W_1 + W_2) + W_3 = W_1 + (W_2 + W_3)$;
(2) $W_1 + W_2 = W_2 + W_1$;
(3) $W_3 \cap [W_2 + (W_1 \cap W_3)] = (W_1 \cap W_3) + (W_2 \cap W_3)$;
(4) *(The modular law of subspaces) If $W_1 \subseteq W_3$ then $W_3 \cap (W_2 + W_1) = W_1 + (W_2 \cap W_3)$.*

PROOF. Parts (1) and (2) follow immediately from the definition.

(3) If $v \in W_3 \cap [W_2 + (W_1 \cap W_3)]$ then we can write $v = w + y$, where $w \in W_2$ and $y \in W_1 \cap W_3$. Thus $w \in W_2 \cap W_3$ and $y \in W_1 \cap W_3$ so $v \in (W_1 \cap W_3) + (W_2 \cap W_3)$. This shows that $W_3 \cap [W_2 + (W_1 \cap W_3)] \subseteq (W_1 \cap W_3) + (W_2 \cap W_3)$. Conversely, assume that $v \in (W_1 \cap W_3) + (W_2 \cap W_3)$. Then surely $v \in W_3$ and we can write $v = w + y$, where $w \in W_2 \cap W_3$ and $y \in W_1 \cap W_3$. Thus $v \in W_3 \cap [W_2 + (W_1 \cap W_3)]$ and so we have equality.

(4) This is a special case of (3). $\square$

Problems

1. Let $V = \mathbb{Q}^2$ on which we have the usual operation of componentwise addition: $[c, d] + [c', d'] = [c + c', d + d']$. If $a + b\sqrt{2} \in \mathbb{Q}(\sqrt{2})$, define

$$(a + b\sqrt{2})[c, d] = [ac + 2bd, bc + ad].$$

With this definition of scalar multiplication, is V a vector space over $\mathbb{Q}(\sqrt{2})$?

2. Which of the following are vector spaces over the field $F = \mathbb{Z}/(2)$?

(i) $\mathbb{Z}/(4)$;
(ii) The set $V = \{0, 1\}$ on which we define addition and scalar multiplication by setting $v + v' = max\{v, v'\}$, $1_F v = v$, and $0_F v = 0_V$ for all $v, v' \in V$;
(iii) $\{[a_1, a_2, a_3, a_4, a_5] \in F^5 \mid \sum_{i=1}^5 a_i = 0_F\}$;
(iv) $\{[a_1, a_2, a_3, a_4, a_5, a_6] \in F^6 \mid a_1 + a_2 + a_3 = a_4 + a_5 + a_6\}$.

3. Let V be the set of all continuous functions from $\mathbb{R}$ to the closed interval $[0, 1]$ on which we define the operation $\vee$ by $(f \vee g)(t) = max\{f(t), g(t)\}$ for all $t \in \mathbb{R}$. Is V, together with this operation and the usual scalar multiplication, a vector space over $\mathbb{R}$?

4. Let ∞ be an element not in $\mathbb{R}$ and assume we have defined an operation $+$ on the set $V = \mathbb{R} \cup \{\infty\}$ satisfying $a + \infty = \infty + a = \infty + \infty = \infty$ for all $a \in \mathbb{R}$. Is it possible to define a scalar product in such a way that V is a vector space over $\mathbb{R}$?

5. Let ∞ and $-\infty$ be distinct elements not in $\mathbb{R}$ and assume that we have define an operation $+$ on the set $V = \mathbb{R} \cup \{-\infty, \infty\}$ satisfying $a + \infty = \infty + a = \infty + \infty = \infty$ and $a + (-\infty) = (-\infty) + a = (-\infty) + (-\infty) = -\infty$ for all $a \in \mathbb{R}$. Is it possible to define a scalar product in such a way that V is a vector space over $\mathbb{R}$?

6. Give an example of a vector space having precisely 125 elements.

7. Which of the following are subspaces of $\mathbb{R}^{\mathbb{R}}$ over $\mathbb{R}$?

(i) The set of all functions f from $\mathbb{R}$ to itself which either satisfy $f(x) = 0$ for all $x \in \mathbb{R}$ or $f(x) = 0$ for at most finitely-many elements x of $\mathbb{R}$;
(ii) The set of all functions f from $\mathbb{R}$ to itself which either satisfy $f(x) = 0$ for all $x \in \mathbb{R}$ or $f(x) \neq 0$ for all $x \in \mathbb{R}$;
(iii) The set of all functions f from $\mathbb{R}$ to itself which satisfy $f(\pi) > f(-\pi)$;
(iv) The set of all functions f from $\mathbb{R}$ to itself for which there exists a positive real number c (which depends on f) satisfying $|f(x)| \leq c|x|$ for all $x \in \mathbb{R}$;
(v) The set of all functions f from $\mathbb{R}$ to itself for which there exist real numbers a and b (which depend on f) satisfying $|f(x)| \leq a|sin(x)| + b|cos(x)|$ for all $0 \geq x \in \mathbb{R}$;
(vi) The set of all piecewise-constant functions from $\mathbb{R}$ to itself;
(vii) The set of all functions from $\mathbb{R}$ to itself which are continuous at all but at most 5 points;
(viii) The set of all functions from $\mathbb{R}$ to itself which are continuous at all but finitely-many points;
(ix) The set of all functions f from $\mathbb{R}$ to itself satisfying $f(2) = f(\frac{1}{2})$;
(x) The set of all functions f from $\mathbb{R}$ to itself for which there exists a positive real number c (which depends on f) such that $f(x) = 0$ for all $-c \leq x \leq c$;
(xi) The set of all continuous functions from $\mathbb{R}$ to itself satisfying $|f(x)| \leq 1$ for all $-1 \leq x \leq 1$;
(xii) The set of all monotonically nondecreasing functions from $\mathbb{R}$ to itself;
(xiii) The set of all functions f from $\mathbb{R}$ to itself satisfying $f(x) = f(x + \pi)$ for all $x \in \mathbb{R}$.

8. Is the set of all vectors $[a, b, c, d, e] \in \mathbb{R}^5$ having an odd number of components equal to 0 a subspace of $\mathbb{R}^5$?

9. Let F be a field. Which of the following are subspaces of $F[X]$ over F?

(i) The set of all polynomials $f(X) \in F[X]$ satisfying $f(-a) = -f(a)$ for all $a \in F$;
(ii) The set of all monic polynomials in $F[X]$;
(iii) The set of all polynomials in $F[X]$ which are not monic;
(iv) The set of all polynomials in $F[X]$ having even degree;
(v) The set of all polynomials in $F[X]$ having degree at most 12;
(vi) The set of all polynomials of the form $(X^2 + X)q(X)$ for some $q(X) \in F[X]$.

10. Let W be a subspace of a vector space V over a field F. Is $(V \setminus W) \cup \{0_V\}$ necessarily a subspace of V?

11. Let $\mathbb{N}$ be the set of nonnegative integers. Let $a_1, \ldots, a_n$ be given elements

of a field F and let

$$W = \{f \in F^{\mathbb{N}} \mid f(k+n+1) = \sum_{i=1}^{n} a_i f(k+i) \text{ for all } k \geq 0\}.$$

Is W a subspace of the vector space $F^{\mathbb{N}}$ over F?

12. Let F be a field and let $1 \leq h \leq n$ be positive integers. Let W be a subspace of F^n over F containing a vector $v = [a_1, \ldots, a_n]$ in which $a_h \neq 0$. Show that for each scalar $c \in F$ there exists a vector $[b_1, \ldots, b_n] \in W$ satisfying $b_h = c$.

13. Let V be a vector space over a field F having subspaces W and W' which satisfy $V = W \cup W'$. Show that $V = W$ or $V = W'$.

14. Let V be a vector space over a field F and let Y and Z be two subspaces of $V \times V$. Let P be the set of all vectors $[v, v'] \in V \times V$ satisfying the condition that there exists an element v'' of V for which $[v, v''] \in Y$ and $[v'', v'] \in Z$. Is P a subspace of $V \times V$?

15. Let V be a vector space over a field F and let v and w be distinct vectors in V. Set $U = \{(1-a)v + aw \mid a \in F\}$. Show that there exists a vector $y \in V$ such that $\{y + u \mid u \in U\}$ is a subspace of V.

16. Let $V = \{v_1, \ldots, v_n\}$ be a vector space over a field F having only finitely-many elements. If $1 \leq h \leq n$, show that the vector v_h appears in each row of the "addition table" of V.

17. Let V be a vector space over the field $\mathbb{Z}/(2)$. Show that $v = -v$ for each $v \in V$.

18. Let W be the subspace of $\mathbb{R}^{\mathbb{R}}$ generated by the functions $f: x \mapsto 7$ and $g: x \mapsto \sin^2(2x)$. Does the function $h: x \mapsto 8\cos(4x)$ belong to W?

19. Express the vector $[4, 1, 7] \in \mathbb{Q}^3$ as a linear combination of

$$\{[1, 0, 2], [3, 2, 6], [4, 2, 10]\}.$$

20. Express the vector $X^2 \in \mathbb{Q}[X]$ as a linear combination of

$$\{X - 1, X + 1, X^2 + 1\}.$$

21. Find a real number c such that the vector $[4, 3, 1] \in \mathbb{R}^3$ is a linear combination of $\{[3, 1, c], [-1, 2, 1]\}$.

22. Let $F = \mathbb{Z}/(3)$ and let $A = \{[1, 1, 0], [1, 0, 2]\} \subseteq F^3$. For which values of c is the vector $[0, 1, c]$ a linear combination of the elements of A?

23. Let F be a field and let $D = \{[a, b, c] \in F^3 \mid a - b + c = 1\}$. Find FD.

24. Let $F = \mathbb{Z}/(2)$ and let A be a nonempty set. If $\{\chi_{B_i} \mid i \in \Omega\}$ is a generating set of the vector space F^A, show that $A = \bigcup_{i \in \Omega} B_i$.

25. Let V be a vector space over a field F in which there exist vectors u, v, and w satisfying $au + bv + cw = 0_V$ for nonzero scalars a, b, and c. Show that $F\{u, v\} = F\{v, w\}$.

26. Let A be a nonempty subset of a vector space V over a field F and let $v, w \in V$ be vectors satisfying $v \in F(A \cup \{w\})$ but $v \notin FA$. Show that $w \in F(A \cup \{v\})$.

CHAPTER III

LINEAR INDEPENDENCE AND DIMENSION

In this chapter we shall see how a small set of vectors can totally determine the structure of an entire vector space, and we will consider the consequences of this observation.

Let V be a vector space over a field F. A nonempty subset D of V is *linearly dependent* if and only if there exist distinct elements $v_1, \ldots, v_n$ of D and elements $a_1, \ldots, a_n$ of F, not all of which are equal to 0_F, satisfying $a_1 v_1 + \cdots + a_n v_n = 0_V$. If no such elements exist then D is *linearly independent*. That is to say, the set D is linearly independent if and only if for every finite subset $\{v_1, \ldots, v_n\}$ of D the only scalars $a_1, \ldots a_n$ in F which satisfy $a_1 v_1 + \cdots + a_n v_n = 0_V$ are $a_1 = \cdots = a_n = 0_F$. The empty subset $\varnothing$ of V is always linearly independent since this condition is vacuously true. The set $\{0_V\}$ is always linearly dependent, since $1_F 0_V = 0_V$.

As immediate consequence of the definitions, we note that any subset of a linearly-independent set of vectors is linearly independent, while any set of vectors having a linearly-dependent subset is itself linearly dependent.

EXAMPLE

The subset $\{[1, 2, 1], [-1, 3, 4], [-4, 7, 11]\}$ of the vector space $\mathbb{R}^3$ over $\mathbb{R}$ is linearly dependent, since $(-1)[1, 2, 1] + 3[-1, 3, 4] + (-1)[-4, 7, 11] = [0, 0, 0]$. The subset $\{[1, 0, 0], [1, 1, 0], [1, 1, 2]\}$ of the same space is linearly independent. To see this, assume that $a, b, c \in \mathbb{R}$ satisfy

$$a[1, 0, 0] + b[1, 1, 0] + c[1, 1, 2] = [0, 0, 0].$$

The lefthand side of this equation is just the vector $[a + b + c, b + c, 2c]$ and so we must have

$$a + b + c = 0$$
$$b + c = 0$$
$$2c = 0$$

the only solution of which, as one sees at once, is $a = b = c = 0$.

24

EXAMPLE

Let F be a field and let Ω be a nonempty set. For each $i \in \Omega$, let V_i be a vector space over F and let $f_i \in V = \prod_{i \in \Omega} V_i$ be a function satisfying the condition that $f_i(i) \neq 0_{V_i}$, while $f_i(j) = 0_{V_j}$ for all $j \neq i$. Then the set $\{f_i \mid i \in \Omega\}$ is a linearly-independent subset of V. To see this, assume that there are elements $i(1), \ldots i(n)$ of Ω and scalars $c_1, \ldots, c_n$ in F such that $\sum_{h=1}^{n} c_h f_{i(h)} = 0_V$. Then for each $1 \leq k \leq n$ we have $0_{V_{i(k)}} = [\sum_{h=1}^{n} c_h f_{i(h)}](i(k)) = c_k f_k(i(k))$. Since $f_k(i(k)) \neq 0_{V_{i(k)}}$, this implies that $c_k = 0_F$.

EXAMPLE

For any field F the subset $\{1_F, X, X^2, X^3, \ldots\}$ of $F[X]$ is linearly independent, since if $\sum_{i=0}^{n} c_i X^i = 0_{F[X]}$ then, by definition, it follows that $c_i = 0$ for each i.

EXAMPLE

The subset $D = \{e^{ax} \mid a \in \mathbb{R}\}$ of $\mathbb{R}^{\mathbb{R}}$ is linearly independent. To see this, assume that there exist distinct real numbers $a_1, \ldots, a_n$ and scalars $c_1, \ldots c_n$ in $\mathbb{R}$ such that the function $f: x \mapsto \sum_{i=1}^{n} c_i e^{a_i x}$ is the 0-function. If $n = 1$ then we must have $c_1 = 0$ since $e^{a_1 x}$ is not the 0-function. We now proceed by mathematical induction. Assume that $n > 1$ and that any subset of D having at most $n - 1$ elements is linearly independent. If we multiply f by $e^{-a_n x}$, we see that

$$g: x \mapsto c_1 e^{b_1 x} + \ldots c_{n-1} e^{b_{n-1} x} + c_n$$

is the 0-function where, for each $1 \leq i \leq n - 1$, $b_i = a_i - a_n \neq 0$. If we differentiate g we get the function $x \mapsto c_1 b_1 e^{b_1 x} + \cdots + c_{n-1} b_{n-1} e^{b_{n-1} x}$, which is also the 0-function. By the induction hypothesis, we see that $c_i b_i = 0$ for all $1 \leq i \leq n - 1$ and, since $b_i \neq 0$ for each such i, this implies that $c_1 = \cdots = c_{n-1} = 0$ whence also $c_n = 0$.

As an immediate consequence of the definition, note that if V is a vector space over a field F and if D is a subset of V containing 0_V then D is linearly dependent.

(3.1) PROPOSITION. *Let V be a vector space over a field F. A nonempty subset D of V is linearly dependent if and only if there exists a vector in D which is a linear combination of other elements of D.*

PROOF. Assume that D is linearly dependent. Then there exists a finite subset $\{v_1, \ldots v_n\}$ of D and there exist scalars $a_1, \ldots a_n$, not all of which equal 0_F, satisfying $a_1 v_1 + \cdots + a_n v_n = 0_V$. If a_h is one of the nonzero scalars then we have $v_h = -a_h^{-1} \sum_{i \neq h} a_i v_i$ and so v_h is a linear combination of other vectors in D.

Conversely, assume that there are vectors $v_1, \ldots, v_n$ in D such that v_1 is a linear combination of $\{v_2, \ldots, v_n\}$. Then there exist scalars $b_2, \ldots, b_n$ in F such that

$v_1 = b_2 v_2 + \cdots + b_n v_n$ and so $(-1)v_1 + b_2 v_2 + \cdots + b_n v_n = 0_V$, proving that D is linearly dependent.　□

EXAMPLE

For each real number a, let $f_a : \mathbb{R} \to \mathbb{R}$ be the function defined by $f_a : x \mapsto |x - a|$. We claim that $D = \{f_a \mid a \in \mathbb{R}\}$ is a linearly-independent subset of the vector space $\mathbb{R}^{\mathbb{R}}$ over $\mathbb{R}$. Indeed, assume that this is not the case. Then D has a finite subset which is linearly dependent and so, by Proposition 3.1, there exist distinct real numbers $b, a_1, \ldots, a_n$ and well as real numbers $c_1, \ldots, c_n$ such that $f_b = \sum_{i=1}^{n} c_i f_{a_i}$. But the function on the righthand side of this equation is differentiable at b (since $b \neq a_i$ for all i) while the function on the lefthand side is not. From this contradiction, we conclude that D must be linearly independent.

We now need another general concept. If A is a nonempty set then a relation $\preceq$ among the elements of A is called a *partial order relation* if and only if the following conditions are satisfied:

(1) $a \preceq a$ for all $a \in A$;
(2) If $a \preceq b$ and $b \preceq a$ then $a = b$;
(3) If $a \preceq b$ and $b \preceq c$ then $a \preceq c$.

If A is a nonempty set on which we have a given partial order relation $\preceq$ then an element a_0 of A is *minimal* with respect to this relation if and only if $a \preceq a_0$ implies $a = a_0$; it is *maximal* with respect to this relation if and only if $a_0 \preceq a$ implies $a = a_0$.

Partial order relations are fundamental in all areas of mathematics. Keep an eye out to see where they occur in various contexts you encounter.

EXAMPLE

The most important example of a partial order relation is the following: if $\mathcal{A}$ is a nonempty family of subsets of a set B then the inclusion relation $\subseteq$ is a partial order relation on $\mathcal{A}$. Whenever we talk about a minimal or maximal element of a family of subsets of a given set, we will be referring to this partial order.

It is important to emphasize that a set may have many minimal or maximal elements, or it may not have any. Thus, for example, if $\mathcal{A}$ is the family of all subsets of $\{1, 2, 3\}$ having at least one element but no more than two elements then $\mathcal{A}$ has three minimal elements: $\{1\}$, $\{2\}$, and $\{3\}$. It also has three maximal elements: $\{1, 2\}$, $\{1, 3\}$, and $\{2, 3\}$. The family $\mathcal{B}$ of all finite nonempty subsets of $\mathbb{Z}$ has infinitely-many minimal elements and no maximal elements; the family $\mathcal{C}$ of all infinite subsets of $\mathbb{Z}$ has one maximal element, namely $\mathbb{Z}$ itself, and no minimal elements.

(3.2) PROPOSITION. *For a subset D of a vector space V over a field F the following conditions are equivalent:*

(1) *D is a minimal generating set of V;*
(2) *D is a maximal linearly-independent subset of V;*

(3) D *is a linearly-independent generating set of* V.

PROOF. (1) $\Rightarrow$ (2): Assume that D is a minimal generating set of V which is linearly dependent. Then, by Proposition 3.1, there exists an element v_0 of D which is a linear combination of elements of the set $D' = D \setminus \{v_0\}$, say $v_0 = a_1 u_1 + \cdots + a_n u_n$, where the u_i are elements of D' and the a_i are scalars. If we choose an arbitrary element v of V then there exist elements $v_1, \ldots, v_t$ of D and scalars $b_0, \ldots, b_t$ in F such that $v = b_0 v_0 + \cdots + b_t v_t$, since D is a generating set of V. But then $v = b_0 a_1 u_1 + \ldots b_0 a_n u_n + b_1 v_1 + \cdots + b_t v_t$ and so D' is also a generating set of V, which contracts the assumption that D is a minimal generating set. Hence D must in fact be linearly independent. If v is a vector in $V \setminus D$ then it is a linear combination of elements of D and so the set $D \cup \{v\}$ is linearly dependent. Hence D is a maximal linearly-independent subset of V.

(2) $\Rightarrow$ (3): Let D be a maximal linearly-independent subset of V and consider a vector $v_0 \in V \setminus D$. Then $D \cup \{v_0\}$ is linearly dependent and so there exist elements $v_1, \ldots v_n$ of D and scalars $a_0, \ldots, a_n$ in F, not all of them equal to 0_F, satisfying $a_0 v_0 + \cdots + a_n v_n = 0_V$. If $a_0 = 0_F$ then D would be linearly dependent, which is a contradiction. Thus we must have $a_0 \neq 0_F$. and so $v_0 = -a_0^{-1} a_1 v_1 - \cdots - a_0^{-1} a_n v_n$. This proves that D is a generating set for V.

(3) $\Rightarrow$ (1): Let D be a linearly-independent generating set for V and let D' be a generating set for V properly contained in D. If $v_0 \in D \setminus D'$ then there exist elements $v_1, \ldots v_n$ of D' and scalars $a_1, \ldots a_n$ in F such that $v_0 = a_1 v_1 + \cdots + a_n v_n$. By Proposition 3.1, this implies that D is linearly dependent, contradicting our assumption. Hence no proper subset of D is a generating set for V, so D is minimal. $\square$

(3.3) PROPOSITION. *Let* V *be a vector space over a field* F *and let* D *be a linearly-independent subset of* V. *If* $v \in V \setminus FD$ *then the set* $D \cup \{v\}$ *is also linearly independent.*

PROOF. Assume that the set $D \cup \{v\}$ is linearly dependent. Then there exist elements $v_1, \ldots, v_n$ of D and scalars $b, a_1, \ldots, a_n$ in F such that

$$bv + a_1 v_1 + \cdots + a_n v_n = 0_V.$$

The scalar b must be nonzero, for otherwise the set D would be linearly dependent. Therefore $v = -b^{-1} a_1 v_1 - \cdots - b^{-1} a_n v_n$, proving that $v \in FD$, which contradicts the choice of v. Therefore our basic assumption is false and $D \cup \{v\}$ must be linearly independent. $\square$

A linearly-independent generating set of a vector space V over a field F is called a *basis* for V over F. Caution! We have not yet proven that a vector space must necessarily have such a basis. Moreover, as we shall see immediately, when such bases exist they need not be unique.

EXAMPLE

If F is a field then the set $\{1_F, X, X^2, \ldots\}$ is a basis for $F[X]$ over F, as is the set $\{1_F, 1_F + X, 1_F + X + X^2, \ldots\}$. Indeed, if $p_i(X) \in F[X]$ is a polynomial of degree i for each $i \geq 0$ then $\{p_i(X) \mid i \geq 0\}$ is a basis for $F[X]$ over F.

EXAMPLE

If F is an arbitrary field then the sets

$$\{[1_F, 0_F, 0_F], [0_F, 1_F, 0_F], [0_F, 0_F, 1_F]\}$$

and

$$\{[1_F, 0_F, 0_F], [1_F, 1_F, 0_F], [1_F, 1_F, 1_F]\}$$

are bases for the vector space F^3 over F. The set

$$\{[0_F, 1_F, 1_F], [1_F, 0_F, 1_F], [1_F, 1_F, 0_F]\}$$

is a basis for F^3 if the characteristic of the field F is not equal to 2. If the characteristic of F equals 2, then $[1_F, 1_F, 0_F] = [0_F, 1_F, 1_F] + [1_F, 0_F, 1_F]$ so this set is linearly dependent and hence not a basis.

EXAMPLE

Let F be a field and let k and n be positive integers. For all $1 \le t \le k$ and all $1 \le s \le n$ let H_{st} be the matrix $[a_{ij}]$ in $\mathcal{M}_{k \times n}(F)$ defined by

$$a_{ij} = \begin{cases} 1_F, & \text{if } i = s \text{ and } j = t \\ 0_F, & \text{otherwise.} \end{cases}$$

Then $\{H_{st} \mid 1 \le s \le k; 1 \le t \le n\}$ is a basis for $\mathcal{M}_{k \times n}(F)$ over F.

EXAMPLE

Neither the set $\{[1, 2, 3], [3, 2, 1]\}$ nor the set

$$\{[0, 1, 1], [1, 2, 1], [3, 0, 3], [-1, -1, -1]\}$$

is a basis for $\mathbb{R}^3$ over $\mathbb{R}$, since the first of these is not a generating set for the space and the second is linearly dependent.

From the above examples we see that a vector space may (and as a rule does) have many bases and so it is natural to ask if there is, in general, a "preferred" basis to choose from among them. For vector spaces of the form F^n for some field F, it is usually convenient to prefer the basis

$$\{[1_F, 0_F, \dots, 0_F], [0_F, 1_F, 0_F, \dots, 0_F], \dots, [0_F, \dots, 0_F, 1_F]\}$$

while for vector spaces of the form $F[X]$ for some field F it is often convenient to prefer the basis $\{1_F, X, X^2, \dots\}$. These bases are therefore called the *canonical bases* of the respective spaces. However, as we shall see later, there are occasions when other bases will better serve specific purposes we have at hand, so that one should beware not to place too great an emphasis on the canonical bases.

There is also another subtle point which should be emphasized here. When we introduced the basic notation and terminology of this book, we pointed out that a set the elements of which are specifically enumerated comes equipped with an implied order given by that enumeration. When dealing with bases — and especially finite bases — this order will often play an important role. This should be kept in mind in all of the discussion which follows.

(3.4) PROPOSITION. *Let V be a vector space over a field F. A nonempty subset D of V is a basis if and only if every element of V can be written in precisely one way as a linear combination of elements of D.*

PROOF. First, let us assume that D is a basis for V and let $v \in V$. Since D is a generating set for V, we know that v can be written in at least one way as a linear combination of elements of D. Assume that this can be done in more than one way. That is to say, assume that there is a finite subset $\{v_1, \ldots v_n\}$ of elements of D and two different sets of scalars, $\{a_1, \ldots a_n\}$ and $\{b_1, \ldots b_n\}$, such that $v = \sum_{i=1}^{n} a_i v_i = \sum_{i=1}^{n} b_i v_i$. Then

$$0_V = v - v = \sum_{i=1}^{n} a_i v_i - \sum_{i=1}^{n} b_i v_i = \sum_{i=1}^{n} (a_i - b_i) v_i,$$

where at least one of the scalars $a_h - b_h$ is nonzero since $a_h \neq b_h$ for at least one index h. But this contradicts the assumption that D is linearly independent, and so we have shown that the representation of each element of v as a linear combination of elements of D must in fact be unique.

Conversely, assume that D satisfies the condition that every element of V can be written in precisely one way as a linear combination of elements of D. Then surely D is a generating set for V. If $\{v_1, \ldots, v_n\}$ is a finite subset of D and if $a_1, \ldots, a_n$ are scalars in F satisfying $a_1 v_1 + \cdots + a_n v_n = 0_V$ then, since we also have $0_F v_1 + \cdots + 0_F v_n = 0_V$, we must have $a_i = 0_F$ for all $1 \leq i \leq n$, proving that D is linearly independent. Therefore D is a basis. $\square$

Let us look at the result of Proposition 3.4 from another angle. Let D be a nonempty subset a vector space V over a field F and recall that $F^{(D)}$ is the set of all functions $f \colon D \to F$ satisfying the condition that $f(u) \neq 0_F$ for only finitely-many elements u of D. We can define a function $\theta \colon F^{(D)} \to V$ by $f \mapsto \sum_{u \in D} f(u) u$. (Note that this sum is well-defined, since only finitely-many of the summands $f(u) u$ are nonzero.) Then:

(1) The function θ is monic if and only if D is linearly independent;
(2) The function θ is epic if and only if D is a generating set for V;
(3) The function θ is bijective if and only if D is a basis for V.

We now, finally, begin addressing the question of whether bases in fact always exist.

(3.5) PROPOSITION. *If V is a vector space finitely generated over a field F then any finite generating set of V contains a basis.*

PROOF. Let D be a finite generating set of V. If D is a minimal generating set then, by Proposition 3.2, it is a basis and we are done. Otherwise, there are generating sets for V properly contained in D. Since D is finite, we know that among all such generating sets for V properly contained in D there is one, call it D', having a minimal number of elements. (There may, of course, be several having this number of elements; in that case, arbitrarily pick one of them.) Then no proper subset of D' can be a generating set for V, so by Proposition 3.2 this is a basis for V. $\square$

COROLLARY. *Every vector space finitely generated over a field F has a basis.*

(3.6) PROPOSITION. *If V is a vector space finitely generated over a field F then any linearly-independent subset of V is contained in a basis for V over F.*

PROOF. By assumption, V has a finite generating set $\{v_1, \ldots, v_n\}$. Let D be a linearly-independent subset of V. If $v_i \in FD$ for all $1 \le i \le n$ then $FD = V$ and so D is itself a basis for V. Otherwise, let $h = min\{i \mid v_i \notin FD\}$. By Proposition 3.3, the set $D' = D \cup \{v_h\}$ is linearly independent. If $FD' = V$, then D' is a basis for V containing D and we are done. Otherwise, let $k = min\{i \mid v_i \notin FD'\}$ and repeat the process. We can go on in this way for only finitely-many steps, at the end of which we must end up with a basis for V which contains D. $\square$

We now want to show that vector spaces which are not finitely generated over their field of scalars also have bases. To do this, we must make use of an axiom of set theory known as variously as the *Hausdorff Maximum Principle* or *Zorn's Lemma*. This axiom is logically equivalent to the Axiom of Choice, which, as we noted at the beginning of this book, was necessary in order to insure that the direct product of an arbitrary family of nonempty sets is nonempty. In order to state this axiom we first need some definitions: Let A be a nonempty set on which we have a partial order $\preceq$ defined. A subset B of A is called a *chain* in A if and only if for all $b, b' \in B$ either $b \preceq b'$ or $b' \preceq b$ holds. A subset C of A is *bounded* if and only if there exists an element $a \in A$ such that $c \preceq a$ for all $c \in C$. The Hausdorff Maximum Principle then states that if every chain in A is bounded, the set A must have a maximal element. Again, let us emphasize that this is to be taken as an axiom of set theory. Given this axiom, we can extend Proposition 3.6 to the case of vector spaces which are not necessarily finitely generated.

(3.7) PROPOSITION. *If V is a vector space over a field F then any linearly-independent subset D of V is contained in a basis for V over F.*

PROOF. Let D be a linearly-independent subset of V and let $\mathcal{A}$ be the collection of all linearly-independent subsets of V containing D, partially ordered by inclusion. Then $\mathcal{A} \ne \varnothing$ since $D \in \mathcal{A}$. Let $\mathcal{C} \subset \mathcal{A}$ be a chain. We want to show that $\mathcal{C}$ is bounded and so we must find an element D^* of $\mathcal{A}$ which contains every member of $\mathcal{C}$. Indeed, choose D^* to be the union of all of the members of $\mathcal{C}$.

We must show that $D^* \in \mathcal{A}$, i.e. that D^* is linearly independent. To do this, it suffices to show that every finite subset of D^* is linearly independent. And, indeed, if $\{v_1, \ldots, v_n\} \subseteq D^*$ then for each $1 \le i \le n$ there exists a member D_i of $\mathcal{C}$ with $v_i \in D_i$. Since $\mathcal{C}$ is a chain, there exists an integer $1 \le h \le n$ such that $D_i \subseteq D_h$ for all $1 \le i \le n$. Hence $\{v_1, \ldots, v_n\} \subseteq D_h$, which suffices to show that $\{v_1, \ldots, v_n\}$

is linearly independent, since D_h is. Thus $D^* \in \mathcal{A}$. Since every element of $\mathcal{C}$ is contained in D^*, this shows that $\mathcal{C}$ is bounded.

We have thus shown that every chain in $\mathcal{A}$ is bounded and so we can apply the Hausdorff Maximum Principle to deduce that $\mathcal{A}$ has a maximal element. In other words, there exists in V a maximal linearly-independent subset containing D, and this subset is a basis for V by Proposition 3.2. $\square$

COROLLARY. *Every vector space over a field F has a basis.*

PROOF. Apply Proposition 3.7 to the special case $D = \varnothing$. $\square$

EXAMPLE

As a consequence of Proposition 3.7 we see that $\mathbb{R}$ has a basis as a vector space over $\mathbb{Q}$. Such a basis is known in analysis as a *Hamel basis*. Mathematicians are not able to give a concrete description of a particular Hamel bases, though we know that such bases exist. Let C be the set of all real numbers which can be represented as a series of the form $\sum_{k \geq 1} a_k 3^{-k}$, where each a_k is either 0 or 2. This set is called the *Cantor set* and it is known to be "sparse" in the unit interval $[0, 1]$ (in a very technical sense of the word which we will not go into here). Nonetheless, it is possible to prove that there exists a Hamel basis for $\mathbb{R}$ contained in C.

We have already seen that a vector space V can have many different bases over its field of scalars F. We now want to show that if V is finitely generated over F then any two such bases have the same number of elements. To do this, we must first proof a supplementary result.

(3.8) PROPOSITION. *If V is a vector space over a field F having a finite generating set $D = \{v_1, \ldots, v_n\}$ then any linearly-independent subset of V has at most n elements.*

PROOF. Assume that the desired result is false. Then there exists a linearly-independent subset $E = \{w_1, \ldots, w_{n+1}\}$ of V having $n + 1$ elements, none of which is 0_V (for otherwise E would be linearly dependent). Since D is a generating set for V, there exist scalars $a_1, \ldots, a_n$, not all equal to 0_F, satisfying $w_1 = \sum_{i=1}^{n} a_i v_i$. By renumbering the elements of D if necessary, we can in fact assume that $a_1 \neq 0_F$. Then

$$v_1 = a_1^{-1} w_1 - a_1^{-1} a_2 v_2 - \cdots - a_1^{-1} a_n v_n$$

and so $D \subseteq FD'$, where $D' = \{w_1, v_2, \ldots, v_n\}$. But $D' \subseteq V = FD$ and so $V = FD'$. Now assume that $k < n$ and that we have already shown

$$V = F\{w_1, \ldots, w_k, v_{k+1}, \ldots, v_n\}.$$

Then there exist scalars $b_1, \ldots, b_n$, not all equal to 0_F, satisfying

$$w_{k+1} = b_1 w_1 + \cdots + b_k w_k + b_{k+1} v_{k+1} + \cdots + b_n v_n.$$

If the scalars $b_{k+1}, \ldots, b_n$ are all equal to 0_F, then we would have shown that E is linearly dependent, which is a contradiction. Therefore at least one of these scalars is nonzero and, by renumbering $\{v_{k+1}, \ldots, v_n\}$ if necessary, we can assume that $b_{k+1} \neq 0_F$. As above, we then conclude that

$$v_{k+1} = -b_{k+1}^{-1} b_1 w_1 - \cdots - b_{k+1}^{-1} b_k w_k + b_{k+1}^{-1} w_{k+1} - b_{k+1}^{-1} b_{k+2} v_{k+2} - \cdots - b_{k+1}^{-1} b_n v_n$$

and so, as above, $V = F\{w_1, \ldots, w_{k+1}, v_{k+2} \ldots v_n\}$. Thus we can continue in this manner and, after a finite number of steps, we get $V = F\{w_1, \ldots w_n\}$. But this implies that w_{n+1} is a linear combination of $\{w_1, \ldots, w_n\}$, contradicting the assumption that E is linearly independent. Thus no such set E can exist. $\square$

(3.9) PROPOSITION. *Any two bases of a vector space V finitely generated over its field of scalars F have the same number of elements.*

PROOF. By hypothesis and Proposition 3.5, we know that V has at least one finite basis D. If D' is another basis of V then, by Proposition 3.8, we see that
(1) Since D is a generating set for V over F and D' is linearly independent, the number of elements of D' is at most that of D; and
(2) Since D' is a generating set for V over F and D is linearly independent, the number of elements of D is at most that of D'.
Therefore D and D' are finite and have the same number of elements. $\square$

We remark parenthetically that the equivalent of Proposition 3.9 for linearly-dependent sets is false. That is to say, a finite linearly-dependent set of vectors may have two minimal linearly-dependent subsets of different sizes. Indeed, there is no efficient algorithm for finding the smallest possible size of a linearly-dependent subset of a given finite linearly-dependent set of vectors.

We should also remark that Proposition 3.9 is a special case of a more general result: If V is an arbitrary vector space over a field F then there exists a bijective correspondence between any two bases of V. The proof of this important fact uses transfinite induction and is therefore beyond the scope of the present work.

We define the *dimension* of a vector space V finitely generated over a field F to be the number of elements in a basis of V over F. Proposition 3.9 shows that this concept is well-defined, since it does not matter which particular basis we happen to select. This number will be denoted by $dim(V)$ or by $dim_F(V)$, when it is necessary to emphasize the field of scalars. If V is not finitely generated over F, we say that it is *infinite dimensional* and write $dim(V) = \infty$.

EXAMPLE

If F is a field then $dim(F^n) = n$ for each positive integer n, since the canonical basis for F^n has n elements.

Since the vector space F^n can be constructed for each field F and each positive integer n, the previous example allows us to make an important existential claim: for each field F and for each positive integer n, there exists a vector space over F having dimension n.

EXAMPLE

If F is a field and if k and n are positive integers then we have seen that the vector space $\mathcal{M}_{k \times n}(F)$ over F has a basis $\{H_{st} \mid 1 \le s \le k; 1 \le t \le n\}$ having kn elements, and so $dim(\mathcal{M}_{k \times n}(F)) = kn$.

EXAMPLE

Let F be a field. We have already seen that the vector space $F[X]$ has an infinite basis over F, and hence it cannot have a finite basis over F. Thus $dim(F[X]) = \infty$. Similarly, if Ω is an infinite set then the vector spaces $F^{(\Omega)}$ and F^{Ω} are not finitely generated over F and so the dimension of each is infinite.

In particular, as a consequence of the definition of dimension, we see that if V is a vector space finitely generated over a field F having finite dimension n then:
(1) Every subset of V having more than n elements must be linearly dependent.
(2) There exists a linearly independent subset D of V having n elements.
(3) If D is such a subset then $V = FD$.

(3.10) PROPOSITION. *Let V be a vector space finitely generated over a field F and let W be a subspace of V. Then*

(1) *W is finitely generated over F;*
(2) *Any basis of W is contained in a basis of V;*
(3) *$dim(W) \le dim(V)$, with equality if and only if $W = V$.*

PROOF. Let $n = dim(V)$.
(1) Since any subset of W is a subset of V, we see that W can have no linearly-independent subset having more than n elements. Therefore W is finitely generated over F.
(2) Any basis of W is surely a linearly-independent subset of V and so, by Proposition 3.7, it is contained in a basis of V.
(3) By (2), we surely have $dim(W) \le dim(V)$, with equality if $W = V$. Conversely, if $dim(W) = dim(V)$ then W has a basis D having n elements. This basis is contained in a basis for V over F, which also has n elements, and so must be D itself. Therefore $W = FD = V$. $\square$

We now want to extend the notion of linear independence. Suppose that W_1 and W_2 are subspaces of a vector space V over a field F. We know that any vector $v \in W_1 + W_2$ can be written in the form $w_1 + w_2$, where $w_1 \in W_1$ and $w_2 \in W_2$. This representation of v is not unique in general. It will be unique if $W_1 \cap W_2 = \{0_V\}$ for then, if $w_1 + w_2 = w_1' + w_2'$ for $w_1, w_1' \in W_1$ and $w_2, w_2' \in W_2$ we have $w_1 - w_1' = w_2' - w_2 \in W_1 \cap W_2 = \{0_V\}$ and so $w_1 = w_1'$ and $w_2 = w_2'$. This consideration suggests the need to emphasize those situations in which the intersection of two subspaces of a given vector space is the trivial subspace. Indeed, if W_1 and W_2 satisfy $W_1 \cap W_2 = \{0_V\}$ we will write $W_1 \oplus W_2$ instead of $W_1 + W_2$ and say that the set $\{W_1, W_2\}$ is *independent*. The subspace $W_1 \oplus W_2$ of V is called the *direct sum* of W_1 and W_2. Note that $W = W \oplus \{0_V\}$ for any subspace W of V.

We next extend the notion of direct sum to cover more than two subspaces of a given vector space. Let V be a vector space over a field F. A collection $\{W_i \mid i \in \Omega\}$ of subspaces of V is *independent* whenever the following condition is satisfied: if Λ is a finite subset of Ω and if $w_i \in W_i$ for each $i \in \Lambda$ then $\sum_{i \in \Lambda} w_i = 0_V$ only when $w_i = 0_V$ for all $i \in \Lambda$. Thus we see that a nonempty subset D of V is linearly independent if and only if the collection of subspaces $\{Fw \mid w \in D\}$ is independent.

EXAMPLE

If $V = \mathbb{R}^3$ then $\{W_1, W_2, W_3\}$ is an independent set of subspaces of V, where $W_1 = \{[a, 0, 0] \mid a \in \mathbb{R}\}$, $W_2 = \{[0, b, 0] \mid b \in \mathbb{R}\}$, and $W_3 = \{[0, 0, c] \mid c \in \mathbb{R}\}$.

(3.11) PROPOSITION. *Let V be a vector space over a field F. Then the following conditions on a finite set $\{W_1, \ldots, W_n\}$ of subspaces of V are equivalent:*

(1) $\{W_1, \ldots, W_n\}$ *is independent;*
(2) *Each element of $\sum_{i=1}^n W_i$ can be written in a unique way as $w_1 + \cdots + w_n$, where $w_i \in W_i$ for all $1 \le i \le n$;*
(3) *For each $1 \le h \le n$ we have $W_h \cap \sum_{i \ne h} W_i = \{0_V\}$.*

PROOF. (1) $\Rightarrow$ (2): Assume (1) and suppose that we have elements w_i and w_i' of W_i for each $1 \le i \le n$ satisfying $w_1 + \cdots + w_n = w_1' + \cdots + w_n'$. Then $(w_1 - w_1') + \cdots + (w_n - w_n') = 0_V$ and so, by (1), $w_i - w_i' = 0_V$ for all $1 \le i \le n$, i.e. $w_i = w_i'$ for all such i. This proves (2).

(2) $\Rightarrow$ (3): Assume (2) and suppose that $0_V \ne w \in W_h \cap \sum_{i \ne h} W_h$. Then for each $i \ne h$ there exists an element $w_i \in W_i$ such that $w = \sum_{i \ne h} w_i$. Since $w \in W_h$, this contradicts (2).

(3) $\Rightarrow$ (1): Assume (3) and suppose that we can write $w_1 + \cdots + w_n = 0_V$, where $w_i \in W_i$ for each $1 \le i \le n$ and not all of these vectors equal 0_V. If we choose h such that $w_h \ne 0$ then $w_h = \sum_{i \ne h} -w_i \in W_h \cap \sum_{i \ne h} W_i$, and this contradicts (3). $\square$

If $\{W_i \mid i \in \Omega\}$ is an independent set of subspaces of a vector space V over a field F, we write $\bigoplus_{i \in \Omega} W_i$ instead of $\sum_{i \in \Omega} W_i$; if $\Omega = \{1, \ldots, n\}$ we write $\bigoplus_{i=1}^n W_i$ instead of $\sum_{i=1}^n W_i$. A representation of a vector space as a direct sum of two or more of its subspaces is called a *direct-sum decomposition* of the space. Direct-sum decompositions will play a major role in much of what follows.

EXAMPLE

If V is a vector space finitely generated over a field F and if $\{v_1, \ldots, v_n\}$ is a basis for V over F, then the set $\{Fv_1, \ldots, Fv_n\}$ is independent by Proposition 3.11 and so we have a direct-sum decomposition $V = \bigoplus_{i=1}^n Fv_i$ of V.

(3.12) PROPOSITION. *Let V be a vector space over a field F and let $\{W_1 \ldots W_n\}$ be a set of subspaces of V. Then the following conditions are equivalent:*

(1) $V = \bigoplus_{i=1}^n W_i$;
(2) *If D_i is a basis for W_i for each $1 \le i \le n$ then $D = \bigcup_{i=1}^n D_i$ is a basis for V.*

PROOF. Assume (1). Let $v \in V$. By (1) we know that v can be written in a unique manner as $w_1 + \cdots + w_n$, where $w_i \in W_i$ for all $1 \le i \le n$. Moreover, each w_i can be written in a unique manner as a linear combination of elements of D_i. Therefore v can be written in a unique manner as a linear combination of elements of D. By Proposition 3.4, this shows that D is a basis for V, and so we have (2). Conversely, if we assume (2) then every element of V can be written in a unique manner as a linear combination of elements of D and hence in a unique manner as $w_1 + \cdots + w_n$, with $w_i \in W_i$ for all $1 \le i \le n$. Thus, by Proposition 3.11, we have (1). $\square$

COROLLARY. *If $W_1, \ldots, W_n$ are subspaces of a vector space V over a field F satisfying $V = W_1 \oplus \cdots \oplus W_n$ then $dim(V) = dim(W_1) + \cdots + dim(W_n)$.*

Let W be a subspace of a vector space V over a field F. A subspace W' of V is a *complement* of W in V if and only if $V = W \oplus W'$.

(3.13) PROPOSITION. *Every subspace of vector space V over a field F has complement in V.*

PROOF. If $W = V$ then the trivial subspace $\{0_V\}$ is a complement of W in V, and if $W = \{0_V\}$ then V itself is a complement of W in V. Otherwise, let D be a basis for W. By Proposition 3.7 we know that there exists a linearly-independent subset D' of V such that $D \cup D'$ is a basis for V. Then $W' = FD'$ is a complement of W in V. $\square$

EXAMPLE

In general, a subspace of a vector space V may have more than one complement in V. For example, if $V = \mathbb{R}^2$ then each of the following subspaces of V is a complement of each of the others:

(1) $W_1 = \{[a, 0] \mid a \in \mathbb{R}\}$;
(2) $W_2 = \{[0, b] \mid b \in \mathbb{R}\}$;
(3) $W_3 = \{[c, c] \mid c \in \mathbb{R}\}$;
(4) $W_4 = \{[d, 2d] \mid d \in \mathbb{R}\}$.

EXAMPLE

Let $V = \mathcal{M}_{n \times n}(\mathbb{Q})$, which is a vector space of dimension n^2 over the field $\mathbb{Q}$ of rational numbers. Let $W_1 = \{A = [a_{ij}] \in V \mid a_{ij} = a_{ji} \text{ for all } 1 \le i, j \le n\}$ and let $W_2 = \{A = [a_{ij}] \in V \mid a_{ij} = -a_{ij} \text{ for all } 1 \le i, j \le n\}$. Both W_1 and W_2 are subspaces of V, respectively called the *space of symmetric matrices* over $\mathbb{Q}$ and the *space of skew-symmetric matrices* over $\mathbb{Q}$, and their intersection is the trivial subspace. Any matrix $A = [a_{ij}] \in V$ can be written as $B + C$, where $B = [b_{ij}]$ is the matrix defined by $b_{ij} = \frac{1}{2}(a_{ij} + a_{ji})$ and $C = [c_{ij}]$ is the matrix defined by $c_{ij} = \frac{1}{2}(a_{ij} - a_{ji})$. Moreover, $B \in W_1$ and $C \in W_2$. Thus we see that $V = W_1 \oplus W_2$.

As we have seen, a subspace of a vector space may have many complements in it. In fact, as the following result shows, it may have very many indeed.

(3.14) PROPOSITION. *Let F be a field with infinitely-many elements and let V be a vector space of dimension at least 2 over F. Then any proper nontrivial subspace W of V has infinitely-many complements in V.*

PROOF. By Proposition 3.13 we know that W has at least one complement in V. Choose such a complement W' and let D be a basis for W'. If $0_V \neq w \in W$ then, by Proposition 2.2(9) we see that Fw is an infinite subset of W, since the field F is infinite, and so we know that the set W is infinite. For each $w \in W$ let $Y_w = F\{u + w \mid u \in D\}$.

Our first claim is that each of the subspaces Y_w of V is a complement of W in V. Indeed, if $v \in W \cap Y_w$ then there exist elements $u_1, \ldots, u_n$ of D and scalars $c_1, \ldots, c_n$ in F such that $v = \sum_{i=1}^{n} c_i(u_i + w) \in W$. But then $\sum_{i=1}^{n} c_i u_i = v - (\sum_{i=1}^{n} c_i)w \in W \cap W' = \{0_V\}$ and since the set $\{u_1, \ldots, u_n\}$ is linearly independent, it follows that $c_i = 0$ for all $1 \leq i \leq n$. This proves that $v = 0_V$, and such we conclude that $W \cap Y_w = \{0_V\}$ for all $w \in W$. If v is an arbitrary vector in V we can write $v = x + (\sum_{i=1}^{n} c_i u_i)$, where x is in W, the vectors $u_1, \ldots u_n$ are in D, and $c_1, \ldots, c_n$ are scalars in F. Then

$$v = \left[x - \left(\sum_{i=1}^{n} c_i \right) w \right] + \sum_{i=1}^{n} c_i (u_i + w) \in W + Y_w$$

and so we have shown that $V = W + Y_w$ for each $w \in W$. Thus each subspace Y_w is indeed a complement of W in V.

We now must show that these complements are indeed distinct, i.e. that $Y_w \neq Y_x$ for elements $w \neq x$ of W. Indeed, let w and x be elements of W satisfying $Y_w = Y_x$. If $u \in D$ then there exist elements $u_1, \ldots u_n$ of D and scalars $c_1, \ldots, c_n$ such that $u + w = \sum_{i=1}^{n} c_i(u_i + x)$. From this we conclude that

$$u - \sum_{i=1}^{n} c_i u_i = \left(\sum_{i=1}^{n} c_i \right) x - w \in W \cap Y_w = \{0_V\}.$$

But the set D is linearly independent and so u must be equal to u_h for one of the indices $1 \leq h \leq n$, and we must have

$$c_i = \begin{cases} 0_F, & i \neq h \\ 1_F, & i = h. \end{cases}$$

Thus $x - w = 0_V$, i.e. $x = w$, and we have shown that if x and w are distinct elements of W then $Y_x \neq Y_w$. $\square$

We have already noted that if $\{W, W'\}$ is an independent set of subspaces of V then $dim(W \oplus W') = dim(W) + dim(W')$. This result can be generalized to the case of an arbitrary pair of subspaces of a vector space.

(3.15) PROPOSITION. **(Grassman's Theorem)** *If W and W' are subspaces of a vector space V over a field F then $dim(W + W') = dim(W) + dim(W') - dim(W \cap W')$.*

PROOF. Let $U = W \cap W'$. This is a subspace both of W and of W' and so, in particular, U has a complement U' in W and a complement U'' in W'. Clearly,

$W + W' = U + U' + U''$. We claim that, in fact, $W + W' = U \oplus U' \oplus U''$. Indeed, assume that $u + u' + u'' = 0_V$, where $u \in U$, $u' \in U'$, and $u'' \in U''$. Then $u' = -u'' - u \in W' \cap W = U$. But $u' \in U'$ and the intersection of U' and U is $\{0_V\}$. Therefore $u' = 0_V$ and so $u = -u''$. But then this is in $U \cap U'' = \{0_V\}$, which implies that $u = u'' = 0_V$. We have thus shown that the set $\{U, U', U''\}$ is independent and hence have established our claim. Moreover, in this case we have

$$\begin{aligned}
dim(W + W') &= dim(U \oplus U' \oplus U'') \\
&= dim(U) + dim(U') + dim(U'') \\
&= dim(W) + dim(U'') \\
&= dim(W) + dim(U'') + dim(U) - dim(U) \\
&= dim(W) + dim(W') - dim(U) \\
&= dim(W) + dim(W') - dim(W \cap W')
\end{aligned}$$

as desired $\square$

EXAMPLE

Consider the subspaces

$$W_1 = \mathbb{R}\{[1, 0, 2], [1, 2, 2]\}$$

and

$$W_2 = \mathbb{R}\{[1, 1, 0], [0, 1, 1]\}$$

of $\mathbb{R}^3$. Each of these subspaces has dimension 2 over $\mathbb{R}$ and so $2 \leq dim(W_1 + W_2) \leq 3$. By Proposition 3.14 we see that this implies that $1 \leq dim(W_1 \cap W_2) \leq 2$. In order to ascertain the exact dimension of $W_1 \cap W_2$ we have to find a basis for it. If $v \in W_1 \cap W_2$ then there exist real numbers a, b, c, d satisfying

$$v = a[1, 0, 2] + b[1, 2, 2] = c[1, 1, 0] + d[0, 1, 1]$$

and so $a + b = c$, $2b = c + d$, and $2a + 2b = d$. This can happen only when $b = -3a$, $c = -2a$, and $d = -4a$ and so we see that v must be of the form

$$a[-2, -6, 4] = (-2a)[1, 1, 0] + (-4a)[0, 1, 1].$$

From this we conclude that $\{[-2, -6, -4]\}$ is a basis for $W_1 \cap W_2$, and so the dimension of $W_1 \cap W_2$ is 1.

Problems

1. Let V be a vector space over a field F. Let v_1, v_2, v_3 be elements of V and let c_1, c_2, c_3 be scalars in F. Show that the set of vectors

$$\{c_2 v_3 - c_3 v_2, c_1 v_2 - c_2 v_1, c_3 v_1 - c_1 v_3\}$$

is linearly dependent.

2. Find rational numbers a and b such that the subset $\{[2, a - b, 1], [a, b, 3]\}$ of $\mathbb{Q}^3$ is linearly dependent.

3. Is the subset $\{[1+i, 3+8i, 5+7i], [1-i, 5, 2+i], [1+i, 3+2i, 4-i]\}$ of $V = \mathbb{C}^3$ linearly dependent over $\mathbb{C}$? Is it linearly dependent when we consider V as a vector space over $\mathbb{R}$?

4. For each nonnegative integer n let $f_n : \mathbb{R} \to \mathbb{R}$ be the function defined by $f_n : t \mapsto \sin^n(t)$. Is the subset $\{f_n \mid n \geq 0\}$ of $\mathbb{R}^\mathbb{R}$ linearly independent?

5. Let V be the vector space over $\mathbb{R}$ consisting of all continuous functions from the interval $[-1, 1]$ to $\mathbb{R}$. Let $f, g \in V$ be the functions defined by $f : x \mapsto x^2$ and $g : x \mapsto x|x|$. Is the set $\{f, g\}$ linearly independent over $\mathbb{R}$?

6. Let V be a vector space over the field $F = \mathbb{Z}/(5)$ and let v_1, v_2, v_3 be vectors in V. Is the set of vectors $\{v_1 + v_2, v_1 - v_2 + v_3, 2v_2 + v_3, v_2 + v_3\}$ linearly independent?

7. Let F be a field of characteristic not equal to 2 and let V be a vector space over F containing a linearly-independent set of vectors $\{v_1, v_2, v_3\}$. Show that the set $\{v_1 + v_3, v_2 + v_3, v_1 + v_2\}$ is also linearly independent.

8. Let $F = \mathbb{Z}/(3)$. Is the subset $\{[1, 1, 2, 2], [1, 2, 1, 2], [1, 1, 1, 2], [0, 2, 2, 0]\}$ of F^4 linearly independent?

9. Let $t \leq n$ be positive integers and for all $1 \leq i \leq t$ let $v_i = [a_{i1}, \ldots, a_{in}]$ be a vector in $\mathbb{R}^n$ satisfying the condition $2|a_{jj}| > \sum_{i=1}^t |a_{ij}|$ for all $1 \leq j \leq n$. Show that the set of vectors $\{v_1, \ldots, v_t\}$ is linearly independent.

10. Let $F = \mathbb{Z}/(2)$ and let A be a nonempty set. If $B \neq C$ are nonempty subsets of A, show that $\{\chi_B, \chi_C\}$ is a linearly-independent subset of F^A.

11. Let A be a subset of $\mathbb{R}$ having at least three elements. Let f_1, f_2, and f_3 be the elements of $\mathbb{R}^A$ defined by $f_1 : t \mapsto 2^{t-1}$, $f_2 : t \mapsto t2^{t-1}$, and $f_3 : t \mapsto t^2 2^{t-1}$. Is the set $\{f_1, f_2, f_3\}$ linearly independent?

12. Let $F = \mathbb{Z}/(5)$ and let $V = F^F$, which is a vector space over F. Let $f : x \mapsto x^2$ and $g : x \mapsto x^3$ be elements of V. Find another element h of V such that the set $\{f, g, h\}$ is linearly independent.

13. Find a basis for the subspace of $\mathbb{R}^4$ generated by

$$\{[4, 2, 6, -2], [1, -1, 3, -1], [1, 2, 0, 0], [1, 5, -3, 1]\}.$$

14. For each real number a let $f_a \in \mathbb{R}^\mathbb{R}$ be the function defined by

$$f_a(r) = \begin{cases} 1, & \text{when } r = a, \\ 0, & \text{otherwise.} \end{cases}$$

Is $\{f_a \mid a \in \mathbb{R}\}$ a basis for $\mathbb{R}^\mathbb{R}$ over $\mathbb{R}$?

15. Let $F = \mathbb{Z}/(2)$ and let A be a nonempty finite set. Is $\{\chi_{\{a\}} \mid a \in A\}$ a basis for the vector space for F^A over F?

16. Let a, b, c be elements of a field F. Show that

$$\{[1_F, a, b], [0_F, 1_F, c], [0_F, 0_F, 1_F]\}$$

is a basis for F^3 over F.

17. Let $F = \mathbb{Z}/(p)$, where p is a prime integer, and let V be a vector space of dimension n over F. How many distinct bases are there for V over F?

18. Let F be a field and let V be the subspace of $F[X]$ consisting of all polynomials of degree no greater than 9. Is the set $\{1_F, X - 1_F, \ldots, (X - 1_F)^9\}$ a basis for V over F?

19. Let $\{v_1, v_2, v_3\}$ be a given basis for a vector space V over a field F. Is the set $\{v_1 + v_2, v_2 + v_3, v_1 - v_3\}$ also a basis for V over F?

20. Let V be a vector space finite dimensional over $\mathbb{C}$ having a basis $D = \{v_1, \ldots, v_n\}$. Show that $\{v_1, \ldots, v_n, iv_1, \ldots, iv_n\}$ is a basis for V as a vector space over $\mathbb{R}$.

21. Let V be the subspace of $\mathbb{Q}[X]$ consisting of all polynomials having dimension at most 5. Extend the set $\{X^5 + X^4, X^5 - 7X^3, X^5 - 4X^2, X^5 + 3X\}$ to a basis for V over $\mathbb{Q}$.

22. Let $W = \mathbb{R}\{[-1, 1, 1, 1], [1, 2, 1, 0]\}$ and $Y = \mathbb{R}\{[2, -1, 0, 1], [-5, 6, 0]\}$ be subspaces of $\mathbb{R}^4$. Calculate the dimensions of $W \cap Y$ and $W + Y$.

23. Let F be a subfield of a field K satisfying the condition that the dimension of K as a vector space over F is finite and equal to $r > 0$. Let V be a vector space over K having finite dimension $n > 0$. What is the dimension of V as a vector space over F?

24. Let V be a vector space over a field F and let n be a positive integer. A matrix $A = [a_{ij}] \in M_{n \times n}(V)$ is called a *Toeplitz matrix* if and only if the entries along each diagonal parallel to the principal diagonal are equal. Thus, for example, the matrix

$$A = \begin{bmatrix} u & v & w \\ x & u & v \\ y & x & u \end{bmatrix}$$

is a Toeplitz matrix of size 3×3. Show that the set of all Toeplitz matrices of size $n \times n$ is a subspace of $M_{n \times n}(V)$. Find the dimension of this subspace for the case $V = F$.

25. Let V be a vector space of dimension 6 over a field F and let W and Y be distinct subspaces of V of dimension 4. What are the possible dimensions of $W \cap Y$?

26. Let V be a vector space which is not finite dimensional over a field F and let W be a proper subspace of V. Show that there exists an infinite set $\{Y_1, Y_2, \ldots\}$ of subspaces of V satisfying the condition that $\bigcap_{i=1}^{n} Y_i \not\subseteq W$ for each $n \geq 1$ but $\bigcap_{i=1}^{\infty} Y_i \subseteq W$.

27. Let V be a vector space which is not finite dimensional over a field F. Show that there exists a countable set of proper subspaces of V the union of which equals V.

28. Let V be the space of all continuous functions from $\mathbb{R}$ to itself and consider the subspaces Y and W of V defined by

$$Y = \{f \in V \mid f(a) = f(-a) \text{ for all } a \in \mathbb{R}\}$$

and
$$W = \{f \in V \mid -f(a) = f(-a) \text{ for all } a \in \mathbb{R}\}.$$

Show that $V = W \oplus Y$.

29. Let F be a field and let W be the subspace of $F[X]$ consisting of all polynomials of degree no greater than 4. Find a complement for W in $F[X]$.

30. Let A be a nonempty set and let B be a subset of A. Let F be a field and let W be the subspace of F^A consisting of all those functions $f: A \to F$ satisfying $f(b) = 0_F$ for all $b \in B$. Find a complement for W in F^A.

31. Let W a be a subspace of a vector space V over a field F and let W' be a complement of W in V. Use W' to construct a complement for W^2 in V^2.

CHAPTER IV

LINEAR TRANSFORMATIONS

Let V and W be vector spaces over a field F. A function $\alpha: V \to W$ is called a *linear transformation* or *homomorphism* if and only if for all $v, v' \in V$ and all $c \in F$ the following conditions are satisfied:

(1) $\alpha(v + v') = \alpha(v) + \alpha(v')$;
(2) $\alpha(cv) = c\alpha(v)$.

Note that as an immediate consequence of condition (2) we see that $\alpha(0_V) = \alpha(0_F 0_V) = 0_F \alpha(0_V) = 0_W$. By an easy induction argument we see that if $v = \sum_{i=1}^{n} c_i v_i$ in V then $\alpha(v) = \sum_{i=1}^{n} c_i \alpha(v_i)$.

EXAMPLE

Let V be a vector space over a field F. Any scalar $c \in F$ defines a linear transformation $\sigma_c: V \to V$ by $\sigma_c: v \mapsto cv$ for all $v \in V$. In particular, we have the linear transformation σ_0 defined by $\sigma_0 : v \mapsto 0_V$ and the linear transformation σ_1 defined by $\sigma_1: v \mapsto v$.

EXAMPLE

Let F be a field and let $c_1, \ldots c_6$ be elements of F. The function $\alpha: F^2 \to F^3$ defined by

$$\alpha: [a, b] \mapsto [ac_1 + bc_2, ac_3 + bc_4, ac_5 + bc_6]$$

is a linear transformation.

This example can be generalized in an important way. Let k and n be positive integers. Any matrix $[a_{ij}] \in \mathcal{M}_{k \times n}(F)$ defines a function $\alpha: F^n \to F^k$ by

$$[b_1, \ldots, b_n] \mapsto \left[\sum_{j=1}^{n} a_{1j} b_j, \ldots, \sum_{j=1}^{n} a_{kj} b_j \right]$$

and, as we shall see later, every linear transformation from F^n to F^k arises in this manner.

EXAMPLE

Let F be a field of characteristic 0. Then there exist linear transformations α and β from $F[X]$ to itself defined by

$$\alpha: \sum_{i \geq 0} a_i X^i \mapsto \sum_{i > 0} (i 1_F) a_i X^{i-1}$$

and

$$\beta: \sum_{i \geq 0} a_i X^i \mapsto \sum_{i \geq 0} [(i+1)1_F]^{-1} a_i X^{i+1}.$$

These linear transformations are known respectively as *formal differentiation* and *formal integration* of polynomials.

EXAMPLE

Let V and W be vector spaces over a field F and let k, n be positive integers. For all $1 \leq i \leq k$ and all $1 \leq j \leq n$, let $\alpha_{ij}: V \to W$ be a linear transformation. Then there exists a linear transformation $\alpha: \mathcal{M}_{k \times n}(V) \to \mathcal{M}_{k \times n}(W)$ defined by $\alpha: [v_{ij}] = [\alpha_{ij}(v_{ij})]$.

EXAMPLE

Linear transformations are considered to be "nice" from an algebraic point of view, but that does not mean that they are also "nice" from the point of view of analysis. To see this, let D be a Hamel basis for $\mathbb{R}$ as a vector space over $\mathbb{Q}$. Then for each $r \in \mathbb{R}$ there exists a unique finite subset $\{u_1(r), \ldots, u_{n(r)}(r)\}$ of D and a finite set $\{a_1(r), \ldots, a_{n(r)}(r)\}$ of scalars in $\mathbb{Q}$ such that

$$r = \sum_{j=1}^{n(r)} a_j(r) u_j(r).$$

The function from $\mathbb{R}$ to itself defined by $r \mapsto \sum_{j=1}^{n(r)} a_j(r)$ is a linear transformation which is not continuous at any point of $\mathbb{R}$ and is therefore, from the point of view of analysis, very "ugly".

EXAMPLE

Let $\alpha: V \to W$ be a linear transformation between vector spaces over a field F and if V' is a subspace of V then the restriction of α to V' is easily seen to be a linear transformation from V' to W.

EXAMPLE

Let V_1 and V_2 be vector spaces over a field F. The *graph* of an arbitrary function $f: V_1 \rightarrow V_2$ is the subset $gr(f) = \{[v, f(v)] \mid v \in V_1\}$ of the vector space $V = V_1 \times V_2$. It is easy to establish that f is a linear transformation from V_1 to V_2 if and only if $gr(f)$ is a subspace of V. Indeed, assume that f is a linear transformation. If $v, v' \in V$ and $c \in F$ then

$$[v, f(v)] + [v', f(v')] = [v + v', f(v) + f(v')] = [v + v', f(v + v')] \in gr(f)$$

and

$$c[v, f(v)] = [cv, cf(v)] = [cv, f(cv)] \in gr(f)$$

so $gr(f)$ is a subspace of V. Conversely, assume that $gr(f)$ is a subspace of V. If $v, v' \in V$ then $[v + v', f(v) + f(v')] = [v, f(v)] + [v', f(v')] \in gr(f)$ and so we must have $f(v) + f(v') = f(v + v')$. Similarly, if $v \in V$ and $a \in F$ then $[cv, cf(v)] \in gr(f)$ and so we must have $cf(v) = f(cv)$. This proves that f is a linear transformation.

Let V and W be vector spaces over a field F. If α and β are linear transformations from V to W then, in particular, they are elements of the set W^V of all functions from V to W and so we can define the function $\alpha + \beta$ from V to W by $v \mapsto \alpha(v) + \beta(v)$. If $v, v' \in V$ and if $c \in F$ then

$$
\begin{aligned}
(\alpha + \beta)(v + v') &= \alpha(v + v') + \beta(v + v') \\
&= \alpha(v) + \alpha(v') + \beta(v) + \beta(v') \\
&= \alpha(v) + \beta(v) + \alpha(v') + \beta(v') \\
&= (\alpha + \beta)(v) + (\alpha + \beta)(v')
\end{aligned}
$$

and

$$
\begin{aligned}
(\alpha + \beta)(cv) &= \alpha(cv) + \beta(cv) \\
&= c\alpha(v) + c\beta(v) \\
&= c[\alpha(v) + \beta(v)] \\
&= c(\alpha + \beta)(v)
\end{aligned}
$$

and so we see that $\alpha + \beta$ is again a linear transformation from V to W. Similarly, if $d \in F$ then the function $d\alpha: V \rightarrow W$ is defined by $v \mapsto d\alpha(v)$. Again, if $v, v' \in V$ and $c \in F$ we see that

$$
\begin{aligned}
[d\alpha](v + v') &= d[\alpha(v + v')] \\
&= d[\alpha(v) + \alpha(v')] \\
&= d\alpha(v) + d\alpha(v') \\
&= [d\alpha](v) + [d\alpha](v')
\end{aligned}
$$

and

$$[d\alpha](cv) = d[\alpha(cv)]$$
$$= dc\alpha(v)$$
$$= c[d\alpha(v)]$$
$$= c[d\alpha](v)$$

and so $d\alpha$ is again a linear transformation from V to W. Thus we conclude that the set of all linear transformations from V to W is a subspace of the vector space W^V over F. We denote this subspace by $Hom(V, W)$.

The next proposition will provide us with an important tool for building linear transformations.

(4.1) PROPOSITION. *Let V and W be vector spaces over a field F. Let D be a basis for V over F and let $f: D \to W$ be an arbitrary function. Then there exists a unique linear transformation $\alpha: V \to W$ satisfying the condition $\alpha(u) = f(u)$ for all $u \in D$*

PROOF. Since D is a basis for V over F, we know that every vector v in V can be represented uniquely as a linear combination $v = \sum_{i=1}^{n} a_i u_i$ of elements of D. Define the function $\alpha: V \to W$ by $\alpha: v \mapsto \sum_{i=1}^{n} a_i f(u_i)$. This function is well-defined by the uniqueness of the representation of v, and is easily verified to be a linear transformation. It also satisfies the condition that $\alpha(u) = f(u)$ for all $u \in D$. Conversely, if $\beta: V \to W$ is a linear transformation satisfying this condition then $\beta(\sum_{i=1}^{n} a_i u_i) = \sum_{i=1}^{n} a_i \beta(u_i) = \sum_{i=1}^{n} a_i f(u_i) = \alpha(\sum_{i=1}^{n} a_i u_i)$ and so $\beta = \alpha$, establishing uniqueness. $\square$

(4.2) PROPOSITION. *Let V, W, and Y be vector spaces over a field F and let $\alpha: V \to W$ and $\beta: W \to Y$ be linear transformations. Then the composed function $\beta\alpha: V \to Y$ is also a linear transformation.*

PROOF. If $v, v' \in V$ and if $c \in F$ then

$$\beta\alpha(v + v') = \beta(\alpha(v + v'))$$
$$= \beta(\alpha(v) + \alpha(v'))$$
$$= \beta(\alpha(v)) + \beta(\alpha(v'))$$
$$= \beta\alpha(v) + \beta\alpha(v')$$

and

$$\beta\alpha(cv) = \beta(\alpha(cv))$$
$$= \beta(c\alpha(v))$$
$$= c\beta(\alpha(v))$$
$$= c(\beta\alpha)(v)$$

and this proves the proposition. $\square$

EXAMPLE

It is often very important and enlightening to be able to decompose a given linear transformation into a product of linear transformations of prescribed types. To illustrate this, let $a < b$ be real numbers and let V be the subspace of $\mathbb{R}^{[a,b]}$ consisting of all differentiable functions. Let $\delta: V \to \mathbb{R}^{[a,b]}$ be the linear transformation which assigns to each function $f \in V$ its derivative f', and, for each $a < c < b$, let $\epsilon_c: \mathbb{R}^{[a,b]} \to \mathbb{R}$ be the linear transformation given by $g \mapsto g(c)$. Then the Mean Value Theorem of calculus says that the linear transformation $\beta: V \to \mathbb{R}$ defined by

$$\beta: f \mapsto \frac{f(b) - f(a)}{b - a}$$

is of the form $\epsilon_c \delta$ for some $c \in (a, b)$.

Let V and W be vector spaces over a field F and let $\alpha: V \to W$ be a linear transformation. For each $w \in W$, we denote the subset $\{v \in V \mid \alpha(v) = w\}$ of V by $Inv(\alpha, w)$. The set $Inv(\alpha, 0_W)$ is called the *kernel* of α, and is denoted by $ker(\alpha)$.

EXAMPLE

Let W be the subspace of $\mathbb{R}^{\mathbb{R}}$ consisting of all everywhere-differentiable functions on $\mathbb{R}$ and let $\delta: W \to \mathbb{R}^{\mathbb{R}}$ be the linear transformation which assigns to each $f \in W$ its derivative f'. The elements of $ker(\delta)$ are the solutions of the *homogeneous differential equation* $f' = 0$. If $0_W \neq g \in W$ the elements of $Inv(\delta, g)$ are the solutions of the *nonhomogeneous differential equation* $f' = g$.

If n is an integer greater than 1 then the linear transformation δ also defines a linear transformation $\delta^n: W^n \to (\mathbb{R}^{\mathbb{R}})^n$ by $\delta^n: [f_1, \ldots, f_n] \mapsto [\delta(f_1), \ldots, \delta(f_n)]$. An element $[f_1, \ldots, f_n]$ of $ker(\delta^n)$ is a solution of the *homogeneous system of differential equations* $f'_i = 0$ and if $g_1, \ldots, g_n$ are elements of W not all equal to 0_W then an element of $Inv(\delta, [g_1, \ldots, g_n])$ is a solution of the *nonhomogeneous system of differential equations* $f'_i = g_i$.

(4.3) PROPOSITION. *Let V and W be vector spaces over a field F and let $\alpha: V \to W$ be a linear transformation. Then*

(1) *$ker(\alpha)$ is a subspace of V; and*
(2) *the function α is monic if and only if $ker(\alpha) = \{0_V\}$.*

PROOF. (1) Let v and v' be elements of $ker(\alpha)$ and let c be a scalar in F. Then $\alpha(v + v') = \alpha(v) + \alpha(v') = 0_W + 0_W = 0_W$ and so $v + v' \in ker(\alpha)$. Also, $\alpha(cv) = c\alpha(v) = c0_W = 0_W$ and so $cv \in ker(\alpha)$. This proves that $ker(\alpha)$ is a subspace of V.

(2) We know that $0_V \in ker(\alpha)$ and so if α is monic we must have $ker(\alpha) = \{0_V\}$. Conversely, assume that $ker(\alpha) = \{0_V\}$ and let $v, v' \in V$ satisfy $\alpha(v) = \alpha(v')$. Then $\alpha(v - v') = \alpha(v) - \alpha(v') = 0_W$ and so $v - v' \in ker(\alpha) = \{0_V\}$. This implies that $v = v'$ and hence we have shown that α is monic. $\square$

A monic linear transformation is called a *monomorphism* between vector spaces.

Let V and W be vector spaces over a field F and let $\alpha: V \to W$ be a linear transformation. The *image* of α is the subset $im(\alpha) = \{\alpha(v) \mid v \in V\}$ of W. This set is always nonempty since we always have $\alpha(0_V) \in im(\alpha)$.

(4.4) PROPOSITION. *Let V and W be vector spaces over a field F and let $\alpha: V \to W$ be a linear transformation. Then*

   (1) *$im(\alpha)$ is a subspace of W; and*

   (2) *the function α is epic if and only if $im(\alpha) = W$.*

PROOF. (1) If $\alpha(v)$ and $\alpha(v')$ are elements of $im(\alpha)$ and if $c \in F$ is a scalar then $\alpha(v) + \alpha(v') = \alpha(v + v') \in im(\alpha)$ and $c\alpha(v) = \alpha(cv) \in im(\alpha)$, thus proving that $im(\alpha)$ is a subspace of W.

(2) This follows immediately from the definition of an epic function. $\square$

An epic linear transformation is called an *epimorphism* between vector spaces. A linear transformation which is both a monomorphism and an epimorphism (i.e. which is bijective) is called an *isomorphism* of vector spaces.

EXAMPLE

Let V be a vector space over a field F. Any linear transformation from V to F (considered as a vector space over itself) other than the 0-map is an epimorphism. We will return to consider linear transformations of this type in Chapter 12.

EXAMPLE

Let $\mathcal{M}_{k \times n}(F)$ be the vector space of all $k \times n$ matrices over a field F. If $A = [a_{ij}]$ is an element of this space then the *transpose* of A is the matrix $A^T = [a_{ji}] \in \mathcal{M}_{n \times k}(F)$. The function $A \mapsto A^T$ is an isomorphism from $\mathcal{M}_{k \times n}(F)$ to $\mathcal{M}_{n \times k}(F)$.

EXAMPLE

We have already noted after Proposition 3.4 that if D is a basis for a vector space V over a field F then there exists a function $\theta: F^{(D)} \to V$ which is monic and epic. It is easy to verify that this function is in fact a linear transformation and so is an isomorphism.

EXAMPLE

Let V and W be vector spaces over a field F and let D be a basis for V over F. Then there exists a function $\phi: Hom(V, W) \to W^D$ defined by restriction: if $u \in D$ and if $\alpha \in Hom(V, W)$ then $\phi(\alpha)(u) = \alpha(u)$. By the definitions of addition and scalar multiplication in both vector spaces it is clear that this is a linear transformation. Moreover, by Proposition 4.1 we know that every function $f \in W^D$ is of the form $\phi(\alpha)$ for one and only one element α of $Hom(V, W)$. Therefore ϕ is an isomorphism.

EXAMPLE

Let $\mathbb{N}$ be the set of all nonnegative integers and let $V = F^{(\mathbb{N})}$. Then the function $\alpha: V \to F[X]$ defined by $\alpha: f \mapsto \sum_{i \geq 0} f(i)X^i$ is an isomorphism of vector spaces over F.

If V and W are vector spaces over a field F and if $\alpha: V \to W$ is a linear transformation then $Inv(\alpha, w)$ is not a subspace of V if $0_W \neq w \in W$. Indeed, if α is not epic, this set may in fact be empty should it be the case that $w \notin im(\alpha)$. Nonetheless, it is close to being a subspace in the following sense.

(4.5) PROPOSITION. *Let* $\alpha: V \to W$ *be a linear transformation between vector spaces over a field* F *and let* $w \in im(\alpha)$. *If* $v_0 \in V$ *is a particular vector satisfying* $\alpha(v_0) = w$ *then* $Inv(\alpha, w) = \{v + v_0 \mid v \in ker(\alpha)\}$.

PROOF. If $v \in ker(\alpha)$ then $\alpha(v + v_0) = \alpha(v) + \alpha(v_0) = 0_W + w = w$ so $v + v_0 \in Inv(\alpha, w)$. Conversely, let $v' \in Inv(\alpha, w)$ and let $v = v' - v_0$. Then $v' = v + v_0$, where $\alpha(v) = \alpha(v') - \alpha(v_0) = w - w = 0_W$ so $v \in ker(\alpha)$. $\square$

EXAMPLE

Let W be the subspace of $\mathbb{R}^{[0,1]}$ consisting of all everywhere differentiable functions and let $\delta: W \to \mathbb{R}^{[0,1]}$ be the linear transformation which assigns to each function $f \in W$ its derivative f'. Then $ker(\delta)$ consists of all the constant functions in $\mathbb{R}^{[0,1]}$. If $g \in im(\delta)$ then $g = \delta(f)$, where f is the function defined by $f(x) = \int_0^x g(t)dt$ for all $x \in [0,1]$. Therefore $Inv(\delta, g)$ consists of all functions of the form $\int_0^x g(t)dt + c$, where c is a constant function.

(4.6) PROPOSITION. *A linear transformation* $\alpha: V \to W$ *is an isomorphism between vector spaces over a field* F *if and only if there exists a linear transformation* $\beta: W \to V$ *satisfying* $\beta\alpha(v) = v$ *for all* $v \in V$ *and* $\alpha\beta(w) = w$ *for all* $w \in W$.

PROOF. Let α be an isomorphism and define β by the condition that $\beta(w) = v$ if and only if $\alpha(v) = w$. This function is well-defined since every element of W is $\alpha(v)$ for precisely one element $v \in V$. It is straightforward to check that β is in fact a linear transformation satisfying the desired conditions. Conversely, assume that such a linear transformation β exists. Then α is an epimorphism since $w = \alpha(\beta(w)) \in im(\alpha)$ for all $w \in W$. Moreover, α is a monomorphism since $\alpha(v) = 0_W$ implies that $v = \beta\alpha(v) = \beta(0_W) = 0_V$. Thus α is an isomorphism. $\square$

If $\alpha: V \to W$ is an isomorphism between vector spaces over a field F then the isomorphism β defined in Proposition 4.6 is denoted by α^{-1} and is called the *inverse* of α.

If there exists an isomorphism between vector spaces V and W over a field F, we write $V \cong W$. We immediately see that if V, W, and Y are vector spaces over F then:

(1) $V \cong V$;

(2) If $V \cong W$ then $W \cong V$;

(3) If $V \cong W$ and $W \cong Y$ then $V \cong Y$.

We also note that if D is a basis for a vector space V over a field F and if $\alpha\colon V \to W$ is an isomorphism then $D' = \{\alpha(u) \mid u \in D\}$ is a basis for W over F. As an immediate consequence of this observation, we see that $V \cong W$ implies $dim(V) = dim(W)$. The next result states that the converse of this is also true in the case of finite-dimensional vector spaces.

(4.7) PROPOSITION. *If V and W are vector spaces both of finite dimension n over field F then $V \cong W$.*

PROOF. Let $D = \{v_1, \ldots, v_n\}$ be a basis for V over F and let $D' = \{w_1, \ldots, w_n\}$ be a basis for W over F. Let $f \in W^D$ be the function defined by $f\colon v_i \mapsto w_i$ for all $1 \leq i \leq n$. By Proposition 4.1 we know that there exists a linear transformation $\alpha\colon V \to W$ satisfying $\alpha(v_i) = w_i$ for all $1 \leq i \leq n$. Then $im(\alpha)$ is a subspace of W containing D' and so by Proposition 2.5 we have $W = FD' \subseteq im(\alpha) \subseteq W$, which implies that $im(\alpha) = W$. Therefore α is an epimorphism. If $v = \sum_{i=1}^{n} c_i v_i$ is an element of $ker(\alpha)$ then

$$0_W = \alpha(v) = \alpha\left(\sum_{i=1}^{n} c_i v_i\right) = \sum_{i=1}^{n} c_i \alpha(v_i) = \sum_{i=1}^{n} c_i w_i$$

and so $c_i = 0_F$ for all i since the set $\{w_1, \ldots, w_n\}$ is linearly independent. Thus $v = 0_V$ and so we have shown that $ker(\alpha) = \{0_V\}$, which implies that α is also a monomorphism and hence an isomorphism. $\square$

As a consequence we have the following corollary.

COROLLARY. *If V is a vector space having finite dimension n over a field F then $V \cong F^n$.*

We can also generalize Proposition 4.7 as follows.

(4.8) PROPOSITION. *Let V and W be vector spaces finitely generated over a field F. Then*

(1) *There exists a monomorphism from V to W if and only if $dim(V) \leq dim(W)$;*

(2) *There exists a epimorphism from V to W if and only if $dim(V) \geq dim(W)$.*

PROOF. (1) If there exists a monomorphism $\alpha\colon V \to W$ then $V \cong im(\alpha)$ and so, by Proposition 3.10, $dim(W) \geq dim(im(\alpha)) = dim(V)$. Conversely, if $dim(W) \geq dim(V)$ then there exists a basis $D = \{v_1, \ldots v_n\}$ for V over F and a basis $\{w_1, \ldots, w_t\}$ for W over F, with $n \leq t$. By Proposition 4.1, the function $f \in W^D$ given by $f(v_i) = w_i$ for all $1 \leq i \leq n$ defines a linear transformation $\alpha\colon V \to W$ satisfying $\alpha(v_i) = w_i$ for all $1 \leq i \leq n$. As in the proof of Proposition 4.7, this is in fact a monomorphism.

(2) If there exists an epimorphism $\alpha\colon V \to W$ and if $\{v_1, \ldots, v_n\}$ is a basis for V over F then $\{\alpha(v_i) \mid 1 \leq i \leq n\}$ is a generating set for $im(\alpha) = W$ and so contains a basis for W. Hence $dim(W) \leq n = dim(V)$. Conversely, assume that

$dim(W) \leq dim(V)$. Choose a basis $\{w_1, \ldots, w_t\}$ for W and for each $1 \leq i \leq t$ let $v_i \in V$ be a vector satisfying $\alpha(v_i) = w_i$. If $\sum_{i=1}^{t} c_i v_i = 0_V$ then

$$0_W = \alpha\left(\sum_{i=1}^{t} c_i v_i\right) = \sum_{i=1}^{t} c_i \alpha(v_i) = \sum_{i=1}^{t} c_i w_i$$

and so $c_i = 0_F$ for each $1 \leq i \leq t$. Therefore $\{v_1, \ldots, v_t\}$ is a linearly-independent subset of V and hence contained in a basis of V over F. This implies that $dim(W) = t \leq dim(V)$. $\square$

The following proposition will play an important role in later chapters.

(4.9) PROPOSITION. *Let V and W be vector spaces finitely generated over a field F and let $\alpha: V \to W$ be a linear transformation. Then*

$$dim(V) = dim(im(\alpha)) + dim(ker(\alpha)).$$

PROOF. Let V'' be a complement of $ker(\alpha)$ in V. Then

$$dim(V) = dim(ker(\alpha) \oplus V'') = dim(ker(\alpha)) + dim(V'')$$

and so it suffices to show that $V'' \cong im(\alpha)$. Indeed, the restriction α'' of α to V'' is just a linear transformation from V'' to $im(\alpha)$. If $v'' \in ker(\alpha'')$ then $v'' \in V'' \cap ker(\alpha) = \{0_V\}$ and so the linear transformation α'' is a monomorphism. If $w \in im(\alpha)$ then there exists a vector $v \in V$ satisfying $\alpha(v) = w$. The vector v can be written as $v' + v''$, where $v' \in ker(\alpha)$ and $v'' \in V''$. But then $w = \alpha(v) = \alpha(v') + \alpha(v'') = \alpha(v') + \alpha(v'') = \alpha(v'') = \alpha''(v'')$ and so $w \in im(\alpha'')$. Therefore α'' is the desired isomorphism between V'' and $im(\alpha)$. $\square$

If $\alpha: V \to W$ is a linear transformation between vector spaces finitely generated over a field F then the number $dim(im(\alpha))$ is called the *rank* of α and the number $dim(ker(\alpha))$ is called the *nullity* of α.

(4.10) PROPOSITION. (**Sylvester's Theorem**) *Let V, W, and Y be vector spaces finitely generated over a field F and let $\alpha: V \to W$ and $\beta: W \to Y$ be linear transformations. Then:*
(1) $dim(im(\beta\alpha)) \leq min\{dim(im(\alpha)), dim(im(\beta))\}$;
(2) $dim(ker(\beta\alpha)) \leq dim(ker(\alpha)) + dim(ker(\beta))$;
(3) $dim(im(\alpha)) + dim(im(\beta)) - dim(W) \leq dim(im(\beta\alpha))$.

PROOF. (1) Clearly $im(\beta\alpha)$ is a subspace of $im(\beta)$ and so $dim(im(\beta\alpha)) \leq dim(im(\beta))$. Moreover, the space $im(\beta\alpha)$ is just the image of the restriction of β to $im(\alpha)$ and so $dim(im(\beta\alpha)) \leq dim(im(\alpha))$, proving (1).
(2) Let β' be the restriction of β to $im(\alpha)$. Then $ker(\beta')$ is a subspace of $ker(\beta)$. By Proposition 4.9 we see that

$$\begin{aligned}
dim(ker(\beta\alpha)) &= dim(V) - dim(im(\beta\alpha)) \\
&= [dim(V) - dim(im(\alpha))] + [dim(im(\alpha)) - dim(im(\beta\alpha))] \\
&= dim(ker(\alpha)) + dim(ker(\beta')) \\
&\leq dim(ker(\alpha)) + dim(ker(\beta))
\end{aligned}$$

and so we have (2).

(3) By Proposition 4.9 we have

$$dim(im(\alpha)) = dim(V) - dim(ker(\alpha)),$$
$$dim(im(\beta)) = dim(W) - dim(ker(\beta)),$$
$$dim(im(\beta\alpha)) = dim(V) - dim(ker(\beta\alpha)),$$

and so (3) is true if and only if $dim(V) - dim(ker(\beta\alpha)) \geq dim(V) - dim(ker(\alpha)) + dim(W) - dim(ker(\beta)) - dim(W)$, which follows directly from (2). $\square$

(4.11) PROPOSITION. *Let V, W, and Y be vector spaces over a field F. Let $\alpha: V \to W$ be a linear transformation and let $\beta: W \to Y$ be a monomorphism. For each $w \in W$ then $Inv(\alpha, w) = Inv(\beta\alpha, \beta(w))$.*

PROOF. If $v \in Inv(\alpha, w)$ then $\alpha(v) = w$ and so $\beta\alpha(v) = \beta(w)$, showing that $v \in Inv(\beta\alpha, \beta(w))$. Conversely, if $v \in Inv(\beta\alpha, \beta(w))$ then $\beta\alpha(v) = \beta(w)$ and, since β is monic, this implies that $\alpha(v) = w$ and so $v \in Inv(\alpha, w)$. $\square$

Problems

1. Let $\alpha: \mathbb{R}^3 \to \mathbb{R}^3$ be a linear transformation over $\mathbb{R}$ satisfying $\alpha([1, 0, 1]) = [-1, 3, 4]$, $\alpha([1, -1, 1]) = [0, 1, 0]$, and $\alpha([1, 2, -1]) = [3, 1, 4]$. What is $\alpha([1, 0, 0])$?

2. Let $\alpha: \mathbb{R}^3 \to \mathbb{R}^3$ be a linear transformation over $\mathbb{R}$ satisfying $\alpha([1, 1, 0]) = [1, 2, -1]$, $\alpha([1, 0, -1]) = [0, 1, 1]$, and $\alpha([0, -1, 1]) = [3, 3, 3]$. Find a vector $v \in \mathbb{R}^3$ satisfying $\alpha(v) = [1, 0, 0]$.

3. Does there exist a real number d such that the function $\mathbb{R}^2 \to \mathbb{R}^2$ defined by $[a, b] \mapsto [a + b + d^2 + 1, a]$ is a linear transformation over $\mathbb{R}$?

4. Does there exist a real number d such that the function $\mathbb{R}^2 \to \mathbb{R}^2$ defined by $[a, b] \mapsto [5da - db, 8d^2 - 8d - 6]$ is a linear transformation over $\mathbb{R}$?

5. Let V be a vector space over a field F and let W, W' be subspaces of F. Let Y be a vector space over F and assume that $\alpha: W \to Y$ and $\beta: W' \to Y$ are linear transformations satisfying $\alpha(v) = \beta(v)$ for all $v \in W \cap W'$. Find a linear transformation $\theta: W + W' \to Y$ the restriction of which to W equals α and the restriction of which to W' equals β.

6. Let F be a field and let α be the function from $\mathcal{M}_{k \times n}(F) \to F^{k+n}$ defined by $\alpha: A \mapsto [b_1, \ldots, b_k, c_1, \ldots, c_n]$ where

   (i) For each $1 \leq i \leq k$, the scalar b_i is the sum of the entries in the ith row of A; and

   (ii) For each $1 \leq j \leq n$, the scalar c_j is the sum of the entries in the jth column of A.

Is α a linear transformation?

7. Let $\alpha: \mathbb{R} \to \mathbb{R}$ be a continuous function satisfying $\alpha(a + b) = \alpha(a) + \alpha(b)$ for all $a, b \in \mathbb{R}$. Show that α is a linear transformation.

8. Let $F = \mathbb{Z}/(3)$ and let $g: F \to F$ be the function defined by

$$g: a \mapsto \begin{cases} 0, & \text{if } a = 0; \\ 2, & \text{if } a = 1; \\ 1, & \text{if } a = 2. \end{cases}$$

Let n be a positive integer and let $\alpha: F^n \to F^n$ be the function defined by

$$\alpha: [a_1, \ldots, a_n] \mapsto [g(a_1), \ldots, g(a_n)].$$

Is α a linear transformation?

9. Let F be a field and let V be the subspace of $F[X]$ composed of all polynomials of degree at most 2. Let $\alpha: V \to F[X]$ be a linear transformation satisfying $\alpha(1_F) = X$, $\alpha(1_F + X) = X^3 + X^5$, and $\alpha(1_F + X + X^2) = 1_F - X^2 + X^4$. Calculate $\alpha(p)$ for an arbitrary element $p(X)$ of V.

10. Does there exist a linear transformation $\alpha: \mathbb{Q}^4 \to \mathbb{Q}[X]$ satisfying the conditions $\alpha([1, 3, 0, -1]) = 2$, $\alpha([-1, 1, 1, 1]) = X$, and $\alpha([-1, 5, 2, 1]) = X + 1$?

11. Let $F = \mathbb{Z}/(3)$ and let $\alpha: F^4 \to F^4$ be the function defined by $[a, b, c, d] \mapsto [a^3, b^3, d, c]$. Is α a linear transformation of vector spaces over F?

12. Let F be a field and let V be the set of infinite series $[a_1, a_2, \ldots]$ of elements of F. This is a vector space over F. Given such a series and given $n \geq 1$, let $s_n = \sum_{i=1}^{n} a_i$ be its nth partial sum. Is the function $\alpha: V \to V$ defined by $[a_1, a_2, \ldots] \mapsto [s_1, s_2, \ldots]$ a linear transformation?

13. Let V be a vector space over $\mathbb{Q}$ having a countably-infinite basis $B = \{v_1, v_2, v_3, \ldots\}$. We then know that each $v \in V$ can be uniquely represented in the form $v = \sum_{i=1}^{\infty} a_i(v)v_i$, where only finitely-many of the scalars $a_i(v)$ are nonzero. Define the function $\alpha: V \to \mathbb{Q}$ by $\alpha: v \mapsto \sum_{i=1}^{\infty} i^2 a_i(v)$. Is α a linear transformation?

14. Let V be the subspace of $\mathbb{R}^{\mathbb{R}}$ composed of all functions which are everywhere differentiable. For each $f \in V$ define the function $Df: \mathbb{R} \times \mathbb{R} \to \mathbb{R}$ by $Df: (a, b) \mapsto f'(a)b$. (This function is called the *differential* of f.) Is the function $D: V \to \mathbb{R}^{\mathbb{R} \times \mathbb{R}}$ defined by $f \mapsto Df$ a linear transformation?

15. Let W be the subspace of $\mathbb{R}^{\mathbb{R}}$ consisting of all twice-differentiable functions and let $\alpha: W \to \mathbb{R}^{\mathbb{R}}$ be the linear transformation defined by $\alpha: f \mapsto f''$. If $g \in \mathbb{R}^{\mathbb{R}}$ is the function defined by $g: x \mapsto x + 1$, find $Inv(\alpha, g)$.

16. Let $\alpha: \mathbb{R}^5 \to \mathbb{R}^3$ be the linear transformation defined by

$$[a, b, c, d, e] \mapsto [b + c - 2d + e, a + 2b + 3c - 4d + e, 2a + 2c - 2e].$$

Find $ker(\alpha)$.

17. Let $F = \mathbb{Z}/(3)$ and let α be the linear transformation from the space F^3 to itself defined by $\alpha: [a, b, c] \mapsto [a + b, 2b + c, 0]$. Find $ker(\alpha)$.

18. Let $\alpha: \mathcal{M}_{3\times 3}(\mathbb{R}) \to \mathbb{R}$ be the function defined by

$$\alpha: [a_{ij}] \mapsto \sum_{i=1}^{3} \sum_{j=1}^{3} a_{ij}.$$

Show that α is a linear transformation and find its kernel.

19. Let $F = \mathbb{Z}/(2)$ and let n be a positive integer. Let W be the set of all vectors $[a_1, \ldots, a_n]$ in F^n having an even number of nonzero entries. Show that W is a subspace of F^n by finding a linear transformation $F^n \to Y$ having kernel equal to W.

20. Set $F = \mathbb{Z}/(2)$. Let A and B be nonempty sets and let $\phi: A \to B$ be a given function. Define the function $\alpha: F^B \to F^A$ as follows: If $f \in F^B$ then

$$\alpha(f): a \mapsto \begin{cases} 1, & \text{if } f(\phi(a)) = 1; \\ 0, & \text{otherwise.} \end{cases}$$

Show that α is a linear transformation and find its kernel.

21. Let V be a vector space over a field F and let $\alpha: V^3 \to V$ be the function defined by $\alpha: [v, v', v''] \mapsto v + v' + v''$. Show that α is a linear transformation and find its kernel.

22. Let $F = \mathbb{Z}/(2)$ and let $\alpha: F^7 \to F^3$ be the linear transformation defined by

$$\alpha: [a_1, \ldots, a_7] \mapsto [a_4 + a_5 + a_6 + a_7, a_2 + a_3 + a_6 + a_7, a_1 + a_3 + a_5 + a_7].$$

Show that if $[0,0,0,0,0,0,0] \neq v \in ker(\alpha)$ then v has at least three components equal to 1.

23. Let V be a vector space over a field F having subspaces W and W'. Let $Y = \{[w, w'] \mid w \in W, w' \in W'\}$, which is a subspace of V^2. Let $\alpha: Y \to V$ be the linear transformation $\alpha: [w, w'] \mapsto w + w'$. Find the kernel of α and show that it is isomorphic to $W \cap W'$.

24. Let W be the subspace of $\mathbb{R}^{\mathbb{R}}$ consisting of all differentiable functions and let $\alpha: W \to \mathbb{R}^{\mathbb{R}}$ be the linear transformation defined by $\alpha(f): x \mapsto f'(x) + cos(x)f(x)$ for all $x \in \mathbb{R}$. Find $ker(\alpha)$.

25. Calculate the kernel of the linear transformation $\alpha: \mathbb{Q}[X] \to \mathbb{Q}[\sqrt{3}]$ (between vector spaces over $\mathbb{Q}$) defined by $\alpha: p(X) \mapsto p(\sqrt{3})$.

26. Let V and W be vector spaces over a field F and let $\alpha, \beta: V \to W$ be monomorphisms. Is $\alpha + \beta$ necessarily a monomorphism?

27. Let $F = \mathbb{Z}/(7)$. How many distinct monomorphisms are there from F^2 to F^4?

28. Let W be the subspace of $\mathbb{R}^6$ composed of all vectors $[a_1, a_2, a_3, a_4, a_5, a_6]$ satisfying $\sum_{i=1}^{6} a_i = 0$. Does there exist a monomorphism from W to $\mathbb{R}^4$?

29. Let n be a positive integer and let V be the subspace of $\mathbb{R}[X]$ consisting of all polynomials having degree less than or equal to n. Let $\alpha: V \to V$ be the linear transformation defined by $\alpha: p(X) \mapsto p(X + 1) - p(X)$. Find $ker(\alpha)$ and $im(\alpha)$.

30. Let $\alpha: \mathbb{R}^3 \to \mathbb{R}^3$ be the linear transformation defined by

$$[a, b, c] \mapsto [a + b + c, -a - c, b].$$

Find $ker(\alpha)$ and $im(\alpha)$.

31. Find a linear transformation $\alpha: \mathbb{Q}^3 \to \mathbb{Q}^4$ satisfying the condition that $im(\alpha) = \mathbb{Q}\{[\frac{2}{3}, -1, 3, 0], [2, 1, 1, -4]\}$.

32. Let V be a vector space over a field F and let $\mathbb{N}$ be the set of nonnegative integers. Let $W = \{f \in V^{\mathbb{N}} \mid f(i) = 0_V \text{ when } i \text{ is odd}\}$ and let Let $W' = \{f \in V^{\mathbb{N}} \mid f(i) = 0_V \text{ when } i \text{ is even}\}$. Find a linear transformation α from $V^{\mathbb{N}}$ to itself having kernel equal to W and image equal to W'.

33. Let V and W be vector spaces over a field F, where $dim(V) > 0$. Show that $W = \sum\{im(\alpha) \mid \alpha \in Hom(V, W)\}$.

34. Let V, W, and Y be vector spaces finitely generated over a field F and let $\alpha: V \to W$ be a linear transformation. Show that the set of all linear transformations $\beta: W \to Y$ satisfying the condition that $im(\alpha) \subseteq ker(\beta)$ is a subspace of $Hom(W, Y)$ and calculate its dimension.

35. Let V be a vector space over $\mathbb{C}$ and let n be a positive integer. Does there exist a linear transformation $\alpha: V \to \mathbb{C}^n$ other than the 0-map satisfying the condition that $im(\alpha) \subseteq \mathbb{R}^n$?

36. Let V and W be vector spaces over a field F and let V' be a proper subspace of V. Decide which are the following sets are subspaces of $Hom(V, W)$:

 (i) $\{\alpha \in Hom(V, W) \mid ker(\alpha) = V'\}$;
 (ii) $\{\alpha \in Hom(V, W) \mid ker(\alpha) \subseteq V'\}$;
 (iii) $\{\alpha \in Hom(V, W) \mid ker(\alpha) \supseteq V'\}$.

37. Let V and W be vector spaces over a field F and let $\alpha, \beta: V \to W$ be epimorphisms. Is $\alpha + \beta$ necessarily an epimorphism?

38. Let W be a subspace of a vector space V over a field F. If $v \in V$ then we will denote the set $\{v + w \mid w \in W\}$ by v/W. The collection of all subsets of V of this form will be denoted by V/W. Define operations of addition and scalar multiplication on V/W as follows:

 (i) $v/W + v'/W = (v + v')/W$ for all $v, v' \in W$; and
 (ii) $a(v/W) = (av)/W$ for all $v \in V$ and all $a \in F$.

Show that these operations are well-define and induce on V/W the structure of a vector space over F. Moreover, show that the function from V to V/W given by $\alpha: v \mapsto v/W$ is an epimorphism of vector spaces over F, the kernel of which is W.

39. Let k be a positive integer and let V be a vector space having finite dimension $2k$ over a field F. Show that there exists an isomorphism $\alpha: V \to V$ satisfying the condition $\alpha^2(v) = -v$ for all $v \in V$.

40. Let V and W be vector spaces over a field F and let $\alpha: V \to W$ be a linear transformation which satisfies the condition that $\alpha\beta\alpha$ is not the 0-function for any linear transformation $\beta: W \to V$ which is not the 0-function. Show that α is an isomorphism.

41. Let F be a field and let $\mathbb{N}$ be the set of nonnegative integers. Let $V = \{f \in F^{\mathbb{N}} \mid f(i) = 0_F$ when i is even$\}$ and let $W = \{f \in F^{\mathbb{N}} \mid f(i) = 0_F$ when i is odd$\}$. Show that $V \cong F^{\mathbb{N}} \cong W$.

42. Let F be a field and let $\alpha: F^3 \to F[X]$ be the linear transformation defined by $[a, b, c] \mapsto (a + b)X + (a + c)X^5$. Find the rank and nullity of α.

43. Let W be a subspace of a vector space V over a field F and let W' be a complement of W in V. If $\alpha \in Hom(W, W')$, show that $Y = \{w + \alpha(w) \mid w \in W\}$ is a subspace of V isomorphic to W.

44. Show that there is no vector space over any field which has precisely 15 elements.

45. Let B be a Hamel basis for $\mathbb{R}$ and let $1 \neq a \in \mathbb{R}$. Show that there exists an element $b \in B$ satisfying $ab \notin B$.

46. Let V and W be vector spaces over a field F, Let $\alpha, \beta \in Hom(V, W)$ satisfy the condition that for each $v \in V$ there exists a scalar $c_v \in F$ (depending on v) such that $\beta(v) = c_v \alpha(v)$. Show that there exists a scalar $c \in F$ such that $\beta = c\alpha$.

CHAPTER V

ENDOMORPHISM RINGS OF VECTOR SPACES

Let V be a vector space over a field F. A linear transformation from V to itself over F is called an *endomorphism* of V. The set of all endomorphisms of V will be denoted by $End(V)$. This set is nonempty since, as we have already noted, it contains the 0-endomorphism $\sigma_0: v \mapsto 0_V$; and the identity endomorphism $\sigma_1: v \mapsto v$. If V is nontrivial these two functions are distinct. In addition to these, we have also seen that every scalar $c \in F$ defines an endomorphism σ_c of V given by $\sigma_c: v \mapsto cv$ for all $v \in V$.

If α and β are endomorphisms of a vector space V over a field F then we have seen that $\alpha + \beta$ and $\alpha\beta$ are also endomorphisms of V. Similarly, if α is an endomorphism of V over F and if $c \in F$ then $c\alpha$ is again an endomorphism of V. Indeed, $c\alpha = \sigma_c\alpha$. Putting all of these together, and deciding once and for all that $\alpha^0 = \sigma_1$ for any $\alpha \in End(V)$, we note that if $p(X) = \sum_{i \geq 0} a_i X^i \in F[X]$ then it makes sense to talk about the endomorphism $p(\alpha) = \sum_{i \geq 0} a_i \alpha^i$ of V. (Recall that there are really only finitely-many nonzero summands here, since only finitely-many of the coefficients a_i are nonzero.)

We also note that σ_0 is a neutral element of $End(V)$ with respect to addition of endomorphisms and σ_1 is a neutral element with respect to composition. Is $End(V)$, together with the operations of addition and composition of functions, a field? Unfortunately, the answer is no. Indeed, we immediately see that at least two conditions in the definition of a field are not satisfied:

(1) First of all, if $\alpha, \beta \in End(V)$ then it is not necessarily true that $\alpha\beta = \beta\alpha$. To see this, consider the following example.

EXAMPLE

Let F be a field and consider the endomorphisms α and β of the vector space F^3 defined by $\alpha: [a, b, c] \mapsto [a, 0_F, 0_F]$ and $\beta: [a, b, c] \mapsto [b, a, c]$. Then $\alpha\beta: [a, b, c] \mapsto [b, 0_F, 0_F]$ while $\beta\alpha: [a, b, c] \mapsto [0_F, a, 0_F]$, so $\alpha\beta \neq \beta\alpha$.

(2) Secondly, if $\sigma_0 \neq \alpha \in End(V)$ there does not necessarily exist an element $\alpha^{-1} \in End(V)$ satisfying $\sigma_1 = \alpha\alpha^{-1} = \alpha^{-1}\alpha$. Consider the following example.

EXAMPLE

Let F be a field and let α be the endomorphism of F^3 defined in the previous example. Then $\alpha \neq \sigma_0$ but $ker(\alpha) \neq \{[0_F, 0_F, 0_F]\}$. This means that α is not a monomorphism and so surely not an isomorphism. Therefore no such α^{-1} can exist.

We also note that another property of fields is not satisfied by $End(V)$: if α and β are elements of $End(V)$ not equal to σ_0, it is nonetheless possible that $\alpha\beta = \sigma_0$. Consider the following example.

EXAMPLE

Let F be a field and let α be the endomorphism of F^3 considered in the previous two examples. Let γ be the endomorphism of F^3 defined by $\gamma \colon [a, b, c] \mapsto [0_F, 0_F, c]$. Then neither α nor γ equals σ_0, but $\alpha\gamma = \sigma_0 = \gamma\alpha$.

While $End(V)$ is not a field, it is an example of another important structure in abstract algebra. A nonempty set R on which we have defined two operations, $(a, b) \mapsto a + b$ (called *addition*), and $(a, b) \mapsto ab$ (called *multiplication*), is a *ring* if and only if the following conditions are satisfied:

(1) *(associativity of addition and multiplication)* $a + (b + c) = (a + b) + c$ and $a(bc) = (ab)c$ for all $a, b, c \in R$.

(2) *(commutativity of addition)* $a + b = b + a$ for all $a, b \in R$.

(3) *(existence of neutral elements with respect to addition and multiplication)* There exist distinct elements of $0_R \neq 1_R$ in R having the property that $a + 0_R = a$ and $a1_R = a$ for all $a \in R$.

(4) *(existence of additive inverses)* For each $a \in R$ there exists an element $-a \in R$ satisfying $a + (-a) = 0_R$.

(5) *(distributivity of multiplication over addition)* $a(b + c) = ab + ac$ and $(a + b)c = ac + bc$ for all $a, b, c \in R$.

In other words, a ring differs from a field in precisely the two aspects we pointed out: multiplication need not be commutative and elements which differ from the neutral element with respect to addition need not have multiplicative inverses.

EXAMPLE

Fields and integral domains are clearly rings. However, as we have seen, $End(F^3)$ is neither a field nor an integral domain for any field F.

A subset of a ring R is a *subring* if it contains 0_R and 1_R and is closed under addition, multiplication, and taking additive inverses.

EXAMPLE

If V is a vector space over a field F and if $\alpha \in End(V)$ then the set $F[\alpha] = \{p(\alpha) \mid p(X) \in F[X]\}$ is a subring of $End(V)$.

EXAMPLE

Let V be a nontrivial vector space over a field F. Then we have a function $\sigma: F \to End(V)$ defined by $c \mapsto \sigma_c$. We claim that this function is monic. Indeed, assume that $\sigma_c = \sigma_d$. If $0_V \neq v \in V$ then $cv = \sigma_c(v) = \sigma_d(v) = dv$ and so $0_V = (c - d)v$. By Proposition 2.2(9), this implies that $c - d = 0$ and so $c = d$. Moreover, if $c, d \in F$ then, as an immediate consequence of the definitions, we have $\sigma_{c+d} = \sigma_c + \sigma_d$ and $\sigma_{cd} = \sigma_c \sigma_d$. As a consequence of this, we see that if $d = c^{-1}$ in F then $\sigma_c \sigma_d = \sigma_1 = \sigma_d \sigma_c$ and so $\sigma_d = \sigma_c^{-1}$. Thus $\{\sigma_c \mid c \in F\}$ is a subring of $End(V)$ which in fact is a subfield, which we identify with F.

Endomorphisms of the form σ_c also commute with all other endomorphisms. Indeed, if $c \in F$ and if $\alpha \in End(V)$ then $\sigma_c \alpha(v) = c\alpha(v) = \alpha(cv) = \alpha\sigma_c(v)$ for all $v \in V$ and so $\sigma_c \alpha = \alpha \sigma_c$.

We have seen that while not every element of $End(V)$ which differs from σ_0 need have a multiplicative inverse, some such elements – such as those of the form σ_c for $0_F \neq c \in F$ – do have such an inverse. Such endomorphisms are called *automorphisms* of V, and, by Proposition 4.6, they are precisely the endomorphisms of V which are isomorphisms. The set of all automorphisms of V is denoted by $Aut(V)$. This set is always nonempty since $\sigma_1 \in Aut(V)$. If $\alpha, \beta \in Aut(V)$ then $\alpha\beta \in Aut(V)$ since $(\alpha\beta)(\beta^{-1}\alpha^{-1}) = \alpha(\beta\beta^{-1})\alpha^{-1} = \alpha\sigma_1\alpha^{-1} = \alpha\alpha^{-1} = \sigma_1$. Moreover, if $\alpha \in Aut(V)$ then surely $\alpha^{-1} \in Aut(V)$. Thus $Aut(V)$ is closed under multiplication and taking multiplicative inverses.

EXAMPLE

Let V be a vector space over a field F and let n be a positive integer. We will define three special sets of automorphisms of V^n, together known as the *elementary automorphisms*:

(1) If $1 \leq h \neq k \leq n$, we define the endomorphism α_{hk} of V^n to be the function which acts on a vector $[v_1, \dots, v_n]$ by interchanging the hth and kth components, while leaving all of the other components fixed. Clearly $\alpha_{hk}^{-1} = \alpha_{hk}$.

(2) If $1 \leq h \leq n$ and if $0 \neq c \in F$, we define the endomorphism $\alpha_{h;c}$ of V^n to be the function which acts on a vector $[v_1, \dots, v_n]$ by multiplying the hth component by c and leaving all of the other components fixed. Clearly $\alpha_{h;c}^{-1} = \alpha_{h;c^{-1}}$.

(3) If $1 \leq h \neq k \leq n$ and if $c \in F$, we define the endomorphism $\alpha_{hk;c}$ of V^n to be the function which acts on a vector $[v_1, \dots, v_n]$ by adding cv_k to v_h while leaving all of the other components fixed. Clearly $\alpha_{hk;c}^{-1} = \alpha_{hk;-c}$.

In a similar way, we can define the elementary automorphisms of the space of column vectors $\mathcal{M}_{n \times 1}(V)$.

EXAMPLE

Let V be a vector space finitely generated over a field F and let $\{v_1, \ldots, v_n\}$ be a basis for V over F. Any permutation τ of the set $\{1, \ldots, n\}$ defines an automorphism of V by $\sum_{i=1}^{n} a_i v_i \mapsto \sum_{i=1}^{n} a_i v_{\tau(i)}$.

We have seen that one can think of the ring $End(V)$ as an extension of the field of scalars F. In a like manner, we can extend the notion of scalar multiplication to include all elements of $End(V)$: if $v \in V$ and $\alpha \in End(V)$ then αv is just $\alpha(v)$. We see immediately that all of the conditions demanded of scalar multiplication still hold:

(1) $\alpha(v + v') = \alpha v + \alpha v'$ for all $v, v \in$ and all $\alpha \in End(V)$;
(2) $(\alpha + \beta)v = \alpha v + \beta v$ for all $v \in V$ and all $\alpha, \beta \in End(V)$;
(3) $(\alpha\beta)v = \alpha(\beta v)$ for all $v \in V$ and all $\alpha, \beta \in End(V)$;
(4) $\sigma_1 v = v$ for all $v \in V$.

A "vector space" with scalars coming from a ring rather than from a field is called a *module* over the ring. The theory of modules is very important in algebra, but is beyond the scope of this book.

By definition, an endomorphism of a vector space V over a field F is an automorphism if and only if it is both monic and epic. However, if V is finite-dimensional, each of these conditions implies the other, as the following result shows.

(5.1) PROPOSITION. *The following conditions on an endomorphism α of a vector space V of finite dimension n over a field F are equivalent:*

(1) *α is an automorphism of V;*
(2) *α is monic;*
(3) *α is epic.*   .

PROOF. By definition, (1) implies (2). Now assume (2). By Proposition 4.3, $ker(\alpha)$ is trivial and so has dimension 0. Therefore, by Proposition 4.9, $dim(V) = dim(im(\alpha))$ and so, by Proposition 3.10, $V = im(\alpha)$, proving (3). Finally, assume (3). Then, by Proposition 4.9, $dim(ker(\alpha)) = dim(V) - dim(im(\alpha)) = 0$ and so $ker(\alpha) = \{0_V\}$. Therefore α is a monomorphism as well as an epimorphism, proving (1). $\square$

In general, in order to prove that an element b of a ring R is the multiplicative inverse of an element a, we must show that both $ab = 1_R$ and $ba = 1_R$. Again, in the case of endomorphism rings of finitely-generated vector spaces, each of these conditions implies the other.

(5.2) PROPOSITION. *Let α be an endomorphism of a vector space V finitely generated over a field F and let β be an endomorphism of V satisfying the condition that either $\alpha\beta$ or $\beta\alpha$ equals σ_1. Then $\beta = \alpha^{-1}$.*

PROOF. If $\beta\alpha = \sigma_1$ then $ker(\alpha) \subseteq ker(\beta\alpha) = \{0_V\}$ and so, by Proposition 5.1, α is an automorphism of V. Moreover, in this case, $\alpha^{-1} = (\beta\alpha)\alpha^{-1} = \beta(\alpha\alpha^{-1}) = \beta$. Similarly, if $\alpha\beta = \sigma_1$ then $im(\alpha) \supseteq im(\alpha\beta) = V$ and so, by Proposition 5.1, α is an automorphism of V. Moreover, in this case, $\alpha^{-1} = \alpha^{-1}(\alpha\beta) = (\alpha^{-1}\alpha)\beta = \beta$. $\square$

COROLLARY. *Let α and β be endomorphisms of a vector space V finitely generated over a field F. Then the following conditions are equivalent:*

(1) $\alpha, \beta \in Aut(V)$;
(2) $\alpha\beta \in Aut(V)$;
(3) $\beta\alpha \in Aut(V)$.

PROOF. We have already noted that (1) implies (2) and (3) irrespective of whether V is finitely generated or not. Now assume (2) and let $\gamma = (\alpha\beta)^{-1}$. Then $\alpha(\beta\gamma) = \sigma_1$ and so, by Proposition 5.2, $\alpha \in Aut(V)$. Similarly, $(\gamma\alpha)\beta = \sigma_1$ and so, by Proposition 5.2, $\beta \in Aut(V)$. Thus we have (1). A similar argument shows that (3) implies (1). $\square$

EXAMPLE

The condition that V be finitely generated in Propositions 5.1 and 5.2 is essential. To see this, let $\mathbb{N}$ be the set of nonnegative integers and let $V = F^{\mathbb{N}}$, where F is a field. Let $\alpha, \beta \in End(V)$ be defined by

$$\alpha(f): i \mapsto \begin{cases} 0_F, & \text{for } i = 0, \\ f(i-1), & \text{for } i > 0. \end{cases}$$

for all $f \in V$ and $\beta(f): i \mapsto f(i+1)$ for all $f \in V$. Neither of these endomorphisms is an automorphism of V, despite the fact that α is a monomorphism and β is an epimorphism. Moreover, $\beta\alpha = \sigma_1$ while $\alpha\beta \neq \sigma_1$.

Let V be a vector space over a field F and let α be an endomorphism of V. A subspace W of V is *stable* under α if and only if $\alpha(w) \in W$ for all $w \in W$. That is to say, W is stable under α if and only if the restriction of α to W is an endomorphism of W.

EXAMPLE

Clearly $\{0_V\}$ and V itself are stable under every endomorphism of V. Moreover, if $\alpha \in End(V)$ then $im(\alpha)$ and $ker(\alpha)$ are stable under α.

EXAMPLE

Let F be a field of characteristic 0. For each nonnegative integer n, let W_n be the subspace of $F[X]$ consisting of all polynomials of degree at most n. Then each W_n is stable under the formal differentiation endomorphism

$$\sum_{i \geq 0} a_i X^i \mapsto \sum_{i > 0} (i1_F) a_i X^{i-1}$$

of $F[X]$.

EXAMPLE

Let $V = \mathbb{R}^2$ and let α be the endomorphism of V defined by $\alpha: [a, b] \mapsto [b, -a]$. Assume that W is a proper subspace of V stable under α. Since W is proper, its dimension is at most 1 and so $W = \mathbb{R}w_0$ for some $w_0 = [c, d] \in W$. Since $[d, -c] = \alpha(w_0) \in W$, there must be a real number r satisfying $[d, -c] = rw_0 = [rc, rd]$ and so $rc = d$ and $rd = -c$. From this we see that $cr^2 = -c$ and so either $r^2 = -1$ (which is impossible for a real number) or $c = 0$ and hence $d = 0$ as well. Thus $W = \{0_V\}$ and so we see that there are no proper nontrivial subspaces of V stable under α.

An endomorphism α of a vector space V over a field F is *idempotent* if and only if $\alpha^2 = \alpha$.

EXAMPLE

If F is a field then the endomorphism α of F^3 defined by $[a, b, c] \mapsto [a, 0_F, c]$ is idempotent.

EXAMPLE

Let V be a vector space over a field F. Let W be a subspace of V having a complement W' in V. We know that every element v of V can be written uniquely in the form $w + w'$ with $w \in W$ and $w' \in W'$. The function from V to W which assigns to each element v of V its component in W is called a *projection* of V onto the subspace W. Such a projection, composed with the inclusion map $W \rightarrow V$, is an idempotent endomorphism of V.

(5.3) PROPOSITION. *Let V be a vector space over a field F and let α be an idempotent endomorphism of V. Then $V = ker(\alpha) \oplus im(\alpha)$.*

PROOF. If $v \in ker(\alpha) \cap im(\alpha)$ then there exists an element y of V satisfying $v = \alpha(y)$ and so $v = \alpha(y) = \alpha^2(y) = \alpha(\alpha(y)) = \alpha(v) = 0_V$. Thus $ker(\alpha) \cap im(\alpha) = \{0_V\}$. On the other hand, if $v \in V$ then we can write $v = [v - \alpha(v)] + \alpha(v)$ where $\alpha(v) \in im(\alpha)$ and $v - \alpha(v) \in ker(\alpha)$ since $\alpha(v - \alpha(v)) = \alpha(v) - \alpha^2(v) = \alpha(v) - \alpha(v) = 0_V$. Thus $V = ker(\alpha) \oplus im(\alpha)$. $\square$

(5.4) PROPOSITION. *Let V be a vector space over a field F and let α be an endomorphism of V. Then the following conditions on a subspace W of V are equivalent:*

(1) *W is stable under α;*
(2) *If π is a projection of V onto W then $\pi\alpha\pi = \alpha\pi$.*

PROOF. Assume (1) and let π be a projection of V onto W coming from some direct-sum decomposition $V = W \oplus W'$. If $v = w + w'$, where $w \in W$ and $w' \in W'$, then $\pi\alpha\pi(v) = \pi\alpha(w) = \alpha(w) = \alpha\pi(v)$ since $\alpha(w) \in W$, and so we have (2).

Conversely, assume (2). If $w \in W$ and if π is a projection of V onto W then $w = \pi(w)$ and so $\pi\alpha(w) = \pi\alpha\pi(w) = \alpha\pi(w) = \alpha(w)$ and so $\alpha(w) \in W$, proving (1). $\square$

Idempotent endomorphisms are important to us since they are closely tied to the notion of a direct-sum decomposition of a vector space, which we introduced in Chapter III, as the following result shows.

(5.5) PROPOSITION. *Let* V *be a vector space over a field* F. *For a collection* $\{W_1, \ldots, W_n\}$ *of subspaces of* V *the following conditions are equivalent:*

(1) $V = \bigoplus_{i=1}^n W_i$;

(2) *There exist idempotent endomorphisms* $\alpha_1, \ldots, \alpha_n$ *of* V *such that:*
    (i) $\alpha_i \alpha_j = \sigma_0$ *for all* $i \neq j$;
    (ii) $\alpha_1 + \cdots + \alpha_n = \sigma_1$;
    (iii) $W_i = im(\alpha_i)$ *for all* $1 \leq i \leq n$.

PROOF. (1) $\Rightarrow$ (2): By (1) we know that every element v of V can be written in a unique manner as $w_1 + \cdots + w_n$, where $w_i \in W_i$ for all $1 \leq i \leq n$. Let α_i be the function which sends v into the ith component w_i of this representation. Then it is easily seen that each such α_i is an idempotent endomorphism of V and that conditions (i) - (iii) are satisfied.

(2) $\Rightarrow$ (1): From (ii) we see that $V = W_1 + \cdots + W_n$. If $0_V \neq v \in W_h \cap (\sum_{j \neq h} W_j)$ then there must exist an index $i \neq h$ such that $\alpha_i(v) \neq 0_V$. But $\alpha_h(v) = v$ since $v \in W_h$ and so $\alpha_i \alpha_h(v) = \alpha_i(v) \neq 0_V$, contradicting (i). Therefore this sum must be direct. $\square$

(5.6) PROPOSITION. *If* W *is a subspace of a vector space* V *over a field* F *then any two complements of* W *in* V *are isomorphic.*

PROOF. Let W' and W'' be complements of W in V. By Proposition 5.5 we know that there exists an idempotent endomorphism α of V which sends each $v \in V$ into its W'-component in the direct-sum representation $V = W \oplus W'$. In particular, $im(\alpha) = W'$ and $ker(\alpha) = W$. Let $\beta: W'' \rightarrow W'$ be the restriction of α to W''. Then β is a monomorphism since $W'' \cap W = \{0_V\}$. If $w' \in W'$ then we can write $w' = y + y''$, where $y \in W$ and $y'' \in W''$. Moreover, $\beta(y'') = \alpha(y'') = \alpha(y'') + \alpha(y) = \alpha(y'' + y) = \alpha(w') = w'$ and so β is an epimorphism as well. Hence it is the isomorphism we seek. $\square$

Problems

1. Let V be a vector space over the field $\mathbb{Z}/(3)$. Show that there exists an endomorphism α of V satisfying $\alpha(v) + \alpha(v) = v$ for all $v \in V$.

2. Let $F = \mathbb{Z}/(2)$ and let n be a positive integer. Let $\alpha: F^n \rightarrow F^n$ be the function defined by $\alpha: [a_1, \ldots, a_n] \mapsto [a'_1, \ldots, a'_n]$, where $0' = 1$ and $1' = 0$. Is α an endomorphism of F^n?

3. Let α and β be the functions from $\mathbb{Q}[X]$ to itself defined by $\alpha: f(X) \mapsto Xf(X)$ and $\beta: f(X) \mapsto X^2 f(X)$. Show that both of these functions are monic endomorphisms of $\mathbb{Q}[X]$ and that $\alpha - \beta$ is also a monic endomorphism of $\mathbb{Q}[X]$.

4. Let $F = \mathbb{Z}/(3)$. What is the probability that an arbitrarily-chosen function from F^4 to itself be an endomorphism of F^4 as a vector space over F?

5. Let α be an endomorphism of a vector space V over a field F. Show that α is not monic if and only if there exists an endomorphism $\beta \neq \sigma_0$ of V satisfying $\alpha\beta = \sigma_0$.

6. Let α be an endomorphism of a vector space V over a field F. Show that α is not epic if and only if there exists an endomorphism $\beta \neq \sigma_0$ of V satisfying $\beta\alpha = \sigma_0$.

7. Let α be an endomorphism of a vector space V over a field F. Show that $ker(\alpha) = ker(\alpha^2)$ if and only if $ker(\alpha) \cap im(\alpha) = \{0_V\}$.

8. Let α be an endomorphism of a vector space V over a field F. Show that $im(\alpha) = im(\alpha^2)$ if and only if $V = ker(\alpha) + im(\alpha)$.

9. Let V be the subspace of $\mathbb{R}^{[0,1]}$ consisting of all those functions which are everywhere differentiable infinitely-many times and let δ be the endomorphism of V which assigns to each $f \in V$ its derivative f'. Find the image and kernel of δ.

10. Let α be the function from $\mathbb{C}$ to itself defined by $\alpha: a + bi \mapsto -b + ai$. Is α an endomorphism of $\mathbb{C}$ as a vector space over $\mathbb{R}$? Is it an endomorphism of $\mathbb{C}$ as a vector space over itself?

11. Let V be a vector space of finite dimension n over a field F and let $\alpha \in End(V)$. Show that there exists an automorphism β of V satisfying $\alpha\beta\alpha = \alpha$.

12. Let V be the vector space of all functions from $\mathbb{R}$ to itself and let $W = \mathbb{R}\{sin(x), cos(x)\}$. If δ is the endomorphism of W which assigns to each $f \in W$ its derivative f', find a polynomial $p(X) \in \mathbb{R}[X]$ of degree 2 satisfying $p(\delta) = \sigma_0$.

13. If V is a vector space over a field F then we know that $End(V)$ is also a vector space over F. Is $Aut(V)$ a subspace of this space?

14. Give an example of a vector space V over a field F, together with idempotent endomorphisms α and β of V, such that $\alpha + \beta$ is not idempotent.

15. Let F be a field of characteristic different from 2 and let V be a vector space over F. Suppose that α and β are idempotent endomorphisms of V satisfying the condition that $\alpha + \beta$ is also idempotent. Show that $\alpha\beta = \beta\alpha = \sigma_0$.

16. Let $V = \mathcal{M}_{2\times 2}(\mathbb{R})$ and let $\alpha: V \to V$ be the function defined by

$$\alpha: \begin{bmatrix} a_{11} & a_{12} \\ a_{21} & a_{22} \end{bmatrix} \mapsto \begin{bmatrix} |a_{11}| & |a_{12}| \\ |a_{21}| & |a_{22}| \end{bmatrix}.$$

Is α an endomorphism of V as a vector space over $\mathbb{R}$?

17. Consider $\mathbb{R}$ as a vector space over $\mathbb{Q}$. An endomorphism α of this space is *bounded* if and only if there exists a nonnegative real number $m(\alpha)$ satisfying $|\alpha(r)| \leq m(\alpha)|r|$ for all $r \in \mathbb{R}$. Does the set of all bounded endomorphisms of $\mathbb{R}$ form a subring of $End(\mathbb{R})$?

18. Consider $\mathbb{R}$ as a vector space over $\mathbb{Q}$ and let α be an endomorphism of this space satisfying the condition that there exists an element $a_0 \in \mathbb{R}$ at which α is continuous. Show that α is continuous everywhere on $\mathbb{R}$.

19. Let V be a vector space over a field F and let $\alpha \in End(V)$. Set $H = \{p(X) \in F[X] \mid p(\alpha) = \sigma_0\}$.
   (i) Show that H is a subspace of $F[X]$.
   (ii) If $p(X) \in H$ and $f(X) \in F[X]$, does $p(X)f(X)$ belong to H? What about $f(X)p(X)$?

20. Let V be the set of all positive real numbers. If $a, b \in V$ and $r \in \mathbb{R}$ write $a \boxplus b = ab$ and $r \boxdot a = a^r$.
   (i) Show that V is a vector space over $\mathbb{R}$ with $\boxplus$ as vector addition and $\boxdot$ as scalar multiplication.
   (ii) Find all endomorphisms of this vector space.

21. Let V be a vector space of finite dimension n over a field F and let $\{\alpha_{ij} \mid 1 \leq i, j \leq n\}$ be a set of n^2 elements of $End(V)$, not all of which are equal to σ_0, which satisfy the condition that

$$\alpha_{ij}\alpha_{kh} = \begin{cases} \alpha_{ih}, & \text{when } i = k, \\ \sigma_0, & \text{otherwise.} \end{cases}$$

Show that there exists a basis $\{v_1, \ldots, v_n\}$ for V satisfying

$$\alpha_{ij}(v_k) = \begin{cases} v_i, & \text{when } j = k, \\ 0_V, & \text{otherwise.} \end{cases}$$

22. Let $V = \mathcal{M}_{2\times 2}(\mathbb{R})$, which is a vector space over $\mathbb{R}$, and let $\alpha: V \to V$ be the function defined by

$$\alpha: \begin{bmatrix} a & b \\ c & d \end{bmatrix} \mapsto \begin{bmatrix} a + 2b + c + 2d & 2a + 4b + 3c + 5d \\ 3a + 6b + 2c + 5d & a + 2b + c + 2d \end{bmatrix}.$$

Is α an endomorphism of V? Is it an automorphism of V?

23. Let V be the vector space of all continuous functions from $\mathbb{R}$ to itself and let $\alpha: V \to V$ be defined by $\alpha: f(x) \mapsto [x^2 + sin(x) + 2]f(x)$. Show that α is an automorphism of V.

24. Let F be a field and let α be the function from $F[X]$ to itself defined by $p(X) \mapsto p(X + 1)$. Is α an endomorphism of $F[X]$? Is it an automorphism?

25. Let α be the endomorphism of $\mathbb{R}^3$ defined by $\alpha: [a, b, c] \mapsto [a - 2b, c, a - b]$. Is α an automorphism of $\mathbb{R}^3$?

26. Let V be a vector space finitely generated over $\mathbb{R}$ and let α be an endomorphism of V satisfying $\alpha^3 + 4\alpha^2 + 2\alpha + \sigma_1 = \sigma_0$. Show that α is an automorphism of V.

27. Let n be a positive integer and let G be the subset of $V = \mathbb{R}^n$ consisting of all vectors $[a_1, \ldots a_n]$ satisfying $a_i \geq 0$ for all $1 \leq i \leq n$. Let $\alpha \in End(V)$ satisfy the condition that $\alpha(v) \in G$ implies $v \in G$. Show that α is an automorphism of V.

28. Let V be the subspace of $\mathbb{R}^{\mathbb{R}}$ consisting of all functions everywhere continuous on $\mathbb{R}$ and let $\alpha: V \to V$ be the function defined by $\alpha(f): t \mapsto f(\frac{t}{2})$ for all $t \in \mathbb{R}$. Is α an endomorphism of V? Is it an automorphism?

29. Consider the field $F = \mathbb{Z}/(3)$ as a vector space over itself. Does there exist automorphism of F other than σ_1?

30. Let V be a vector space over a field F and let α be an endomorphism of V. Show that $W = \bigcup_{i \geq 1} ker(\alpha^i)$ is a subspace of V stable under α.

31. Let α be an endomorphism of a vector space V over a field F and let W be a subspace of V. Show that in the set of all subspaces of W which are stable under α there is one which contains all of the others.

32. Let $\mathbb{P}$ be the set of positive integers and let W be the subspace of $\mathbb{R}^{\mathbb{P}}$ consisting of all functions f satisfying the condition that the sequence of real numbers $f(1), f(2), \ldots$ converges. Let τ be a permutation of $\mathbb{P}$ and let α be the function from $\mathbb{R}^{\mathbb{P}}$ to itself defined by $\alpha(f): i \mapsto f(\tau(i))$. Is W necessarily stable under α?

33. Let V be a vector space over a field F. Let $0_F \neq c \in F$ and let α be an endomorphism of V. Let $\{x_0, x_1, \ldots, x_n\}$ be a set of vectors in V satisfying the following conditions:

(1) $\alpha(x_0) = cx_0$;
(2) $\alpha(x_i) - cx_i = x_{i-1}$ for all $1 \leq i \leq n$.

Show that the subspace $F\{x_0, x_1, \ldots, x_n\}$ of V is stable under α.

34. Let F be a field having infinitely-many elements and let V be a vector space over F having dimension greater than 1. For each $0_F \neq c \in F$, show that there exist infinitely-many subspaces of V which are stable under the endomorphism σ_c of V

35. Let V be a vector space over a field F and let α and β be endomorphisms of V. Let $\theta \in Aut(V)$ satisfy $\theta\alpha = \beta\theta$. Show that a subspace W of V is stable under α if and only if $W' = \{\theta(w) \mid w \in W\}$ is stable under β.

36. Let α and β be endomorphisms of a vector space V over a field F satisfying $\alpha\beta = \beta\alpha$. Is $ker(\alpha)$ necessarily stable under β?

37. Let α be an idempotent endomorphism of a vector space V over a field F. Show that $\sigma_1 - \alpha$ is also idempotent.

CHAPTER VI

REPRESENTATION OF LINEAR
TRANSFORMATIONS BY MATRICES

Let V and W be vector spaces finitely generated over a field F with $dim(V) = n$ and $dim(W) = k$, and choose fixed bases $B = \{v_1, \ldots, v_n\}$ for V over F and $D = \{w_1, \ldots, w_k\}$ for W over F. Let $\alpha\colon V \to W$ be a linear transformation. By Proposition 3.4 we know that for each $1 \leq j \leq n$ there exists a unique set of scalars $\{a_{1j}, \ldots a_{kj}\}$ such that $\alpha(v_j) = \sum_{i=1}^{k} a_{ij} w_i$. The matrix $[a_{ij}] \in \mathcal{M}_{k \times n}(F)$ totally determines the linear transformation α in the following sense: if $v \in V$ then there exists a unique representation of v as $\sum_{j=1}^{n} b_j v_j$ and so we must have

$$\alpha(v) = \sum_{i=1}^{k} \left(\sum_{j=1}^{n} b_j a_{ij} \right) w_i.$$

Thus we see that, given fixed bases for V and W over F, we can associate with each linear transformation $\alpha\colon V \to W$ a matrix $[a_{ij}] \in \mathcal{M}_{k \times n}(F)$ which totally determines the action of α. The converse is also true: each matrix $[a_{ij}] \in \mathcal{M}_{k \times n}(F)$ uniquely defines a linear transformation $\alpha\colon V \to W$ by

$$\alpha\colon \sum_{j=1}^{n} b_j v_j \mapsto \sum_{i=1}^{k} \left(\sum_{j=1}^{n} b_j a_{ij} \right) w_i.$$

From these definitions, we see that distinct matrices define distinct linear transformations and that distinct linear transformations define distinct matrices. The theory of matrices was first developed and used in this context by the British mathematician Arthur Cayley.

All of the above discussion can be summed up by the following proposition.

(6.1) PROPOSITION. *Let V be a vector space of dimension n over a field F and let W be a vector space of dimension k over F. For each basis B of V over F and each basis D of W over F there exists an isomorphism of vector spaces $\Phi_{BD}\colon Hom(V, W) \to \mathcal{M}_{k \times n}(F)$ given by $\Phi_{BD}(\alpha) = [a_{ij}]$, where $\alpha(v_j) = \sum_{i=1}^{k} a_{ij} w_i$ for all $1 \leq k \leq n$.*

PROOF. We have already seen, above, that the function $\Phi_{BD}\colon Hom(V, W) \to \mathcal{M}_{k \times n}(F)$ is bijective, so all that is left to show is that it is in fact a linear transformation. Indeed, if $\alpha, \beta \in Hom(V, W)$ satisfy $\Phi_{BD}(\alpha) = [a_{ij}]$ and $\Phi_{BD}(\beta) = [b_{ij}]$

then for each $1 \leq j \leq n$ we have

$$(\alpha + \beta)(v_j) = \sum_{i=1}^{k}(a_{ij} + b_{ij})w_i$$

$$= \sum_{i=1}^{k} a_{ij}w_i + \sum_{i=1}^{k} b_{ij}w_i$$

$$= \alpha(v_j) + \beta(v_j)$$

and so $\Phi_{BD}(\alpha+\beta) = \Phi_{BD}(\alpha)+\Phi_{BD}(\beta)$. Similarly, if $c \in F$ then for each $1 \leq j \leq n$ we have

$$(c\alpha)(v_j) = \sum_{i=1}^{k} c a_{ij} w_i = c\left(\sum_{i=1}^{k} a_{ij} w_i\right) = c(\alpha(v_j))$$

and so $\Phi_{BD}(c\alpha) = c\Phi_{BD}(\alpha)$. This shows that Φ_{BD} is indeed a linear transformation and so an isomorphism. $\square$

COROLLARY. *If V is a vector space of dimension n over a field F and if W is a vector space of dimension k over F then the vector space $Hom(V, W)$ has dimension kn over F.*

PROOF. We have already seen that $dim(\mathcal{M}_{k \times n}(F)) = kn$ and so the result follows from Proposition 4.8. $\square$

EXAMPLE

Consider the bases $\{[\frac{1}{2}, -\frac{1}{2}, 0], [\frac{1}{2}, 0, -\frac{1}{2}], [0, \frac{1}{2}, \frac{1}{2}]\}$ of $\mathbb{R}^3$ and $\{[1, 1], [1, 0]\}$ of $\mathbb{R}^2$. If $[r, s, t] \in \mathbb{R}^3$ then there exist real numbers b_1, b_2, b_3 such that

$$[r, s, t] = b_1[\frac{1}{2}, -\frac{1}{2}, 0] + b_2[\frac{1}{2}, 0, -\frac{1}{2}] + b_3[0, \frac{1}{2}, \frac{1}{2}]$$

$$= [\frac{1}{2}(b_1 + b_2), \frac{1}{2}(-b_1 + b_3), \frac{1}{2}(-b_2 + b_3)]$$

and so $r = \frac{1}{2}(b_1 + b_2)$, $s = \frac{1}{2}(-b_1 + b_3)$, and $t = \frac{1}{2}(-b_2 + b_3)$, from which we derive $b_1 = r - s + t$, $b_2 = r + s - t$, and $b_3 = r + s + t$. The matrix $\begin{bmatrix} 3 & 5 & 7 \\ 4 & 8 & 2 \end{bmatrix}$ defines a linear transformation $\alpha: \mathbb{R}^3 \rightarrow \mathbb{R}^2$ which sends a vector

$$b_1[\frac{1}{2}, -\frac{1}{2}, 0] + b_2[\frac{1}{2}, 0, -\frac{1}{2}] + b_3[0, \frac{1}{2}, \frac{1}{2}] \in \mathbb{R}^3$$

to the vector $(3b_1 + 5b_2 + 7b_3)[1, 1] + (4b_1 + 8b_2 + 3b_3)[1, 0] \in \mathbb{R}^2$. That is to say,

$$\alpha: [r, s, t] \mapsto (15r+9s+5t)[1, 1]+(14r+6s-2t)[1, 0] = [29r+15s+3t, 15r+9s+5t].$$

It is important to reemphasize that the matrix representation of a linear transformation depends on the bases which we have chosen! If we choose different bases
of the respective vector spaces we will get a different matrix altogether. It is also
important to notice that the matrix representation depends on the order in which
the basis elements are taken, so it is worth restating at this point the convention
that we made at the beginning of the book: if the elements of a finite set are listed
explicitly, the set is assumed to have an induced order, going from left to right in
the given list.

Later, we will deal in detail with the relations between the matrices representing
a given linear transformation obtained from choosing various bases.

Let V be a vector space finitely-generated over a field F and let α be an endomorphism of V. If W is a subspace of V stable under α then the restriction α' of
α to W is an endomorphism of W. Let $B = \{v_1, \ldots, v_k\}$ be a basis of W, which
we will extend to a basis $D = \{v_1, \ldots, v_n\}$ of V. If $A = [a_{ij}]$ is the matrix $\Phi_{DD}(\alpha)$
then for each $1 \leq j \leq n$ we have $\alpha(v_j) = \sum_{i=1}^n a_{ij} v_i$ and so we must have $a_{ij} = 0_F$
for $1 \leq i \leq k$ and all $k + 1 \leq j \leq n$. Thus we see that the matrix A is of the
form $\begin{bmatrix} A_{11} & O \\ A_{21} & A_{22} \end{bmatrix}$, where $A_{11} = \Phi_{BB}(\alpha')$. The subspace $W' = F\{v_{k+1}, \ldots, v_n\}$ is
a complement of W in V. In case W' is also stable under α, we see that $A_{21} = O$
and we obtain A in the form of two square matrices "strung out" along the diagonal
of A. We will return to this construction later, and generalize it.

In addition to adding matrices of the same size and multiplying a matrix by a
scalar, we can define the product of two matrices, provided their sizes are com-
patible. In particular, let F be a field. If $A = [a_{ij}] \in \mathcal{M}_{k \times n}(F)$ and $B = [b_{jh}] \in \mathcal{M}_{n \times t}(F)$ then we set AB to be the matrix $[c_{ih}] \in \mathcal{M}_{k \times t}(F)$ defined by
$c_{ih} = \sum_{j=1}^n a_{ij} b_{jh}$ for all $1 \leq i \leq k$ and all $1 \leq h \leq t$. Watch out! The product
of two matrices is defined only when the number of columns in the first equals
the number of rows in the second. Thus, if A and B are matrices, it is perfectly
possible that AB may be defined while BA may not be defined. Even if both of
these products are defined, they may well be of different sizes.

EXAMPLE

Over $\mathbb{R}$, we have

$$\begin{bmatrix} 2 & 1 & 1 \\ 4 & 3 & 0 \end{bmatrix} \begin{bmatrix} 1 & 0 & 3 \\ 7 & 1 & 3 \\ 4 & 2 & 0 \end{bmatrix} = \begin{bmatrix} 13 & 3 & 9 \\ 25 & 3 & 21 \end{bmatrix}$$

while

$$\begin{bmatrix} 1 & 0 & 3 \\ 7 & 1 & 3 \\ 4 & 2 & 0 \end{bmatrix} \begin{bmatrix} 2 & 1 & 1 \\ 4 & 3 & 0 \end{bmatrix}$$

is not defined.

EXAMPLE

Over $\mathbb{R}$, we have

$$[1 \quad 2 \quad 3] \begin{bmatrix} 5 \\ 4 \\ 2 \end{bmatrix} = [19]$$

while

$$\begin{bmatrix} 5 \\ 4 \\ 2 \end{bmatrix} [1 \quad 2 \quad 3] = \begin{bmatrix} 5 & 10 & 15 \\ 4 & 8 & 12 \\ 2 & 4 & 6 \end{bmatrix}.$$

This seemingly odd definition of matrix multiplication was not, needless to say, chosen at random, but rather to insure that certain important algebraic conditions are satisfied. Indeed, if F is as field, if $A \in \mathcal{M}_{k \times n}(F)$, if $B, B' \in \mathcal{M}_{n \times t}(F)$, and if $C \in \mathcal{M}_{t \times p}(F)$ then we can check in a by straightforward computation that:

(1) $A(BC) = (AB)C$;
(2) $A(B + B') = AB + AB'$; and
(3) $(B + B')C = BC + B'C$.

From this it follows that if B is a given matrix in $\mathcal{M}_{k \times t}(F)$ then the function $\mathcal{M}_{n \times k}(F) \rightarrow \mathcal{M}_{n \times t}(F)$ defined by $A \mapsto AB$ is a linear transformation.

Actually, the definition of the matrix product is a direct consequence of the connection between matrices and linear transformations, as the following proposition shows.

(6.2) PROPOSITION. *Let V be a vector space of dimension n over a field F, let W be a vector space of dimension k over F, and let Y be a vector space of dimension t over F. Choose a basis B for V over F, a basis D for W over F, and a basis E for Y over F. If $\alpha: V \rightarrow W$ and $\beta: W \rightarrow Y$ are linear transformations then $\Phi_{BE}(\beta\alpha) = \Phi_{DE}(\beta)\Phi_{BD}(\alpha)$.*

PROOF. If $B = \{v_1, \ldots, v_n\}$, $D = \{w_1, \ldots, w_k\}$, and $E = \{y_1, \ldots, y_t\}$ and if $\Phi_{BD}(\alpha) = [a_{ij}]$ and $\Phi_{DE}(\beta) = [b_{hi}]$, then for each $v = \sum_{j=1}^{n} c_j v_j \in V$ we have

$$\alpha(v) = \sum_{i=1}^{k} \sum_{j=1}^{n} c_j a_{ij} w_i$$

and so

$$\beta(\alpha(v)) = \sum_{h=1}^{t} \sum_{i=1}^{k} \sum_{j=1}^{n} c_j b_{hi} a_{ij} y_h = \sum_{h=1}^{t} \sum_{j=1}^{n} c_j \left(\sum_{i=1}^{k} b_{hi} a_{ij} \right) y_h$$

which proves the desired equality. $\square$

We can extend the definition of matrix multiplication in the following manner: Let h, k and n be positive integers and let V be a vector space over a field F. If $A = [a_{ij}] \in \mathcal{M}_{h \times k}(F)$ and if $M = [v_{jt}] \in \mathcal{M}_{k \times n}(V)$ then we can define the matrix $AM \in \mathcal{M}_{h \times n}(V)$ to be the matrix $[u_{it}]$, where $u_{it} = \sum_{j=1}^{k} a_{ij} v_{jt}$ for each

$1 \leq i \leq h$ and each $1 \leq t \leq n$. By straightforward computation, we note that if $A, B \in M_{k \times k}(F)$ and if $M, N \in M_{k \times n}(V)$ then

(1) $A(BM) = (AB)M$;
(2) $A(M + N) = AM + AN$;
(3) $(A + B)M = AM + BM$.

In general, especially when we are dealing with computational linear algebra, it is much easier to work with matrices rather than with linear transformations, and most computer languages are designed to handle matrix arithmetic with ease. Therefore, in dealing with finite-dimensional vector spaces and linear transformations between them one usually begins by fixing bases for each of the spaces and then representing all linear transformations by matrices relative to these bases. How to choose these bases in order to make our computations come out easier is a matter of definite concern, and we shall return to it throughout the rest of this book.

In case we are dealing with linear transformations between vector spaces of the form F^n there is an advantage to using canonical bases. Indeed, let F be a field and let k, n be positive integers. Assume that $\alpha: F^n \to F^k$ is a linear transformation satisfying $\Phi_{BD}(\alpha) = [a_{ij}]$, where B and D are the canonical bases of F^n and F^k respectively. If $v = [b_1, \ldots, b_n] \in F^n$ then, by definition,

$$\alpha(v) = \left[\sum_{j=1}^{n} a_{1j} b_j, \ldots, \sum_{j=1}^{n} a_{kj} b_j \right].$$

That is to say, $\alpha(v)$ is defined by the matrix product $\alpha(v)^T = \Phi_{BD}(\alpha) v^T$.

EXAMPLE

If $\alpha: \mathbb{R}^3 \to \mathbb{R}^2$ is the linear transformation represented with respect to the canonical bases by the matrix $\begin{bmatrix} 1 & 2 & -1 \\ 0 & -1 & -2 \end{bmatrix}$ then

$$\alpha: [1, 0, 1] \mapsto \begin{bmatrix} 1 & 2 & -1 \\ 0 & -1 & -2 \end{bmatrix} \begin{bmatrix} 1 \\ 0 \\ 1 \end{bmatrix} = [0, -2]$$

and

$$\alpha: [1, 2, 3] \mapsto \begin{bmatrix} 1 & 2 & -1 \\ 0 & -1 & -2 \end{bmatrix} \begin{bmatrix} 1 \\ 2 \\ 3 \end{bmatrix} = [2, -8].$$

Multiplication of two matrices over a field F may, in practice, involve a very large amount of computation. Indeed, if $A \in M_{k \times n}(F)$ and $B \in M_{n \times t}(F)$ then the computation of AB involves knt multiplications and $k(n-1)t$ additions so that if the sizes of these matrices are large (say over 1000×1000, which is not unlikely in modern applications of matrix theory), a considerable amount of computational power must be utilized to calculate the product of two matrices. Therefore, much effort has gone into finding computational shortcuts for calculating matrix products. One such shortcut is known as the method of block multiplication.

Let $A = [a_{ij}] \in M_{k \times n}(F)$ and $B = [b_{ij}] \in M_{n \times t}(F)$, where F is a field. Suppose that we pick positive integers $1 = k_1 < k_2 < \cdots < k_{p+1} = k$, $1 = n_1 < n_2 < \cdots < n_{q+1} = n$, and $1 = t_1 < t_2 < \cdots < t_{r+1} = t$. For all $1 \le i \le p$ and all $1 \le j \le q$ let A_{ij} be the matrix

$$\begin{bmatrix} a_{k_i, n_j} & \cdots & a_{k_i, n_{j+1}} \\ & \cdots & \\ a_{k_{i+1}, n_j} & \cdots & a_{k_{i+1}, n_{j+1}} \end{bmatrix}$$

so that we can write A as a matrix of blocks

$$\begin{bmatrix} A_{11} & \cdots & A_{1q} \\ & \cdots & \\ A_{p1} & \cdots & A_{pq} \end{bmatrix}.$$

Similarly, we can write B as a matrix of blocks

$$\begin{bmatrix} B_{11} & \cdots & B_{it} \\ & \cdots & \\ B_{q1} & \cdots & B_{qt} \end{bmatrix}$$

and the matrix AB becomes

$$\begin{bmatrix} C_{11} & \cdots & C_{it} \\ & \cdots & \\ C_{p1} & \cdots & C_{pt} \end{bmatrix}$$

where, for each $1 \le i \le p$ and each $1 \le h \le t$ we have $C_{ih} = \sum_{j=1}^{q} A_{ij} B_{jh}$. Enlightened use of this observation can be of great help in reducing the time needed to compute matrix products, especially for matrices known to contain large blocks of zeros.

EXAMPLE

If

$$A = \begin{bmatrix} 1 & 2 & 3 & 0 & 0 \\ 1 & 1 & 2 & 0 & 0 \\ 2 & 2 & 1 & 1 & 1 \\ 0 & 0 & 0 & 0 & 1 \end{bmatrix}$$

and

$$B = \begin{bmatrix} 1 & 2 & 0 & 1 \\ 3 & 1 & 1 & 0 \\ 4 & 0 & 1 & 1 \\ 1 & 1 & 0 & 0 \\ 1 & 0 & 0 & 0 \end{bmatrix}$$

then

EXAMPLE (Continued)

$$AB = \left[\begin{array}{cc} \begin{bmatrix} 1 & 2 & 3 \\ 1 & 1 & 2 \\ 2 & 2 & 1 \\ 0 & 0 & 0 \end{bmatrix} & \mathbf{O} \\ \begin{bmatrix} 1 & 1 \\ 0 & 1 \end{bmatrix} \end{array} \right] \left[\begin{array}{cc} \begin{bmatrix} 1 & 2 \\ 3 & 1 \\ 4 & 0 \end{bmatrix} & \begin{bmatrix} 0 & 1 \\ 1 & 0 \\ 1 & 1 \end{bmatrix} \\ \begin{bmatrix} 1 & 1 \\ 1 & 0 \end{bmatrix} & \mathbf{O} \end{array} \right]$$

$$= \left[\begin{array}{cc} \begin{bmatrix} 19 & 4 \\ 12 & 3 \\ 14 & 7 \\ 1 & 0 \end{bmatrix} & \begin{bmatrix} 5 & 4 \\ 3 & 3 \\ 3 & 3 \\ 0 & 0 \end{bmatrix} \end{array} \right].$$

Let us now look at another method of fast matrix multiplication. Suppose that F is a field and n is a positive integer. Let $A = [a_{ij}]$ and $B = [b_{ij}]$ belong to $\mathcal{M}_{n \times n}(F)$ and let $C = AB$. In order to compute each of the n^2 entries in C by straightforward application of the definitions, we need to carry out n multiplications and $n-1$ additions, so that in all we need n^3 multiplications and $n^3 - n^2$ additions to calculate C. In other words, the number of arithmetic operations in F needed to compute C is on the order of n^c, where $c = 3$. Could we do better using a more sophisticated approach? The answer is affirmative, and we will illustrate this using a particular method, known as the Strassen-Winograd computational algorithm.

Let us first consider the case $n = 2$. Given A and B as above, we can compute the scalars

$$p_0 = (a_{11} + a_{12})(b_{11} + b_{12}),$$
$$p_1 = (a_{11} + a_{22})b_{11},$$
$$p_2 = a_{11}(b_{12} - b_{22}),$$
$$p_3 = (-a_{11} + a_{21})(b_{11} + b_{12}),$$
$$p_4 = (a_{11} + a_{12})b_{22},$$
$$p_5 = a_{22}(-b_{11} + b_{21}),$$
$$p_6 = (a_{12} - a_{22})(b_{21} + b_{22}).$$

Then

$$c_{11} = p_0 + p_5 - p_4 + p_6,$$
$$c_{12} = p_2 + p_4,$$
$$c_{21} = p_1 + p_5,$$
$$c_{22} = p_0 - p_1 + p_2 + p_3.$$

In this computation we used 7 multiplications and 18 additions, as opposed to 8 multiplications and 4 additions using the ordinary method. Now suppose that

$n = 4$. We can think of A and B as made up of blocks of size 2×2: $A = \begin{bmatrix} A_{11} & A_{12} \\ A_{21} & A_{22} \end{bmatrix}$ and $B = \begin{bmatrix} B_{11} & B_{12} \\ B_{21} & B_{22} \end{bmatrix}$. We now calculate matrices $P_1, \ldots, P_6 \in \mathcal{M}_{2 \times 2}(F)$ as above and use them to compute $C = \begin{bmatrix} C_{11} & C_{12} \\ C_{21} & C_{22} \end{bmatrix} = AB$. To do this we need a total of 49 multiplications and $198 = 7 \cdot 18 + 4 \cdot 18$ additions, as opposed to 64 multiplications and 46 additions using the ordinary method. In general, if $n = 2^h$ then the number of multiplications needed to compute AB by the Strassen-Winograd method is $M(h) = 7^h$ and the number of additions is $A(h) = 6(7^h - 4^h)$. Thus $M(h) + A(h) < 7^{h+1}$ and we see that the number of arithmetic operations in F needed is on the order of n^c, where $c \leq \log_2 7 = 2.807\ldots$, as opposed to an order of n^3 by the ordinary method. If n is large, this can lead to a considerable savings in computational power. More sophisticated algorithms of this sort can reduce the number of arithmetic operations needed to an order of n^c, where $c < 2.4$.

Problems

1. Find the linear transformations from $\mathbb{Q}^4$ to $\mathbb{Q}^3$ represented by the matrix

$$\begin{bmatrix} 1 & 2 & 3 & 4 \\ 1 & 2 & 4 & 3 \\ 2 & 3 & 1 & 4 \end{bmatrix}$$

with respect to each of the following pairs of bases:

(i) $\{[1,0,0,0], [0,1,0,0], [0,0,1,0], [0,0,0,1]\}$ and $\{[1,0,0], [0,1,0], [0,0,1]\}$;

(ii) $\{[1,1,1,1], [0,1,1,1], [0,0,1,1], [0,0,0,1]\}$ and $\{[1,1,1], [0,1,1], [0,0,1]\}$;

(iii) $\{[1,0,0,1], [0,1,0,1], [0,0,1,1], [0,0,0,1]\}$ and $\{[2,0,0], [0,3,1], [0,0,1]\}$;

(iv) $\{[1,2,0,0], [0,1,2,0], [0,0,1,2], [0,0,0,1]\}$ and $\{[1,2,3], [2,1,2], [2,2,2]\}$.

2. Let V be the subspace of $\mathbb{R}[X]$ consisting of all those polynomials having degree at most 2 and let W be the subspace of $\mathbb{R}[X]$ consisting of all those polynomials having degree at most 3. Let $\alpha: V \to W$ be the linear transformation defined by $\alpha: a + bX + cX^2 \mapsto (a+b) + (b+c)X + (a+c)X^2 + (a+b+c)X^3$. Find $\Phi_{BD}(\alpha)$ when

(i) B and D are the canonical bases;

(ii) $B = \{1, X+1, X^2+X+1\}$ and $D = \{X^3 - X^2, X^2 - X, X - 1, 1\}$;

(iii) $B = \{X^2, X, 1\}$ and $D = \{X^3 - 3, X^2 - 3, X - 1, -1\}$.

3. Let $\alpha: \mathbb{R}^3 \to \mathbb{R}^2$ be the linear transformation defined by

$$\alpha: [a, b, c] \mapsto [a + b + c, b + c].$$

Find the matrices which represent α with respect to the following bases:

(i) The canonical bases;

(ii) $\{[-1,0,2], [0,1,1], [3,-1,0]\}$ and $\{[-1,1], [1,0]\}$;

(iii) $\{[3,-1,0], [0,1,1], [-1,0,2]\}$ and $\{[1,0], [-1,1]\}$;

(iv) $\{[0,1,1], [3,-1,0], [-1,0,2]\}$ and $\{[-1,1], [1,0]\}$.

4. Let α be the endomorphism of the vector space $\mathbb{R}^3$ defined by $[a, b, c] \mapsto$ $[3a + 2b, -a - c, a + 3b]$. Find $\Phi_{BB}(\alpha)$ for each of the following bases B of $\mathbb{R}^3$ over $\mathbb{R}$:

(i) $\{[1, -1, 0], [1, 0, -1], [0, 1, 0]\}$;
(ii) $\{[-1, -1, -1], [2, 0, 3], [1, 2, 3]\}$;
(iii) $\{[1, 1, 2], [1, 0, 0], [0, -5, 0]\}$.

5. Let α be the endomorphism of the vector space $\mathbb{R}^3$ over $\mathbb{R}$ represented with respect to some basis by the matrix $\begin{bmatrix} 0 & 2 & -1 \\ -2 & 5 & -2 \\ -4 & 8 & -3 \end{bmatrix}$. Is α idempotent?

6. Let V be the subspace of $\mathbb{R}[X]$ consisting of all polynomials of degree at most 2, and let $B = \{1, X, X^2\}$ and $D = \{1, 1 + X, 3 + 4X + 2X^2\}$ be bases for V. Let α be the endomorphism of V satisfying $\Phi_{BB}(\alpha) = \begin{bmatrix} 1 & 1 & 1 \\ 0 & 2 & 2 \\ 0 & 0 & 3 \end{bmatrix}$. Find $\Phi_{DD}(\alpha)$.

7. Let V be the subspace of $\mathbb{R}^3$ with basis $\{[1, 1, 0], [0, 1, 1]\}$ and let W be the subspace of $\mathbb{R}^5$ with basis $\{[1, 1, 1, 0, 0], [0, 0, 1, 1, 0], [0, 0, 0, 1, 1]\}$. Let $\alpha: V \to \mathbb{R}^5$ be the linear transformation defined by

$$\alpha: [a, b, c] \mapsto [a + b + c, a + b + c, a - b + c, -2b, 0].$$

Show that $im(\alpha) \subseteq W$ and find the matrix which represents α with respect to the given bases.

8. Let W be the subspace of $\mathbb{R}^{\mathbb{R}}$ generated by the linearly-independent set of vectors $D = \{1, x, e^x, xe^x\}$ and let δ be the endomorphism of W defined by $\delta: f \mapsto f'$. Find the matrix represented δ with respect to D.

9. Let $D = \{1 + i, 2 + i\}$ be a basis for $\mathbb{C}$ as a vector space over $\mathbb{R}$ and let α be the endomorphism of $\mathbb{C}$ given by $\alpha: z \mapsto \bar{z}$. Find the matrix representing α with respect to D.

10. Let $F = \mathbb{Z}/(3)$. Let $\alpha: F^3 \to F^2$ be the linear transformation defined by $\alpha: [a, b, c] \mapsto [a - b, 2a - c]$ and let $\beta: F^2 \to F^4$ the linear transformation defined by $\beta: [a, b] \mapsto [b, a, 2b, 2a]$. Find the matrix which represents $\beta\alpha$ with respect to the canonical bases.

11. Calculate $\begin{bmatrix} 1 & 3 & 1 \\ 2 & 1 & 1 \\ 1 & 2 & 3 \end{bmatrix} \begin{bmatrix} 1 & 2 & 2 \\ 4 & 3 & 2 \\ 1 & 4 & 2 \end{bmatrix}$ in $\mathcal{M}_{3\times3}(\mathbb{Z}/(5))$.

12. If $A = \begin{bmatrix} 0 & 1 & 2 & 3 \\ 1 & 3 & 4 & 0 \\ 3 & 2 & 0 & 1 \end{bmatrix} \in \mathcal{M}_{3\times4}(\mathbb{R})$, find all matrices $B \in \mathcal{M}_{4\times3}(\mathbb{R})$ satisfying $AB = \begin{bmatrix} 1 & 0 & 0 \\ 0 & 1 & 0 \\ 0 & 0 & 1 \end{bmatrix}$.

13. If $A = \begin{bmatrix} 1 & -1 & 2 \\ 0 & 1 & 2 \\ -1 & 3 & 1 \\ 2 & 1 & -1 \end{bmatrix} \in \mathcal{M}_{4\times3}(\mathbb{R})$, find all matrices $B \in \mathcal{M}_{3\times4}(\mathbb{R})$ satisfying $BA = \begin{bmatrix} 1 & 0 & 0 \\ 0 & 1 & 0 \\ 0 & 0 & 1 \end{bmatrix}$.

14. Find the set of all matrices $A \in \mathcal{M}_{4\times3}(\mathbb{R})$ satisfying

$$\begin{bmatrix} 0 & 1 & 0 & 0 \\ 0 & 0 & 1 & 0 \\ 0 & 0 & 0 & 1 \\ 0 & 0 & 0 & 0 \end{bmatrix} A = A \begin{bmatrix} 0 & 1 & 0 \\ 0 & 0 & 1 \\ 0 & 0 & 0 \end{bmatrix}.$$

15. Let n be a positive integer. A matrix $[a_{ij}] \in \mathcal{M}_{n\times n}(\mathbb{R})$ is called a *stochastic matrix* if and only if $a_{ij} \geq 0$ for all $1 \leq i, j \leq n$ and $\sum_{i=1}^{n} a_{ij} = 1$ for all $1 \leq j \leq n$. Show that the set of all stochastic matrices is closed under matrix multiplication.

16. Do the elements $A = \begin{bmatrix} 1 & 0 & -1 & 0 \\ 0 & 1 & 0 & -1 \\ 1 & 0 & -1 & 0 \\ 0 & 1 & 0 & -1 \end{bmatrix}$ and $B = \begin{bmatrix} 4 & -1 & -1 & 0 \\ -1 & 4 & 0 & -1 \\ 1 & 0 & 2 & -1 \\ 0 & 1 & -1 & 2 \end{bmatrix}$ of $\mathcal{M}_{4\times4}(\mathbb{Q})$ satisfy $AB = BA$?

17. Let F be a field, let n be a positive integer, and let $H \in \mathcal{M}_{n\times n}(F)$ be a given matrix. Let $\alpha: \mathcal{M}_{n\times n}(F) \to \mathcal{M}_{n\times n}(F)$ be the function defined by $\alpha: A \mapsto AH + HA$. Is α an endomorphism of the vector space $\mathcal{M}_{n\times n}(F)$?

18. Let α be the endomorphism of $\mathbb{R}^4$ represented with respect to the canonical bases by the matrix

$$\begin{bmatrix} 3 & -1 & 0 & 0 \\ -1 & 2 & -1 & 0 \\ 0 & -1 & 2 & -1 \\ 0 & 0 & -1 & 1 \end{bmatrix}.$$

If $v \in \mathbb{R}^4$ is a vector satisfying the condition that all components of $\alpha(v)$ are nonnegative. Show that all components of v are nonnegative.

19. Let V and W be vector spaces over a field F, which are not necessarily finitely generated over F. Pick bases $\{v_i \mid i \in \Omega\}$ and $\{w_j \mid j \in \Lambda\}$ for V and W respectively. Let $p: \Omega \times \Lambda \to F$ be function satisfying the condition that the set $\{j \in \Lambda \mid p(i, j) \neq 0_F\}$ is finite for all $i \in \Omega$ and let $\alpha_p: V \to W$ be the function defined as follows: if $v = \sum_{i \in \Gamma} a_i v_i$, where Γ is a finite subset of Ω and the a_i are scalars in F, then $\alpha_p(v) = \sum\{a_i p(i, j) w_j \mid i \in \Gamma; j \in \Lambda\}$. Show that α_p is a linear transformation and that all of the linear transformations from V to W are of this form.

20. Let k and n be positive integers and let $v \in \mathbb{R}^n$. If $A \in \mathcal{M}_{k\times n}(\mathbb{R})$, show that $Av^T = \begin{bmatrix} 0 \\ \vdots \\ 0 \end{bmatrix}$ if and only if $A^T A v^T = \begin{bmatrix} 0 \\ \vdots \\ 0 \end{bmatrix}$.

21. Let $A \in \mathcal{M}_{3\times 2}(\mathbb{R})$ and $B \in \mathcal{M}_{2\times 3}(\mathbb{R})$ be matrices satisfying

$$AB = \begin{bmatrix} 8 & 2 & -2 \\ 2 & 5 & 4 \\ -2 & 4 & 5 \end{bmatrix}.$$

Calculate BA.

22. Find matrices $A \in \mathcal{M}_{3\times 2}(\mathbb{Q})$ and $B \in \mathcal{M}_{2\times 3}(\mathbb{Q})$ satisfying

$$AB = \begin{bmatrix} 1 & 1 & 1 \\ -2 & 0 & -6 \\ 0 & 1 & -2 \end{bmatrix}.$$

23. Let F be a field and let n be a positive integer. Let W be a nontrivial subspace of $V = \mathcal{M}_{n\times n}(F)$ satisfying the condition that if A and B belong to W then so do AB and BA. Show that $W = V$.

CHAPTER VII

RINGS OF SQUARE MATRICES

Given a positive integer n and a field F, we want to look in more detail at the algebraic structure of the set $\mathcal{M}_{n\times n}(F)$ of all $n \times n$ matrices over F. We already know that this set is closed under addition and multiplication of matrices and it is easily verified (do it!) that it is in fact a ring. The neutral element of is ring with respect to addition is the matrix $\mathbf{O}$ all of the entries of which are 0_F. The neutral element of this ring with respect to multiplication is the matrix $I = [a_{ij}]$ defined by

$$a_{ij} = \begin{cases} 1_F, & \text{for } i = j \\ 0_F, & \text{otherwise,} \end{cases}$$

called the *identity matrix*. All of this, of course, should not be surprising since we have already seen that if V is a vector space of dimension n over F and if D is a basis for V then the function $\Phi_{DD}: End(V) \to \mathcal{M}_{n\times n}(F)$ is bijective and preserves addition and multiplication.

We begin looking at spaces of the form $\mathcal{M}_{n\times n}(V)$, where V is an arbitrary vector space over a field F, and identifying some special matrices in them. By restricting to the special case $V = F$, we then obtain some significant subrings of the ring $\mathcal{M}_{n\times n}(F)$.

(1) A matrix $M \in \mathcal{M}_{n\times n}(V)$ is called a *scalar matrix* if and only if it is of the form Iv for some $v \in V$. In other words, $M = [v_{ij}]$ where $v_{ij} = v$ when $i = j$ and $v_{ij} = 0_V$ otherwise. Such matrices arise very frequently.

EXAMPLE

Let V be a vector space of finite dimension t over a field F and let $\{y_1, \ldots, y_t\}$ be a basis for V over F. If $M = [v_{ij}] \in \mathcal{M}_{n\times n}(V)$ then for each $1 \leq i, j \leq n$ there exist scalars $a_{ij1}, \ldots, a_{ijt}$ such that $v_{ij} = \sum_{h=1}^{t} a_{ijh}y_h$. Therefore $M = A_1Y_1 + \cdots + A_tY_t$, where, for each $1 \leq h \leq t$, $A_h = [a_{ijh}] \in \mathcal{M}_{n\times n}(F)$ and Y_h is the scalar matrix $[u_{ij}] \in \mathcal{M}_{n\times n}(V)$ with $u_{ii} = y_h$ for all $1 \leq i \leq n$ and $u_{ij} = 0$ if $1 \leq i \neq j \leq n$.

For the special case $V = F$ we note that scalar matrices in $\mathcal{M}_{n\times n}(F)$ commute with all other matrices, i.e., if $A = cI$ is a scalar matrix and $B \in \mathcal{M}_{n\times n}(F)$ is arbitrary, then $AB = BA$. Since $(cI)B$ is just the same as the scalar product cB each $c \in F$ and $B \in \mathcal{M}_{n\times n}(F)$, we see that multiplication of a matrix by a scalar is really just a special case of matrix multiplication.

The sum, difference, and product of two scalar matrices in $\mathcal{M}_{n\times n}(F)$ are again scalar matrices as are the matrices $\mathbf{O}$ and I. Therefore, the set of all scalar matrices is a subring of $\mathcal{M}_{n \times n}(F)$. Moreover, if $0_F \neq c \in F$ then the scalar matrix cI is invertible, with $(cI)^{-1} = c^{-1}I$. Therefore this subring is in fact a subfield.

(2) A matrix $M = [v_{ij}] \in \mathcal{M}_{n \times n}(V)$ is a *diagonal matrix* if and only if $v_{ij} = 0_V$ whenever $i \neq j$. Thus we do not demand that all elements of the diagonal of a diagonal matrix be nonzero, nor that they be equal. Scalar matrices are, of course, diagonal matrices, but the converse is not true.

In the ring $\mathcal{M}_{n \times n}(F)$, diagonal matrices do not necessarily commute with all other matrices but do commute with each other: if A and B are diagonal matrices in $\mathcal{M}_{n \times n}(F)$ then $AB = BA$. The sum, difference, and product of two diagonal matrices in $\mathcal{M}_{n \times n}(F)$ are again diagonal matrices as are the matrices O and I. Therefore the set of all diagonal matrices is a subring of $\mathcal{M}_{n \times n}(F)$. This ring will be denoted by $\mathcal{D}_{n \times n}(F)$.

(3) A matrix $M = [v_{ij}] \in \mathcal{M}_{n \times n}(V)$ is an *upper-triangular matrix* if and only if $v_{ij} = 0_V$ when $i > j$. Clearly all diagonal matrices are upper triangular, but the converse is not true. The sum, difference, and product of two upper-triangular matrices in $\mathcal{M}_{n \times n}(F)$ are again upper-triangular matrices as are the matrices O and I. Therefore the set of all upper-triangular matrices is a subring of $\mathcal{M}_{n \times n}(F)$. This ring will be denoted by $\mathcal{U}\mathcal{T}_{n \times n}(F)$.

Similarly, a matrix $M = [v_{ij}] \in \mathcal{M}_{n \times n}(V)$ is a *lower-triangular matrix* if and only if $v_{ij} = 0_V$ when $i < j$. The set of all lower-triangular matrices also forms a subring of $\mathcal{M}_{n \times n}(F)$. This ring will be denoted by $\mathcal{L}\mathcal{T}_{n \times n}(F)$. It is evident that $\mathcal{U}\mathcal{T}_{n \times n}(F) \cap \mathcal{L}\mathcal{T}_{n \times n}(F) = \mathcal{D}_{n \times n}(F)$.

EXAMPLE

In $\mathcal{M}_{3 \times 3}(\mathbb{Q})$:

(1) The matrix $\begin{bmatrix} 2 & 0 & 0 \\ 0 & 2 & 0 \\ 0 & 0 & 2 \end{bmatrix}$ is a scalar matrix.

(2) The matrix $\begin{bmatrix} 2 & 0 & 0 \\ 0 & 0 & 0 \\ 0 & 0 & 3 \end{bmatrix}$ is a diagonal matrix which is not a scalar matrix.

(3) The matrix $\begin{bmatrix} 2 & 1 & 1 \\ 0 & 3 & 0 \\ 0 & 0 & 1 \end{bmatrix}$ is an upper-triangular matrix which is not a diagonal matrix.

(4) The matrix $\begin{bmatrix} 2 & 0 & 0 \\ 1 & 0 & 0 \\ 1 & 3 & 5 \end{bmatrix}$ is a lower-triangular matrix which is not a diagonal matrix.

A matrix $A = [v_{ij}] \in \mathcal{M}_{n \times n}(V)$ is a *symmetric matrix* if and only if it equals its own transpose, i.e. if and only if $v_{ij} = v_{ji}$ for all $1 \leq i, j \leq n$. Clearly diagonal matrices are symmetric and an upper-triangular (or lower-triangular) matrix is symmetric if and only if it is a diagonal matrix. If V is a matrix over a field of characteristic other than 2 and if $A \in \mathcal{M}_{n \times n}(V)$ then the matrix $\frac{1}{2}(A + A^T)$ is symmetric.

For vector spaces V over an arbitrary field F, The set of all symmetric matrices in $\mathcal{M}_{n \times n}(V)$ is clearly a subspace of $\mathcal{M}_{n \times n}(V)$ as a vector space over F, but the set of all symmetric matrices in $\mathcal{M}_{n \times n}(F)$ is not necessarily a subring of $\mathcal{M}_{n \times n}(F)$, since it is not necessarily closed under matrix multiplication. We will denote this subspace by $\mathcal{S}_{n \times n}(V)$. If the characteristic of F is not equal to 2, the dimension of $\mathcal{S}_{n \times n}(V)$ over F is $\frac{1}{2}n(n+1)$, a fact which can be proven by induction on n.

EXAMPLE

The matrices $\begin{bmatrix} 2 & 3 & 1 \\ 3 & 2 & 0 \\ 1 & 0 & 1 \end{bmatrix}$ and $\begin{bmatrix} 1 & 2 & 1 \\ 2 & 0 & 0 \\ 1 & 0 & 3 \end{bmatrix}$ belong to $\mathcal{S}_{3 \times 3}(\mathbb{Q})$ but their product is $\begin{bmatrix} 9 & 4 & 5 \\ 7 & 6 & 3 \\ 2 & 2 & 4 \end{bmatrix} \notin \mathcal{S}_{3 \times 3}(\mathbb{Q})$.

Since not every endomorphism of a finite-dimensional vector space differing from σ_0 has a multiplicative inverse, we see that not every nonzero matrix in $\mathcal{M}_{n \times n}(F)$ has a multiplicative inverse. A matrix which does not have such an inverse is said to be *singular*; a matrix which has a multiplicative inverse is *nonsingular*. Thus, if A is a nonsingular matrix in $\mathcal{M}_{n \times n}(F)$ then there exists a matrix $B \in \mathcal{M}_{n \times n}(F)$ satisfying $BA = AB = I$. This matrix is necessarily unique, since if $CA = AC = I$ then $C = C(AB) = (CA)B = B$. We will denote the matrix B by A^{-1}.

If V is a vector space of dimension n over F, if D is a basis for V over F, and if $\alpha \in End(V)$, then the matrix $\Phi_{BB}(\alpha) \in \mathcal{M}_{n \times n}(F)$ is nonsingular if and only if $\alpha \in Aut(V)$. Note that as a consequence of this observation and of Proposition 5.2, we see that in order to show that $B = A^{-1}$ it suffices to show that *one* of the conditions $AB = I$ or $BA = I$ holds. Calculating the inverse of a nonsingular matrix efficiently is one of the major problems of computational linear algebra.

EXAMPLE

We have already seen that every scalar matrix not equal to $\mathbf{O}$ is nonsingular. If F is a field and $A = [a_{ij}] \in \mathcal{D}_{n \times n}(F)$ satisfies the condition that $a_{ii} \neq 0_F$ for all $1 \leq i \leq n$ then A is nonsingular and A^{-1} is the matrix $[b_{ij}] \in \mathcal{D}_{n \times n}(F)$ with $b_{ii} = a_{ii}^{-1}$ for all $1 \leq i \leq n$.

More generally, a matrix $[a_{ij}] \in \mathcal{UT}_{n \times n}(F) \cup \mathcal{LT}_{n \times n}(F)$ is nonsingular if and only if $a_{ii} \neq 0_F$ for all $1 \leq i \leq n$.

In Chapter 5 we defined the elementary automorphisms $\alpha_{hk}, \alpha_{h;c}, \alpha_{hk;c}$ of the vector space F^n for all $1 \leq h \neq k \leq n$ and $c \in F$. If D is the canonical basis of F^n, then matrices of the form $E_{hk} = \Phi_{DD}(\alpha_{hk})$, $E_{h;c} = \Phi_{DD}(\alpha_{h;c})$, and $E_{hk;c} = \Phi_{DD}(\alpha_{hk;c})$ in $\mathcal{M}_{n \times n}(F)$ are *elementary matrices*. All of these matrices are nonsingular. These matrices can be characterized as follows:

(1) E_{hk} is the matrix formed from I by interchanging the hth and kth columns;

(2) $E_{h;c}$ is the matrix formed from I by multiplying the hth column by the scalar c;

(3) $E_{hk;c}$ is the matrix formed from I by adding c times the kth column to the hth column.

The inverses of these matrices are easily calculated: $E_{hk}^{-1} = E_{hk}$, $E_{h;c}^{-1} = E_{h,c^{-1}}$, and $E_{hk;c}^{-1} = E_{hk;-c}$ for all $1 \leq h \neq k \leq n$ and $c \in F$.

(7.1) PROPOSITION. *Let F be a field, let n be a positive integer, and let $A, B \in \mathcal{M}_{n \times n}(F)$. Then:*

(1) *The matrix A is nonsingular if and only if A^T is nonsingular. In this case, $(A^T)^{-1} = (A^{-1})^T$.*

(2) *The matrices A and B are both nonsingular if and only if AB is nonsingular. Moreover, in this case, $(AB)^{-1} = B^{-1}A^{-1}$.*

PROOF. (1) If A and B are matrices in $\mathcal{M}_{n \times n}(F)$ then $(AB)^T = B^T A^T$ as a direct consequence of the definition. In particular, if $B = A^{-1}$ then $AB = I = I^T = (AB)^T = B^T A^T$ so A^T is nonsingular and $(A^T)^{-1} = (A^{-1})^T$. The reverse direction holds since $A^{TT} = A$.

(2) Assume that both A and B are nonsingular. Then

$$I = B^{-1}IB = B^{-1}(A^{-1}A)B = (B^{-1}A^{-1})(AB)$$

and so AB is nonsingular with $(AB)^{-1} = B^{-1}A^{-1}$. Conversely, if AB is nonsingular and $C = (AB)^{-1}$ then $(CA)B = C(AB) = I = (AB)C = A(BC)$ so both A and B are nonsingular. $\square$

COROLLARY. *Let F be a field, let n be a positive integer, and assume that $A, B \in \mathcal{M}_{n \times n}(F)$. Then AB is nonsingular if and only if BA is nonsingular.*

(7.2) PROPOSITION. *Let F be a field, let n be a positive integer, and let $A, B \in \mathcal{M}_{n \times n}(F)$, with A nonsingular. Then there exist unique elements C and D of $\mathcal{M}_{n \times n}(F)$ satisfying $CA = B = AD$.*

PROOF. If $C = BA^{-1}$ and $D = A^{-1}B$ then surely $CA = B = AD$. Conversely, if $C'A = B$ then $C' = C'AA^{-1} = BA^{-1} = C$ and so C is unique. The uniqueness of D is proven similarly. $\square$

EXAMPLE

The matrices C and D mentioned in Proposition 7.2 need not necessarily be the same. For example, if $A = \begin{bmatrix} 1 & 2 \\ 0 & 1 \end{bmatrix}$ and $B = \begin{bmatrix} 1 & 0 \\ 2 & 1 \end{bmatrix}$ in $\mathcal{M}_{2 \times 2}(\mathbb{Q})$ then $C = \begin{bmatrix} 1 & -2 \\ 2 & -3 \end{bmatrix}$ and $D = \begin{bmatrix} -3 & -2 \\ 2 & 1 \end{bmatrix}$.

(7.3) PROPOSITION. *Let F be a field, let n be a positive integer. For a matrix $C = [c_{ij}] \in \mathcal{M}_{n \times n}(F)$ the following conditions are equivalent:*

(1) *C is nonsingular;*

(2) *If V is a vector space of dimension n over F having a basis D then there exists an automorphism $\alpha \in \mathrm{Aut}(V)$ satisfying $\Phi_{DD}(\alpha) = C$;*

   (3) *The columns of C are distinct and form a linearly-independent subset of $\mathcal{M}_{n\times1}(F)$;*

   (4) *The rows of C are distinct and form a linearly-independent subset of F^n.*

PROOF. (1) $\Leftrightarrow$ (2): Assume (1). Then there exists a matrix $B \in \mathcal{M}_{n\times n}(F)$ satisfying $BC = CB = I$. Let $\alpha, \beta \in End(V)$ be endomorphisms satisfying $\Phi_{DD}(\alpha) = C$ and $\Phi_{DD}(\beta) = B$. Then $\Phi_{DD}(\alpha\beta) = CB = I = \Phi_{DD}(\sigma_1)$ and $\Phi_{DD}(\beta\alpha) = BC = I = \Phi_{DD}(\sigma_1)$. Since Φ_{DD} is monic, this implies that $\alpha\beta = \sigma_1 = \beta\alpha$ and so $\alpha \in Aut(V)$.

Conversely, assume (2). If $\Phi_{DD}(\alpha) = C$ and if there exists an automorphism β of V satisfying $\alpha\beta = \sigma_1 = \beta\alpha$ then $BC = CB = I$, where $B = \Phi_{DD}(\beta)$, and so C is nonsingular.

(2) $\Leftrightarrow$ (3): Let $V = \mathcal{M}_{n\times1}(F)$ and let $D = \{v_1, \ldots, v_n\}$ be the canonical basis of V and let α be the endomorphism of V satisfying $\Phi_{DD}(\alpha) = C$. Assume (2). Then α is an automorphism of V and so, in particular, it is monic. In particular, if $0_V \neq v \in V$ then $\alpha(v) \neq 0_V$. But if C has two equal columns or if the set of columns of C is linearly dependent then there exist scalars $a_1, \ldots, a_n$ not all of which equal 0_F, satisfying

$$\begin{bmatrix} 0_F \\ \vdots \\ 0_F \end{bmatrix} = \sum_{j=1}^{n} a_j \begin{bmatrix} c_{1j} \\ \vdots \\ c_{nj} \end{bmatrix} = C \begin{bmatrix} a_1 \\ \vdots \\ a_n \end{bmatrix} = \alpha\left(\begin{bmatrix} a_1 \\ \vdots \\ a_n \end{bmatrix} \right)$$

and this is a contradiction. Thus we have (3). Conversely, assume (3). Then the set of columns of C is a basis for $\mathcal{M}_{n\times1}(F)$. Moreover, for each $1 \le i \le n$ the ith column of C is just $\alpha(v_i)$ and so we see that α is an automorphism of V.

(2) $\Leftrightarrow$ (4): This follows directly from the equivalence of (2) and (3). $\square$

(7.4) PROPOSITION. *Let F be a field, let n be a positive integer. A matrix in $\mathcal{M}_{n\times n}(F)$ is nonsingular if and only if it is a product of elementary matrices.*

PROOF. We have already seen that every elementary matrix is nonsingular and so, by Proposition 7.1, every product of elementary matrices is nonsingular. Therefore all we have to show is that every nonsingular matrix in $\mathcal{M}_{n\times n}(F)$ is a product of elementary matrices. First, let us check the meaning of multiplication of a matrix on the left by an elementary matrix. Indeed, if $B \in \mathcal{M}_{n\times n}(F)$ then:

   (1) $E_{hk}B$ is the matrix obtained from B be interchanging the hth and kth rows of B;

   (2) $E_{h;c}B$ is the matrix obtained from B obtained by multiplying the hth row of B by the scalar c;

   (3) $E_{hk;c}B$ is the matrix obtained from B by adding c times the hth row of B to the kth row.

Now assume that $A = [a_{ij}] \in \mathcal{M}_{n\times n}(F)$ is a nonsingular matrix. Then, needless to say, $B = A^{-1}$ is also nonsingular. By Proposition 7.3, it follows that the columns of B are distinct and form a linearly-independent set. In particular, the first column must have a nonzero component, say b_{h1}. Therefore the matrix $E_{h1}B$ has a nonzero element as the $(1, 1)$ entry. After multiplying this matrix on the left by $E_{1;c}$ for the appropriate value of c, we obtain a matrix $C = [c_{ij}] \in \mathcal{M}_{n \times n}(F)$ with 1_F as the $(1, 1)$ entry.

If $2 \leq t \leq n$ we now multiply the matrix C on the left by $E_{t1;-c_{t1}}$, we obtain a matrix with 0_F as the $(t, 1)$ entry. Hence, after multiplying B on the left by a number of elementary matrices, we obtain a matrix the first column of which is

$$\begin{bmatrix} 1_F \\ 0_F \\ \vdots \\ 0_F \end{bmatrix}.$$ This matrix is a product of nonsingular matrices and hence is nonsingular.

Moreover, the columns of this new matrix are distinct and still form a linearly-independent set. Therefore the second column must contain a nonzero component in a row other than the first row. Again, by multiplying the matrix on the left by appropriate elementary matrices, we obtain a nonsingular matrix the first two

columns of which look like $\begin{bmatrix} 1_F & 0_F \\ 0_F & 1_F \\ 0_F & 0_F \\ \vdots & \vdots \\ 0_F & 0_F \end{bmatrix}$. Continuing in this manner, we obtain a

matrix E which is a product of elementary matrices and which satisfies $EB = I$. Therefore $A = B^{-1} = E$ is a product of elementary matrices, as desired. $\square$

Actually, we do not need all of these elementary matrices. For instance, consider a matrix E_{hk} where $h < k$. If $k \neq h + 1$ then one easily checks that $E_{hk} = E_{h,h+1} E_{h+1,k} E_{h,h+1}$ and so, continuing in a recursive manner, we see that E_{hk} can itself be written as a product of an odd number of elementary matrices of the form $E_{i,i+1}$.

Proposition 7.4 provides us with a method for calculating A^{-1}, when A is a nonsingular matrix in $\mathcal{M}_{n \times n}(F)$: we form the matrix $[I\ A]$ in $\mathcal{M}_{n \times 2n}(F)$ and multiply it on the left by the succession of elementary matrices which transform A into I. The result will transform $[I\ A]$ into $[A^{-1}\ I]$. Multiplication on the left by an elementary matrix is called a *elementary operation*. Thus the elementary operations are $\psi_{hk} \colon A \mapsto E_{hk} A$; $\psi_{h;c} \colon A \mapsto E_{h;c} A$; $\psi_{hk;c} \colon A \mapsto E_{hk;c} A$.

EXAMPLE

If $A = \begin{bmatrix} 1 & 2 & 3 \\ 2 & 3 & 0 \\ 0 & 1 & 2 \end{bmatrix} \in \mathcal{M}_{3 \times 3}(\mathbb{Q})$, we begin by forming the matrix

$$\begin{bmatrix} 1 & 0 & 0 & 1 & 2 & 3 \\ 0 & 1 & 0 & 2 & 3 & 0 \\ 0 & 0 & 1 & 0 & 1 & 2 \end{bmatrix}.$$

We now perform the following series of elementary operations:

EXAMPLE (Continued)

$$\begin{bmatrix} 1 & 0 & 0 & 1 & 2 & 3 \\ 0 & 1 & 0 & 2 & 3 & 0 \\ 0 & 0 & 1 & 0 & 1 & 2 \end{bmatrix} \xrightarrow{\psi_{12,-2}}$$

$$\begin{bmatrix} 1 & 0 & 0 & 1 & 2 & 3 \\ -2 & 1 & 0 & 0 & -1 & -6 \\ 0 & 0 & 1 & 0 & 1 & 2 \end{bmatrix} \xrightarrow{\psi_{2;-1}}$$

$$\begin{bmatrix} 1 & 0 & 0 & 1 & 2 & 3 \\ 2 & -1 & 0 & 0 & 1 & 6 \\ 0 & 0 & 1 & 0 & 1 & 2 \end{bmatrix} \xrightarrow{\psi_{2,1;-2}}$$

$$\begin{bmatrix} -3 & 2 & 0 & 1 & 0 & -9 \\ 2 & -1 & 0 & 0 & 1 & 6 \\ 0 & 0 & 1 & 0 & 1 & 2 \end{bmatrix} \xrightarrow{\psi_{2,3,-1}}$$

$$\begin{bmatrix} -3 & 2 & 0 & 1 & 0 & -9 \\ 2 & -1 & 0 & 0 & 1 & 6 \\ -2 & 1 & 1 & 0 & 0 & -4 \end{bmatrix} \xrightarrow{\psi_{3,-4}}$$

$$\begin{bmatrix} -3 & 2 & 0 & 1 & 0 & -9 \\ 2 & -1 & 0 & 0 & 1 & 6 \\ 1/2 & -1/4 & -1/4 & 0 & 0 & 1 \end{bmatrix} \xrightarrow{\psi_{3,1;9}}$$

$$\begin{bmatrix} 3/2 & -1/4 & -9/4 & 1 & 0 & 0 \\ 2 & -1 & 0 & 0 & 1 & 6 \\ 1/2 & -1/4 & -1/4 & 0 & 0 & 1 \end{bmatrix} \xrightarrow{\psi_{3,1;-6}}$$

$$\begin{bmatrix} 3/2 & -1/4 & -9/4 & 1 & 0 & 0 \\ -1 & 1/2 & 3/2 & 0 & 1 & 0 \\ 1/2 & -1/4 & -1/4 & 0 & 0 & 1 \end{bmatrix}.$$

Thus $A^{-1} = \frac{1}{4} \begin{bmatrix} 6 & -1 & -9 \\ -4 & 2 & 6 \\ 2 & -1 & -1 \end{bmatrix}.$

However, one must be very careful in using this method on a computer, for in actual practice accumulated roundoff and truncation error can totally destroy the reliability of the answer. To see this, consider the following example:

EXAMPLE

For each positive integer n, the matrix $H_n = [1/(i+j-1)]$ is called the *Hilbert matrix* in $\mathcal{M}_{n \times n}(\mathbb{Q})$, for it was first studied by David Hilbert in 1894. This matrix often serves as a benchmark to test matrix-inversion programs on the computer since, while the entries of H_n are all between 0 and 1, the entries in H_n^{-1} become very large very rapidly. Thus, for example,

EXAMPLE (Continued)

$$H_6 = \begin{bmatrix} 1 & 1/2 & 1/3 & 1/4 & 1/5 & 1/6 \\ 1/2 & 1/3 & 1/4 & 1/5 & 1/6 & 1/7 \\ 1/3 & 1/4 & 1/5 & 1/6 & 1/7 & 1/8 \\ 1/4 & 1/5 & 1/6 & 1/7 & 1/8 & 1/9 \\ 1/5 & 1/6 & 1/7 & 1/8 & 1/9 & 1/10 \\ 1/6 & 1/7 & 1/8 & 1/9 & 1/10 & 1/11 \end{bmatrix}$$

while

$$H_6^{-1} = \begin{bmatrix} 36 & -630 & 3360 & -7560 & 7560 & -2772 \\ -630 & 14700 & -88200 & 211680 & -220500 & 83160 \\ 3360 & -88200 & 564480 & -1411200 & 1512000 & -582120 \\ -7560 & 211680 & -1411200 & 3628800 & -3969000 & 1552320 \\ 7560 & -220500 & 1512000 & -3969000 & 4410000 & -1746360 \\ -2772 & 83160 & -582120 & 15523230 & -1746360 & 698544 \end{bmatrix}.$$

As a result, if we were trying to invert H_6 using the above algorithm on a computer having only 7-digit accuracy, there would be a good likelihood of an error of 100% in our answer.

Let us recapitulate the algorithm we have used. Given a nonsingular matrix $A \in \mathcal{M}_{n \times n}(F)$, we form the matrix $B = [I \ A] = [b_{ij}]$ in $\mathcal{M}_{n \times 2n}(F)$ and then perform the following operations for $1 \leq h \leq n$:

(I) Exchange row h with one of the rows under it so that we obtain $b_{h,n+h} \neq 0_F$. If this cannot be done, it is a sign that A is singular;

(II) Multiply the new row h by $b_{h,n+h}^{-1}$ to obtain a matrix in which $b_{h,n+h} = 1_F$;

(III) For each $i \neq h$, subtract $b_{i,n+h}$ times row h from row i to obtain a matrix in which $b_{i,n+h} = 0_F$ for all $i \neq h$.

At the end of this process, we have transformed B into the matrix $[A^{-1} \ I]$.

The elements $b_{h,n+h}$ are called the *pivots* of the algorithm. One way of avoiding roundoff and truncation errors, when working with $F = \mathbb{R}$ or $F = \mathbb{Q}$, is to insure in step (I), that we choose not just any nonzero element but the one of maximal absolute value available to us. This refinement of the basic algorithm is known as *partial pivoting*, and it usually suffices in most computational situations. There are more sophisticated pivoting techniques available, but a rule of thumb seems to be that when they are necessary it is probably best to consider a different inversion algorithm altogether.

There are other methods of inverting a nonsingular matrix. Certainly it is easy to invert an upper-triangular or lower-triangular matrix having a diagonal composed of nonzero elements (if one of the diagonal elements of such a matrix is zero then the matrix is singular). Let F be a field and let n be a positive integer. Assume that $A \in \mathcal{M}_{n \times n}(F)$ is a nonsingular matrix which we want to invert. If we can write $A = LU$, where $L \in \mathcal{LT}_{n \times n}(F)$ and $U \in \mathcal{UT}_{n \times n}(F)$, then $A^{-1} = U^{-1}L^{-1}$ and we are done. How do we go about finding the matrices L and U? If $A = [a_{ij}]$, $U = [b_{ij}]$, and $L = [c_{ij}]$ then for each $1 \leq i, j \leq n$ we have to find elements b_{ij}

$(i \geq j)$ and elements c_{ij} $(i \leq j)$ such that $a_{ij} = \sum_{k=1}^{n} c_{ik}b_{kj}$. The number of possibly nonzero summands depends on whether i is less than, greater than, or equal to j but in all we have n^2 equations in $n^2 + n$ indeterminates. This allows us a certain amount of leeway and so we can make the additional restriction that the diagonal elements of L all be equal to 1_F. (This is also good for computational reasons – it allows us to store the elements of U and the elements of $L - I$ together as one $n \times n$ matrix in the computer.) Now we have $n^2 + n$ equations in the same number of unknowns, and we can solve them by the following method, known as *Crout's algorithm*:

(I) Set $c_{ii} = 1$ for $1 \leq i \leq n$;

(II) For each $1 \leq j \leq n$ first calculate

$$b_{ij} = a_{ij} - \sum_{k=1}^{i-1} c_{ik}b_{kj}$$

for all $1 \leq i \leq j$ and then

$$c_{ij} = b_{jj}^{-1}\left(a_{ij} - \sum_{k=1}^{j-1} c_{ik}b_{kj}\right)$$

for all $j + 1 \leq i \leq n$.

Sometimes it is also possible to make use of block decompositions of a matrix to calculate its inverse. Let F be a field and let n be a positive integer. Assume that $A \in M_{n \times n}(F)$ is a nonsingular matrix which we can write in the form $\begin{bmatrix} A_{11} & A_{12} \\ A_{21} & A_{22} \end{bmatrix}$, where $A_{11} \in M_{k \times k}(F)$. In order to calculate A^{-1} we must find a matrix of the form $\begin{bmatrix} B_{11} & B_{12} \\ B_{21} & B_{22} \end{bmatrix}$, where $B_{11} \in M_{k \times k}(F)$ and where

$$A_{11}B_{11} + A_{12}B_{21} = I;$$
$$A_{12}B_{12} + A_{12}B_{22} = O;$$
$$A_{21}B_{11} + A_{22}B_{21} = O;$$
$$A_{21}B_{12} + A_{22}B_{22} = I.$$

Suppose that we know that the matrix A_{11} is nonsingular. From the first of these equations, we then conclude that $B_{11} = A_{11}^{-1}(I - A_{12}B_{21})$. If we plug this value into the third equation, we get $B_{21} = -(A_{22} - A_{21}A_{11}^{-1}A_{12})^{-1}A_{21}A_{11}^{-1}$ provided, of course, that the matrix $C = A_{22} - A_{21}A_{11}^{-1}A_{12}$ is in fact nonsingular. Similarly, we get $B_{22} = C^{-1}$ and $B_{12} = -A_{11}^{-1}A_{12}C^{-1}$. In other words, we see that a sufficient condition for A to be nonsingular is that A_{11} and $A_{22} - A_{21}A_{11}^{-1}A_{12}$ to be nonsingular and in this case we know exactly how to compute A^{-1}. Needless to say, this method is of use primarily if we can form a block decomposition of A for which this computation is particularly easy.

Let V be a vector space of finite dimension $n > 0$ over a field F and let $B = \{v_1, \ldots, v_n\}$ and $D = \{w_1, \ldots, w_n\}$ be bases for V over F. For each $1 \leq j \leq n$ there exists elements $q_{1j}, \ldots, q_{nj}$ of F such that $w_j = \sum_{i=1}^{n} q_{ij} v_i$. By Proposition 7.3 the matrix $Q = [q_{ij}]$ is nonsingular. If $v = \sum_{i=1}^{n} a_i v_i = \sum_{j=1}^{n} b_j w_j$ is an arbitrary vector in V, then we see that

$$v = \sum_{j=1}^{n} b_j w_j = \sum_{j=1}^{n} b_j \left(\sum_{i=1}^{n} q_{ij} v_i \right) = \sum_{i=1}^{n} \left(\sum_{j=1}^{n} q_{ij} b_j \right) v_i$$

and so $a_i = \sum_{j=1}^{n} q_{ij} b_j$ for each $1 \leq i \leq n$. Thus we see that

$$\begin{bmatrix} a_1 \\ \vdots \\ a_n \end{bmatrix} = Q \begin{bmatrix} b_1 \\ \vdots \\ b_n \end{bmatrix}.$$

The matrix Q is called the *change of basis matrix* from B to D.

EXAMPLE

Let F be a field and let n be a positive integer. Let V a be the subspace of $F[X]$ consisting of all polynomials of degree less than n. Then $B = \{1, X, X^2, \ldots, X^{n-1}\}$ is a basis for V over F. Let $c_1, \ldots, c_n$ be distinct elements of F and for all $1 \leq i \leq n$ let

$$p_i(X) = \prod_{j \neq i} (c_i - c_j)^{-1} (X - c_j).$$

This is a polynomial in V of degree $n-1$, called a *Lagrange interpolation polynomial* defined by $\{c_1, \ldots, c_n\}$. It is clear that

$$p_i(c_j) = \begin{cases} 1_F, & \text{when } i = j \\ 0_F, & \text{otherwise.} \end{cases}$$

Moreover, the subset $D = \{p_1(X), \ldots, p_n(X)\}$ of V is linearly independent for if $\sum_{i=1}^{n} a_i p_i(X) = 0_F$ then for each $1 \leq h \leq n$ we have $a_h = \sum_{i=1}^{n} a_i p_i(c_h) = 0_F$. Therefore, D is also a basis for V over F. If $q(X) \in V$ then there exist scalars $b_1, \ldots, b_n$ in F satisfying $q(X) = \sum_{i=1}^{n} b_i p_i(X)$ and so for each $1 \leq h \leq n$ we have $q(c_h) = \sum_{i=1}^{n} b_i p_i(c_h) = b_h$. In particular, if $q(X) = X^k$ we see that $X^k = \sum_{i=1}^{n} c_i^k p_i(X)$. Thus we see that the change of basis matrix from B to D equals

$$\begin{bmatrix} 1_F & 1_F & \cdots & 1_F & 1_F \\ c_1 & c_2 & \cdots & c_{n-1} & c_n \\ & & \cdots & & \\ c_1^{n-1} & c_2^{n-1} & \cdots & c_{n-1}^{n-1} & c_n^{n-1} \end{bmatrix}.$$

A matrix of this form is called a *Vandermonde matrix*.

Let F be a field. Let V be a vector space of finite dimension n over F and let Y be a vector space of finite dimension k over F. Suppose that $B = \{v_1, \ldots, v_n\}$ and $D = \{w_1, \ldots, w_n\}$ are bases for V over F and that $E = \{y_1, \ldots, y_k\}$ is a basis for Y over F. Let $Q = [q_{ij}]$ be the change of basis matrix from B to D. If $\alpha \in Hom(V, Y)$ and if $C = [c_{ij}] = \Phi_{BE}(\alpha)$, what is $\Phi_{DE}(\alpha)$? For all $1 \leq j \leq n$ we have $\alpha(w_j) = \alpha\left(\sum_{h=1}^n q_{hj}v_h\right) = \sum_{h=1}^n q_{hj}\alpha(v_h) = \sum_{h=1}^n q_{hj}\left(\sum_{i=1}^k c_{ih}y_i\right) = \sum_{i=1}^k \left(\sum_{h=1}^n c_{ih}q_{hj}\right)y_i$ and so we have $\Phi_{DE}(\alpha) = CQ$. If $G = \{z_1, \ldots, z_k\}$ is also a basis for Y over F and if $P = [p_{ij}]$ is the change of basis matrix from E to G then, similarly, $C = P\Phi_{BG}(\alpha)$ and so $\Phi_{BG}(\alpha) = P^{-1}C$. Therefore we also have $\Phi_{DG}(\alpha) = P^{-1}CQ = P^{-1}\Phi_{BE}(\alpha)Q$.

EXAMPLE

Consider the bases

$$B = \{[1, 0, 0], [0, 1, 1], [0, -1, 0]\}$$

and

$$D = \{[1, 0, 0], [1, 0, 1], [0, 1, -1]\}$$

for $\mathbb{R}^3$ over $\mathbb{R}$. Also, consider the bases $E = \{[1, 0], [0, -1]\}$ and $G = \{[1, 1], [1, -1]\}$ for $\mathbb{R}^2$ over $\mathbb{R}$. Since

$$[1, 0, 0] = 1[1, 0, 0] + 0[0, 1, 1] + 0[0, -1, 0]$$
$$[1, 0, 1] = 1[1, 0, 0] + 1[0, 1, 1] + 1[0, -1, 0]$$
$$[0, 1, -1] = 0[1, 0, 0] + (-1)[0, 1, 1] + (-2)[0, -1, 0]$$

we see that the change of basis matrix from B to D is $Q = \begin{bmatrix} 1 & 1 & 0 \\ 0 & 1 & -1 \\ 0 & 1 & -2 \end{bmatrix}$. Also, since

$$[1, 1] = 1[1, 0] + (-1)[0, -1]$$
$$[1, -1] = 1[1, 0] + 1[0, -1]$$

we see that the change of basis matrix from E to G is $P = \begin{bmatrix} 1 & 1 \\ -1 & 1 \end{bmatrix}$ and it is easy to check that $P^{-1} = \frac{1}{2}\begin{bmatrix} 1 & -1 \\ 1 & 1 \end{bmatrix}$. Consider the linear transformation $\alpha: \mathbb{R}^3 \to \mathbb{R}^2$ defined by $\alpha: [a, b, c] \mapsto [a + b, b + c]$. Then

$$\alpha: [1, 0, 0] \mapsto [1, 0] = 1[1, 0] + 0[0, -1]$$
$$\alpha: [0, 1, 1] \mapsto [1, 2] = 1[1, 0] + (-2)[0, -1]$$
$$\alpha: [0, -1, 0] \mapsto [-1, -1] = (-1)[1, 0] + 1[0, 1]$$

EXAMPLE (Continued)

and so $\Phi_{BE}(\alpha) = \begin{bmatrix} 1 & 2 & -1 \\ 0 & -2 & 1 \end{bmatrix}$. Similarly,

$$\alpha:[1,0,0] \mapsto [1,0] = \frac{1}{2}[1,1] + \frac{1}{2}[1,-1]$$
$$\alpha:[1,0,1] \mapsto [1,1] = 1[1,1] + 0[1,-1]$$
$$\alpha:[0,1,-1] \mapsto [1,0] = \frac{1}{2}[1,1] + \frac{1}{2}[1,-1]$$

and so $\Phi_{DG}(\alpha) = \begin{bmatrix} 1/2 & 1 & 1/2 \\ 1/2 & 0 & 1/2 \end{bmatrix}$ and, indeed, we have $\Phi_{DG}(\alpha) = P^{-1}\Phi_{BE}(\alpha)Q$.

EXAMPLE

Let V be the subspace of the vector space $\mathbb{R}^{\mathbb{R}}$ consisting of all those functions infinitely differentiable at each point in $\mathbb{R}$ and let $\delta: V \to V$ be the linear transformation given by differentiation, $\delta: f \mapsto f'$. Let a and b be distinct real numbers and let W be the subspace of V with basis $D = \{e^{ax}\sin(bx), e^{ax}\cos(bx)\}$. Then W is stable under δ. In fact, the restriction δ' of δ to W is the automorphism of W with $\Phi_{DD}(\delta') = \begin{bmatrix} a & -b \\ b & a \end{bmatrix}$. It is easy to verify that

$$\Phi_{DD}(\delta')^{-1} = \frac{1}{a^2+b^2}\begin{bmatrix} a & b \\ -b & a \end{bmatrix}.$$

Thus we see that δ' is an automorphism of W satisfying

$$(\delta')^{-1}(e^{ax}\sin(bx)) = \frac{e^{ax}}{a^2+b^2}[a\sin(bx) - b\cos(bx)]$$

which is, as one would expect, just $\int e^{ax}\sin(bx)dx$ up to a constant. Similarly,

$$(\delta')^{-1}(e^{ax}\cos(bx)) = \frac{e^{ax}}{a^2+b^2}[b\sin(bx) - a\cos(bx)]$$

and this is $\int e^{ax}\cos(bx)dx$ up to a constant.

Problems

1. Let $A = \begin{bmatrix} 1 & 1 \\ 0 & 1 \end{bmatrix} \in M_{2\times 2}(\mathbb{R})$. Find the set of all matrices $B \in M_{2\times 2}(\mathbb{R})$ satisfying $BA = AB$.

2. Let n be a positive integer and let F be a field of characteristic 0. If $A, B \in M_{n \times n}(F)$, show that $BA - AB \neq I$.

3. Let F be a field. Find the set of all matrices $A \in M_{2 \times 2}(F)$ satisfying $A^2 = O$.

4. Calculate $\begin{bmatrix} 1 & 2 & -3 \\ 0 & 1 & 2 \\ 0 & 0 & 1 \end{bmatrix}^{-1}$ in $M_{3 \times 3}(\mathbb{R})$.

5. Calculate $\begin{bmatrix} 1 & 1 & 1 & 1 \\ 1 & 1 & -1 & -1 \\ 1 & -1 & 1 & -1 \\ 1 & -1 & -1 & 1 \end{bmatrix}^{-1}$ in $M_{4 \times 4}(\mathbb{R})$.

6. Find a matrix $A \in M_{3 \times 3}(\mathbb{R})$ satisfying $A \begin{bmatrix} 1 & 1 & -1 \\ 2 & 1 & 0 \\ 1 & -1 & 1 \end{bmatrix} = \begin{bmatrix} 1 & -1 & 3 \\ 4 & 3 & 2 \\ 1 & -2 & 5 \end{bmatrix}$.

7. Let $A = \begin{bmatrix} a & 1 & 0 \\ 0 & a & 1 \\ 0 & 0 & a \end{bmatrix} \in M_{3 \times 3}(\mathbb{R})$. Show that for each $n > 1$ we have $A^n = \begin{bmatrix} b & c & d \\ 0 & b & c \\ 0 & 0 & b \end{bmatrix}$, where $b = a^n$, $c = na^{n-1}$, and $d = \frac{1}{2}n(n-1)a^{n-2}$.

8. Let F be a field and let S be the set of all matrices $[a_{ij}] \in M_{3 \times 3}(F)$ satisfying $a_{12} = a_{21} = a_{23} = a_{32} = 0_F$. Is S a subring of $M_{3 \times 3}(F)$?

9. Let K be the set of all matrices in $M_{2 \times 2}(\mathbb{Q})$ of the form $\begin{bmatrix} a & b \\ 2b & a \end{bmatrix}$. Show that K is a subfield of $M_{2 \times 2}(\mathbb{Q})$.

10. Find a matrix $A \in M_{2 \times 2}(\mathbb{R})$ satisfying $A^2 = \begin{bmatrix} 1 & 3 \\ 0 & 1 \end{bmatrix}$.

11. The *Fibonnaci sequence* is the sequence $a_0, a_1, \ldots$ of positive integers defined by $a_0 = 0$, $a_1 = 1$, and $a_i = a_{i-1} + a_{i-2}$ for $i > 1$. Show that $\begin{bmatrix} 0 & 1 \\ 1 & 1 \end{bmatrix}^n = \begin{bmatrix} a_{n-1} & a_n \\ a_n & a_{n+1} \end{bmatrix}$ for all $n \geq 1$.

12. Let $A = \begin{bmatrix} a & 1 & 1 \\ 1 & a & 1 \\ 1 & 1 & a \end{bmatrix} \in M_{3 \times 3}(\mathbb{R})$, where $1 \neq a \neq -2$. Calculate A^{-1}.

13. Find all matrices $A \in M_{3 \times 3}(\mathbb{R})$ satisfying

$$A \begin{bmatrix} 1 & 1 & 1 \\ 2 & 2 & 2 \\ 0 & 1 & 1 \end{bmatrix} = O.$$

14. Let $A = [a_{ij}] \in M_{2 \times 2}(\mathbb{R})$ be a matrix satisfying $A^2 = A$. Show that $a_{11} + a_{22}$ equals 0, 1, or 2.

15. Show that $\begin{bmatrix} 1 & 1 \\ 1 & 1 \end{bmatrix}^n = \begin{bmatrix} 2^{n-1} & 2^{n-1} \\ 2^{n-1} & 2^{n-1} \end{bmatrix}$ in $M_{2\times 2}(\mathbb{Q})$ for all $n \geq 1$.

16. Let F be a field and let $b, c \in F$. Find $\begin{bmatrix} 0_F & b \\ c & 0_F \end{bmatrix}^n$ for all $n > 0$.

17. Let S be the set of all matrices $[a_{ij}] \in M_{2\times 2}(\mathbb{Q})$ satisfying the conditions that a_{ij} is an integer for all $1 \leq i, j \leq n$ and $a_{11}a_{22} - a_{12}a_{21} = 1$. Is S closed under taking inverses?

18. Let F be a field and let n be a positive integer. Let $A \in M_{n\times n}(F)$ satisfy $AA^T = A$. Show that $A^2 = A$.

19. Show that every nonsingular element of $M_{2\times 2}(\mathbb{Q})$ is the product of matrices of the form $\begin{bmatrix} 0 & 1 \\ 1 & 0 \end{bmatrix}$, $\begin{bmatrix} 1 & 1 \\ 0 & 1 \end{bmatrix}$, or $\begin{bmatrix} a & 0 \\ 0 & 1 \end{bmatrix}$ for some $a \in \mathbb{Q}$.

20. Find all pairs (A, B) of matrices in $M_{2\times 2}(\mathbb{Q})$ satisfying $(A - B)(A + B) \neq A^2 - B^2$.

21. Let F be a field and let $A = \begin{bmatrix} 1_F & 0_F & 0_F \\ 1_F & 0_F & 1_F \\ 0_F & 1_F & 0_F \end{bmatrix} \in M_{3\times 3}(F)$. Show that $A^{k+2} = A^k + A^2 - I$ for all positive integers k.

22. Let n be a positive integer, let F be a field, and let A and B be matrices in $M_{n\times n}(F)$ satisfying $A + B = I$. Show that $AB = O$ if and only if $A^2 = A$ and $B^2 = B$.

23. Let n be a positive integer and let F be a field. The *Schur product* $*$ in $M_{n\times n}(F)$ is defined by $[a_{ij}] * [b_{ij}] = [a_{ij}b_{ij}]$. Does $M_{n\times n}(F)$, together with ordinary matrix addition and the Schur product, form a ring?

24. Let F be a field.

(i) Find a matrix $A \in M_{3\times 3}(F)$ satisfying $A^2 = \begin{bmatrix} 0_F & 0_F & 1_F \\ 0_F & 0_F & 0_F \\ 0_F & 0_F & 0_F \end{bmatrix}$.

(ii) Does there exist a matrix $B \in M_{3\times 3}(F)$ satisfying $B^2 = \begin{bmatrix} 0_F & 1_F & 0_F \\ 0_F & 0_F & 0_F \\ 0_F & 0_F & 0_F \end{bmatrix}$?

25. Find a matrix $A \in M_{2\times 2}(\mathbb{Q})$ satisfying

$$A \begin{bmatrix} 0 & c \\ c & 0 \end{bmatrix} A^T = \begin{bmatrix} 2c & 0 \\ 0 & -c/2 \end{bmatrix}$$

for all $c \in \mathbb{Q}$.

26. Let a and b be elements of a field F satisfying $ab \neq 1_F$ and let $A = \begin{bmatrix} 1_F & 0_F & b \\ 0_F & 1_F & 0_F \\ a & 0_F & 1_F \end{bmatrix} \in M_{3\times 3}(F)$. Calculate A^{-1}.

27. Let F be a field and let $0_F \neq c \in F$. Calculate A^{-1}, where

$$A = \begin{bmatrix} c & 1_F & 0_F & 0_F \\ 0_F & c & 1_F & 0_F \\ 0_F & 0_F & c & 1_F \\ 0_F & 0_F & 0_F & c \end{bmatrix} \in \mathcal{M}_{4 \times 4}(F).$$

28. Find infinitely-many triples (A, B, C) of elements of $\mathcal{M}_{3 \times 3}(\mathbb{Q})$ having integer entries and satisfying $A^3 + B^3 = C^3$. (This shows that Fermat's Last Theorem cannot be extended from the integers to matrices over the integers.)

29. Let F be a field. Find a matrix $A \in \mathcal{M}_{4 \times 4}(F)$ satisfying $A^4 = I$.

30. Let n be a positive integer and let $F = \mathbb{Z}/(p)$. Show that for each matrix $A \in \mathcal{M}_{n \times n}(F)$ there exist positive integers $k > h$ satisfying $A^k = A^h$. Would this still be true if we took $F = \mathbb{Q}$ instead?

31. Let $A = [a_{ij}] \in \mathcal{M}_{2 \times 2}(\mathbb{C})$ be a matrix satisfying $a_{11} + a_{22} \neq \pm 2d$, where d satisfies $d^2 = a_{11}a_2 - a_{12}a_{21}$. Show that there exist four distinct matrices $B \in \mathcal{M}_{2 \times 2}(\mathbb{C})$ satisfying $B^2 = A$.

32. Given a complex number c, find all matrices $A \in \mathcal{M}_{2 \times 2}(\mathbb{C})$ satisfying $(A - cI)^2 = O$.

33. Let n be a positive integer and let S be the set of all matrices $[a_{ij}]$ in $\mathcal{M}_{n \times n}(\mathbb{Q})$ all of the entries in which are integers. Is S a subring of $\mathcal{M}_{n \times n}(\mathbb{Q})$?

34. Show that there is no matrix $A \in \mathcal{M}_{2 \times 2}(\mathbb{C})$ satisfying $A^2 = \begin{bmatrix} 0 & 1 \\ 0 & 0 \end{bmatrix}$.

35. Let n be a positive integer and let F be a field. How many matrices $A = [a_{ij}] \in \mathcal{M}_{n \times n}(F)$ satisfy the following conditions:
(1) Each a_{ij} is either 0_F or 1_F;
(2) 1_F appears exactly once in each row and each column?

36. Let $n \geq 4$ be an integer and let F be a field. Show that there exist matrices A and B in $\mathcal{M}_{n \times n}(F)$ satisfying $A^2 = B^2 = O$ but $AB = BA \neq O$.

37. Let $A = [a_{ij}] \in \mathcal{M}_{n \times n}(\mathbb{Q})$ be the matrix defined by

$$a_{ij} = \begin{cases} j, & \text{if } i > 1 \text{ and } j = i - 1 \\ 0, & \text{otherwise.} \end{cases}$$

Show that for each positive integer k we have $A^k = [b_{ij}]$ where $b_{ij} = (i-1)!/(j-1)!$ if $i = j + k$ and $b_{ij} = 0$ otherwise.

38. Find a matrix $I \neq A \in \mathcal{M}_{3 \times 3}(\mathbb{Q})$ satisfying $AB = BA$ and $AC = CA$, where

$$B = \begin{bmatrix} 1 & 0 & 0 \\ 1 & 1 & 0 \\ 0 & 0 & 1 \end{bmatrix} \text{ and } C = \begin{bmatrix} 1 & 0 & 0 \\ 0 & 1 & 0 \\ 0 & 1 & 1 \end{bmatrix}.$$

39. Let F be a field and let $A, B \in \mathcal{M}_{2 \times 2}(F)$. Show that $(AB - BA)^2$ is a diagonal matrix.

40. Let F be a field and let n be a positive integer. Let $A_1, \dots, A_n \in \mathcal{UT}_{n \times n}(F)$ satisfying the condition that the (i, i) entry of A_i is 0_F for all $1 \le i \le n$. Show that $A_1 \cdot \ldots \cdot A_n = O$.

41. Find a matrix $A \in \mathcal{LT}_{4 \times 4}(\mathbb{R})$ and a matrix $B \in \mathcal{UT}_{4 \times 4}(\mathbb{R})$ satisfying

$$AB = \begin{bmatrix} 1 & 4 & 1 & 3 \\ 0 & -1 & 3 & -1 \\ 3 & 1 & 0 & 2 \\ 1 & -2 & 5 & 1 \end{bmatrix}.$$

42. Let n be a positive integer and let F be a field. Let $A = [a_{ij}] \in \mathcal{UT}_{n \times n}(F)$ be a matrix satisfying the following conditions:

(1) a_{ii} equals either 0_F or 1_F for all $1 \le i \le n$;
(2) If $a_{hh} = 0_F$ then $a_{ih} = 0_F$ for all $1 \le i \le n$;
(3) If $a_{hh} = 1_F$ then $a_{hj} = 0_F$ for all $j \ne h$.

Show that $A = A^2$.

43. Let a and b be nonzero elements of a field F. Show that there exists a matrix $A \in \mathcal{UT}_{2 \times 2}(F)$ satisfying

$$\begin{bmatrix} 0_F & a \\ 0_F & 0_F \end{bmatrix} A = \begin{bmatrix} 0_F & b \\ 0_F & 0_F \end{bmatrix}.$$

Is A necessarily unique?

44. Let F be a field and let n be a positive integer. If $A \in \mathcal{M}_{n \times n}(F)$, show that $AA^T \in \mathcal{S}_{n \times n}(F)$.

45. Let F be a field and let n be a positive integer. If $A, B \in \mathcal{S}_{n \times n}(F)$, is ABA also symmetric?

46. Write the matrix $\begin{bmatrix} 1 & -2 \\ 2 & 2 \end{bmatrix} \in \mathcal{M}_{2 \times 2}(\mathbb{R})$ as a product of elementary matrices.

47. Let F be a field and n be a positive integer, and let $A \in \mathcal{M}_{2n \times 2n}(F)$ be a matrix which can be represented in the form $\begin{bmatrix} A_{11} & A_{12} \\ A_{21} & A_{22} \end{bmatrix}$, where the A_{ij} are nonsingular matrices in $\mathcal{M}_{n \times n}(F)$. Is A necessarily nonsingular?

48. Let F be a field and let A and B be matrices in $\mathcal{M}_{n \times n}(F)$, where A is nonsingular. Show that $(A + B)A^{-1}(A - B) = (A - B)A^{-1}(A + B)$.

49. Let F be a field and let n be a positive integer. If $A \in \mathcal{M}_{n \times n}(F)$ is a nonsingular matrix and if $v \in F^n$, show that $A - v^T v$ is a nonsingular matrix if and only if $vA^{-1}v^T \ne 1$.

50. Let F be a field, let n be a positive integer, and let $V = \mathcal{M}_{n \times 1}(F)$. If $0_V \ne v \in V$, show that there exists a nonsingular matrix in $\mathcal{M}_{n \times n}(F)$ the first column of which equals v.

51. Let F be a field and let n be a positive integer. Let $A \in \mathcal{M}_{n \times n}(F)$ be a nonsingular matrix satisfying the condition that the sum of the elements of each row of A equals 1. Show that A^{-1} satisfies this same condition.

52. For each real number t, let $A(t)$ be the matrix $\begin{bmatrix} 1 & 0 & t \\ -t & 1 & -t^2/2 \\ 0 & 0 & 1 \end{bmatrix} \in \mathcal{M}_{3 \times 3}(\mathbb{R})$.
Show that each such matrix is nonsingular and that the set of all such matrices is closed under multiplication.

53. Let n be a positive integer, let F be a field, and let A be a matrix in $\mathcal{M}_{n \times n}(F)$ for which there exists a positive integer k satisfying $A^k = O$. Show that the matrix $I - A$ is nonsingular and find its inverse.

54. Let n be a positive integer, let F be a field, and let A be a matrix in $\mathcal{M}_{n \times n}(F)$ for which there exists a matrix B satisfying $I + A + AB = O$. Show that A is nonsingular.

55. Find matrices A and B in $\mathcal{M}_{2 \times 2}(\mathbb{R})$ satisfying $A^2 = B^2 = O$ such that the matrix $A + iB \in \mathcal{M}_{2 \times 2}(\mathbb{C})$ is nonsingular.

56. Let $n > 1$ be an integer and let $B \in \mathcal{M}_{n \times n}(\mathbb{Q})$ be the matrix all of the entries of which equal 1. Show that there exists a matrix $A \in \mathcal{M}_{n \times n}(\mathbb{Q})$ satisfying the condition that $A + cB$ is nonsingular for all $c \in \mathbb{Q}$.

57. For each positive integer n let $A_n = [a_{ij}^{(n)}] \in \mathcal{M}_{n \times n}(\mathbb{R})$ be the matrix defined by $a_{ij}^{(n)} = min\{i, j\}$ for all $1 \leq i, j \leq n$. Show that each of the matrices A_n is nonsingular.

58. Let k be a positive integer and let n be a prime integer. How many distinct nonsingular matrices are there in $\mathcal{M}_{k \times k}(\mathbb{Z}/(n))$?

59. Do there exist matrices A and B in $\mathcal{M}_{2 \times 2}(\mathbb{R})$ such that $A^2 + B^2$ is nonsingular but the matrix $A + iB \in \mathcal{M}_{2 \times 2}(\mathbb{C})$ is singular?

60. Find the change of bases matrix from the canonical basis of $\mathbb{R}^3$ to the basis $\{[1, 1, 1], [1, 1, 0], [1, 0, 0]\}$.

61. Let $F = \mathbb{Z}/(2)$ and let K be a field not containing F as a subfield. Define a function $\phi: \mathcal{M}_{2 \times 2}(K) \to \mathcal{M}_{2 \times 2}(F)$ as follows: if $[a_{ij}] \in \mathcal{M}_{2 \times 2}(K)$ then $\phi([a_{ij}]) = [b_{ij}]$, where

$$b_{ij} = \begin{cases} 1, & \text{if } a_{ij} \neq 0 \\ 0, & \text{otherwise.} \end{cases}$$

Is it true that $\phi(A + A') = \phi(A) + \phi(A')$ holds for all $A, A' \in \mathcal{M}_{2 \times 2}(K)$? What about $\phi(AA') = \phi(A)\phi(A')$?

CHAPTER VIII

SYSTEMS OF LINEAR EQUATIONS

The theory of finite-dimensional vector spaces was created primarily in connection with one problem, and that is the simultaneous solution of a system of k linear equations in n indeterminates over a field F of the form

(*)
$$a_{11}X_1 + \cdots + a_{1n}X_n = b_1$$
$$a_{21}X_1 + \cdots + a_{2n}X_n = b_2$$
$$\cdots$$
$$a_{k1}X_1 + \cdots + a_{kn}X_n = b_k$$

where the a_{ij} and the b_i are elements of F and the X_j are indeterminates taking values in F. Such systems arise in many applications and also in many areas of mathematics.

EXAMPLE

Let $[a, b]$ be a closed interval of $\mathbb{R}$ and let V be the vector space of all continuous functions from this interval to $\mathbb{R}$. If W is an n-dimensional subspace of V then the *interpolation problem* for V via W is the following: given a function $f \in V$ and real numbers $a \leq t_1 < \cdots < t_n \leq b$, find a function $g \in W$ satisfying $g(t_j) = f(t_j)$ for all $1 \leq j \leq n$, if such a function exists. Given a basis $\{g_1, \ldots, g_n\}$ for W over $\mathbb{R}$, then we have to find real numbers $c_1, \ldots, c_n$ such that $\sum_{i=1}^n c_i g_i(t_j) = f(t_j)$ for all $1 \leq j \leq n$. That is to say, we have to find a solution to the system of linear equations

$$g_1(t_1)X_1 + \cdots + g_n(t_1)X_n = f(t_1)$$
$$g_1(t_2)X_1 + \cdots + g_n(t_2)X_n = f(t_2)$$
$$\cdots$$
$$g_1(t_k)X_1 + \cdots + g_n(t_k)X_n = f(t_k).$$

A system of linear equations (*) is *homogeneous* if $b_i = 0_F$ for all $1 \leq i \leq k$. Otherwise it is *inhomogeneous*. At this stage, we do not know the answer to the following questions.

(1) Must a system (*) have a solution?
(2) If a solution exists, is it necessarily unique?
(3) If there are many solutions, what is the structure of the set of all solutions?
(4) If there are solutions, is there an algorithm to find them?

In order to answer these questions, we rephrase the problem in terms of matrices. Historically, the use of matrices for this purpose was first appears in Europe in the 19th century in the work of the British mathematician J. J. Sylvester. However, matrices were already used almost 2,000 years earlier for this purpose in the classical Chinese text "Nine Chapters on the Mathematical Art" ("Jiuzhang Suanshu").

We immediately note that the system (*) can be rewritten as

$$\begin{bmatrix} a_{11} & \cdots & a_{1n} \\ a_{21} & \cdots & a_{2n} \\ & \cdots & \\ a_{k1} & \cdots & a_{kn} \end{bmatrix} \begin{bmatrix} X_1 \\ X_2 \\ \vdots \\ X_n \end{bmatrix} = \begin{bmatrix} b_1 \\ b_2 \\ \vdots \\ b_k \end{bmatrix},$$

where $X = \begin{bmatrix} X_1 \\ X_2 \\ \vdots \\ X_n \end{bmatrix}$ is a vector of indeterminates. The matrix $A = [a_{ij}]$ is the

coefficient matrix of the system (*). If we think of A as representing a linear transformation $\alpha: F^n \to F^k$ with respect to the canonical bases, then the set of all solutions of (*) is precisely $Inv(\alpha, w) = \{v \in F^n \mid \alpha(v) = w\}$, where $w = [b_1, \ldots, b_k]$.

The system $AX = w^T$ is homogeneous precisely when $w = [0_F, \ldots, 0_F]$. In this case the set of all solutions to (*) is just the kernel of α and so, in particular, is a subspace of F^n. This space is called the *solution space* of the homogeneous system. If the system is inhomogeneous then, by Proposition 4.5, the set $Inv(\alpha, w)$ of all solutions to (*) either empty or of the form $\{v + v_0 \mid v \in ker(\alpha)\}$, where v_0 is one particular solution to the system. If the set of solutions is empty, the system is *inconsistent*; otherwise it is *consistent*.

How do we go about finding solutions to systems of the form $AX = w^T$, or showing that no such solutions exist? We note by Proposition 4.11 that if $v \in F^n$ then $\alpha(v) = w$ if and only if $\beta\alpha(v) = \beta(w)$ for any automorphism β of F^k. In particular, this is true if we take β to be one of the elementary automorphisms of F^k:

(1) Interchanging two components;
(2) Multiplying a one of the components by a nonzero scalar;
(3) Adding a scalar multiple of one component to another component.

As we have already seen, this can be accomplished by multiplying the matrix $A \in \mathcal{M}_{k \times n}(F)$ on the left by one of the elementary matrices in $\mathcal{M}_{k \times k}(F)$. This, of course, can be repeated finitely-many times, and so we conclude that the set of solutions to a system of the form $AX = w^T$ is the same as the set of all solutions of the system $(EA)X = Ew^T$, where E is a product of elementary matrices. By Proposition 7.4, of course, these matrices E are precisely the nonsingular matrices in $\mathcal{M}_{k \times n}(F)$.

Our course of action is therefore the following: given a system of linear equations which we write in the form $AX = w^T$, where $A \in \mathcal{M}_{k \times n}(F)$ and $w \in F^k$, we wish to multiply the matrix A and the column vector w^T on the left by a succession of elementary matrices in $\mathcal{M}_{k \times k}(F)$ until we obtain a new system which is easily solvable. Rather than multiplying each of them separately, it is easier to consider the *augmented coefficient matrix* $[A\ w^T] \in \mathcal{M}_{k \times (n+1)}(F)$ obtained from A by adding w^T as an additional column on the right. By successive multiplications on the left by elementary matrices (which, we recall, correspond to the operations of interchanging rows, multiplying a row by a nonzero scalar, or adding a scalar multiple of one row to another row) we wish to bring this augmented matrix to a "nice" form. There are several possible such forms, but the usual one chosen is the following. A matrix $D = [d_{ij}] \in \mathcal{M}_{k \times (n+1)}(F)$ is in *echelon form* whenever it satisfies the following condition: if $1 \leq s \leq k$ and $1 \leq t \leq (n+1)$ are indices such that $d_{sj} = 0_F$ for all $1 \leq j < t$ (i.e., all entries in the matrix to the left of d_{st} equal 0_F) then $d_{it} = 0_F$ for all $s < i \leq k$ (i.e., all entries below d_{st} are equal to 0_F).

We remark that if a matrix $A \in \mathcal{M}_{n \times n}(F)$ is in echelon form then it surely belongs to $\mathcal{UT}_{n \times n}(F)$. Such a matrix is nonsingular precisely when it has no row equal to $[0_F, \ldots, 0_F]$.

EXAMPLE

The matrices
$$\begin{bmatrix} 1 & 6 & 7 & 0 & 2 \\ 0 & 9 & 0 & 1 & 2 \\ 0 & 0 & 0 & 1 & 0 \\ 0 & 0 & 0 & 0 & 7 \end{bmatrix} \text{ and } \begin{bmatrix} 0 & 7 & 0 & 0 & 0 \\ 0 & 0 & 0 & 6 & 1 \\ 0 & 0 & 0 & 0 & 0 \\ 0 & 0 & 0 & 0 & 0 \end{bmatrix}$$
belonging to $\mathcal{M}_{4 \times 5}(\mathbb{Q})$ are in echelon form.

The method of bringing a matrix in $C = [c_{ij}] \in \mathcal{M}_{k \times (n+1)}(F)$ into echelon form can be summarized as follows:

(I) Let $s = 1$.

(II) Let t be the number of the leftmost column of D containing a nonzero entry in a row i for some $s \leq i \leq k$. If there is no such column, stop.

(III) Interchange rows if necessary to obtain a matrix $D = [d_{ij}]$ in which $d_{st} \neq 0_F$.

(IV) For each $s < i \leq k$, multiply the sth row of D by $-d_{it}d_{st}^{-1}$ and add it to the ith row. As a result, we get a new matrix in which all elements of the tth column under d_{st} are 0_F.

(V) If $s < k$, replace s by $s + 1$ and return to step (II). If $s = k$, stop.

EXAMPLE

To illustrate this method, let us consider the matrix

$$\begin{bmatrix} 1 & 2 & 3 & 1 \\ 2 & 1 & 4 & 2 \\ 1 & -1 & 1 & 1 \end{bmatrix} \in \mathcal{M}_{3 \times 3}(\mathbb{Q}).$$

We begin with $s = t = 1$. Multiply the first row by -2 and add it to the second row to obtain

EXAMPLE (Continued)

$$\begin{bmatrix} 1 & 2 & 3 & 1 \\ 0 & -3 & -2 & 0 \\ 1 & -1 & 1 & 1 \end{bmatrix}$$

and then multiply the first row by -1 and add it to the third row to obtain

$$\begin{bmatrix} 1 & 2 & 3 & 1 \\ 0 & -3 & -2 & 0 \\ 0 & -3 & -2 & 0 \end{bmatrix}.$$

Now take $s = 2$ and $t = 2$. Multiply the second row by -1 and add it to the third row to obtain

$$\begin{bmatrix} 1 & 2 & 3 & 1 \\ 0 & -3 & -2 & 0 \\ 0 & 0 & 0 & 0 \end{bmatrix}.$$

Since this matrix is in echelon form, we stop here.

Let us return to the system of linear equations $AX = w^T$. We begin with the matrix $[A\ w^T] \in \mathcal{M}_{k \times (n+1)}(F)$ and bring it to a matrix $D = [A'\ w'^T] \in \mathcal{M}_{k \times (n+1)}(F)$ in echelon form, where $A' = [a'_{ij}]$ and $w' = [b'_1, \ldots, b'_k]$. We know that the set of solutions to $AX = w^T$ is the same as the set of solutions to $A'X = w'^T$. But these can be determined with ease, by a process known as *back substitution*. Indeed, let s be the largest index i such that the ith row of D is not $[0_F, \ldots, 0_F]$. We distinguish between three possible cases:

Case I: $a'_{sj} = 0_F$ for all $1 \leq j \leq n$ but $b'_s \neq 0_F$. In that case, it is clear that there exist no solutions to the system $A'X = w'^T$ and hence no solutions to the system $AX = w^T$.

Case II: $a'_{sj} \neq 0_F$ for precisely one index $1 \leq j \leq n$. In this case, any solution $[d_1, \ldots, d_n]$ to the system $A'X = w'^T$ must satisfy $a'_{sj}d_j = b'_s$ and so $d_j = a'^{-1}_{sj}b'_s$. Replacing X_j in all of the other equations by this value and then deleting the sth equation, we have reduced our problem to one of a system of $k - 1$ linear equations in $n - 1$ indeterminates.

Case III: $a'_{sj} \neq 0_F$ for more than one value of j, say $j_1, \ldots, j_p$. The sth equation then becomes $X_{j_1} = a^{-1}_{sj_1}b_s - a^{-1}_{sj_1}a_{sj_2}X_{j_2} - \cdots - a^{-1}_{sj_1}a_{sj_p}X_{j_p}$ and so we can replace X_{j_1} in all of the other equations by the expression on the right and then deleting the sth equation, we have again reduced our problem to one of a system of $k - 1$ linear equations in $n - 1$ indeterminates.

We now repeat this process with a new value of s, necessarily smaller than the previous one. Eventually, we will either conclude that there is no solution to the system $A'X = w'^T$ (and hence to the system $AX = w^T$) or we will obtain prescribed values for some of the indeterminates and a single linear equation in the remaining indeterminates, the set of solutions to which gives us the set of solutions to the original system.

The process which we have used is known as *Gaussian elimination*. Again, over $\mathbb{R}$ or $\mathbb{Q}$ we try to reduce the errors created by roundoff and/or truncation my means of partial pivoting: in step (III) we choose d_{st} to be the element having maximal absolute value from among all of those we have available to us.

EXAMPLE

Consider the inhomogeneous system of 3 linear equations in 3 indeterminates

$$3X_1 + 2X_2 + X_3 = 0$$
(*) $$-2X_1 + X_2 - X_3 = 2$$
$$2X_1 - X_2 + 2X_3 = -1$$

with coefficients in $\mathbb{R}$. The augmented coefficient matrix of this system is

$$\begin{bmatrix} 3 & 2 & 1 & 0 \\ -2 & 1 & -1 & 2 \\ 2 & -1 & 2 & -1 \end{bmatrix},$$

which reduces to the matrix in echelon form $\begin{bmatrix} 1 & 3 & 0 & 2 \\ 0 & 7 & -1 & 6 \\ 0 & 0 & 1 & 1 \end{bmatrix}$. The last row implies that $X_3 = 1$ and so, replacing X_3 by this value, we obtain the new system of equations

$$X_1 + 3X_2 = 2$$
$$7X_2 = 7$$

from which we conclude that $X_2 = 1$ and $X_1 = -1$. Thus we conclude that the set of solutions to (*) is $\{[-1, 1, 1]\}$.

EXAMPLE

Consider the inhomogeneous system of 3 linear equations in 3 indeterminates

$$X_1 + 2X_2 = 1$$
(*) $$X_1 - X_2 = 3$$
$$-X_1 + 2X_2 = -2$$

with coefficients in $\mathbb{R}$. The augmented coefficient matrix of this system is $\begin{bmatrix} 1 & 1 & 1 \\ 1 & -1 & 3 \\ -1 & 2 & -2 \end{bmatrix}$, which reduces to the matrix in echelon form $\begin{bmatrix} 1 & 1 & 1 \\ 0 & -2 & 2 \\ 0 & 0 & 2 \end{bmatrix}$. Thus the system (*) is inconsistent.

EXAMPLE

Consider the inhomogeneous system of 3 linear equations in 3 indeterminates

$$
\begin{aligned}
X_1 + 2X_2 + X_3 &= -1 \\
2X_1 + 4X_2 + 3X_3 &= 3 \\
3X_1 + 6X_2 + 4X_3 &= 2
\end{aligned}
$$

(*)

with coefficients in $\mathbb{R}$. The augmented coefficient matrix of this system is

$$
\begin{bmatrix}
1 & 2 & 1 & -1 \\
2 & 4 & 3 & 3 \\
3 & 6 & 4 & 2
\end{bmatrix}
$$

which reduces to the matrix in echelon form

$$
\begin{bmatrix}
1 & 2 & 1 & -1 \\
0 & 0 & 1 & 5 \\
0 & 0 & 0 & 0
\end{bmatrix}.
$$

From the second row we see that $X_3 = 5$. Inserting this in the first row, we see that the set of all solutions of (*) is $\{[d_1, d_2, 5] \mid d_1 + 2d_2 = -6\} \subseteq \mathbb{R}^3$. One of the parameters d_1 or d_2 can therefore be written as a function of the other, and so this set of solutions can be variously denoted as $\{[d, -3 - \frac{1}{2}d, 5] \mid d \in \mathbb{R}\}$ or $\{[-6 - 2d, d, 5] \mid d \in \mathbb{R}\}$.

Gaussian elimination can be used to check whether a given finite subset of F^k is linearly independent. Indeed, if F is a field and k is a positive integer, let $\{v_1, \ldots, v_n\} \subseteq F^k$, where $v_j = [a_{1j}, \ldots, a_{kj}]$ for all $1 \leq j \leq n$. In order for this set of vectors to be linearly dependent, we have to find scalars $b_1, \ldots, b_n$ not all equal to 0_F, such that $[0_F, \ldots, 0_F] = \sum_{j=1}^n b_j v_j = [\sum_{j=1}^n b_j a_{1j}, \ldots, \sum_{j=1}^n b_j a_{nj}]$. In other words, we must show that the homogeneous system of linear equations $AX = [0_F, \ldots, 0_F]^T$ has a nontrivial solution, where $A = [a_{ij}]$. If no such solution exists, then the given set of vectors is linearly independent.

EXAMPLE

Let us check whether the subset $\{[1, -1, 3, 4], [3, -3, 6, 4], [-1, 1, 0, 4]\}$ of $\mathbb{Q}^4$ is linearly dependent or linearly independent. Set

$$
A = \begin{bmatrix}
1 & 3 & -1 \\
-1 & -3 & 1 \\
3 & 6 & 0 \\
4 & 4 & 4
\end{bmatrix}.
$$

EXAMPLE (Continued)

This matrix reduces to the matrix in echelon form

$$B = \begin{bmatrix} 1 & 3 & -1 \\ 0 & -1 & 1 \\ 0 & 0 & 0 \\ 0 & 0 & 0 \end{bmatrix}.$$

The system of linear equations

$$X_1 + 3X_2 - X_3 = 0$$
$$-X_2 + X_3 = 0$$

has a nontrival solution $[-2, 1, 1]$. We check to see that

$$-2[1, -1, 3, 4] + 1[3, -3, 6, 4] + 1[-1, 1, 0, 4] = [0, 0, 0, 0]$$

and so the given set of vectors is linearly dependent.

There are methods other than Gaussian elimination for solving systems of linear equations of the form $AX = w^T$, especially when the matrix A is a square matrix known to be nonsingular. However, any method which we use is bounded to entail a variety of computational errors. Indeed, when we think that we have found a vector v satisfying $Av^T = w^T$, we in fact have found some vector $v_1 = v + v_e$, where v_e is a vector of errors which we cannot identify (but hope is very close to $[0, \ldots, 0]$) and where $Av_1^T = (w + w_e)^T$, where w_e is also (hopefully) very near $[0, \ldots, 0]$ and is given by $Av_e^T = w_e^T$. We then note that $Av_e^T - w^T = w_e^T$ and so we can obtain the equation $Av_e^T = Av_1^T - w^T$, in which the righthand side is known to us. Thus, if we replace v_1 by the vector v_2 defined by $v_2^T = v_1^T - A^{-1}(Av_1^T - w^T)$ we are likely to get a better approximation to the true solution v which we seek. Of course, this procedure can be iterated several times if necessary.

These remarks suggest another method of solution of systems of linear equations based on the notion of convergence, which works when the field F is equal to $\mathbb{R}$. Suppose we have a system of linear equations $AX = w^T$, where $w = [b_1, \ldots, b_n]$ and where $A = [a_{ij}]$ is an invertible matrix in $\mathcal{M}_{n \times n}(\mathbb{R})$ which we so arranged (by interchanging rows if necessary) so that all of the diagonal elements a_{ii} are nonzero. By transferring various terms from the lefthand side of the equations to the righthand side, we obtain

$$(*) \qquad X_i = a_{ii}^{-1} \left[b_i - \sum_{j \neq i} a_{ij} X_j \right]$$

for $1 \leq i \leq n$. Now suppose that we have some reason to believe that $v = [d_1, \ldots, d_n]$ is a solution to this system of equations. If we plug in the d_j instead of the X_j on the righthand side of the equations $(*)$, we obtain a new vector

$v' = [d'_1, \ldots, d'_n]$. If $v = v'$ then we have found a solution to the system of equations. If not, we now repeat the process, using v' instead of v. Hopefully, after a reasonable number of steps we will converge on a solution to the system. Each repetition in this method is called an *iteration* and the method itself is known as *Jacobi iteration algorithm.*

If we think of the matrix A and being of the form $L + U + D$, where

(1) L has the same entries as A below the main diagonal and 0 elsewhere;
(2) U has the same entries as A above the main diagonal and 0 elsewhere;
(3) D is the diagonal matrix having the same diagonal entries as A;

then the iteration step in the Jacobi algorithm can be written as

$$v'^T = -D^{-1}(U + L)v^T + w^T.$$

There is another iteration algorithm, called the *Gauss-Seidel iteration algorithm*, in which replace this iteration step by

$$v'^T = -(U + D)^{-1}Lv^T + w^T.$$

Its advantage over the Jacobi method is that we use the new components of v immediately after we compute them and not just at the beginning of the next step.

Of course, we have no guarantee that in a particular case either of these methods – or any of the other various iteration methods – indeed converges. The study of conditions for convergence and estimates on the rate of convergence are the domain of that branch of mathematics known as numerical analysis. There exist examples of systems of linear equations for which the Jacobi method converges while the Gauss-Seidel method does not, and vice versa. In general, one can say that iteration methods work best when the matrix A is sparse (i.e. has many zero entries), as with systems of linear equations arising from partial differential equations. The main advantage of iteration methods over elimination methods (as in Gaussian elimination, with or without partial pivoting) is that the roundoff and truncation errors tend – at least partially – to cancel each other out, rather than accumulate.

We now return to our general discussion.

If $A \in \mathcal{M}_{k \times n}(F)$ is a matrix over a field F, then the *row space* of A is the subspace of F^n generated by the rows of A and the *column space* of A is the subspace of $\mathcal{M}_{k \times 1}(F)$ generated by the columns of A.

(8.1) PROPOSITION. *Let F be a field, let $A \in \mathcal{M}_{k \times n}(F)$, and let $w \in F^k$. Then the system of linear equations $AX = w^T$ has a solution if and only if w^T belongs to the column space of A.*

PROOF. Let $v = [d_1, \ldots, d_n]$ be a solution to the system $AX = w^T$. Then

$$w^T = \left[\sum_{j=1}^{n} a_{1j}d_j, \ldots, \sum_{j=1}^{n} a_{kj}d_j \right]^T = d_1 \begin{bmatrix} a_{11} \\ \vdots \\ a_{k1} \end{bmatrix} + \cdots + d_n \begin{bmatrix} a_{1n} \\ \vdots \\ a_{kn} \end{bmatrix}$$

and so w^T belongs to the column space of A. Conversely, if w^T belongs to the column space of A then there exists scalars $d_1, \ldots, d_n$ such that

$$w^T = d_1 \begin{bmatrix} a_{11} \\ \vdots \\ a_{k1} \end{bmatrix} + \cdots + d_n \begin{bmatrix} a_{1n} \\ \vdots \\ a_{kn} \end{bmatrix}$$

and so $[d_1, \ldots, d_n]^T$ is a solution to $AX = w^T$. $\square$

The *rank* of a matrix $A \in \mathcal{M}_{k \times n}(F)$ is defined to be the dimension of the column space of A. In Proposition 8.4, we will prove on that this also equals the dimension of the row space of A.

EXAMPLE

Let F be a field and let n be a positive integer. If v and w are vectors in F^n then the matrix $v^T w$ is a matrix in $\mathcal{M}_{n \times n}(F)$ having rank at most 1. The converse is also true. If A is a matrix in $\mathcal{M}_{n \times n}(F)$ of rank 1, then there exists a column of A, call it v^T, which is not $[0_F, \ldots, 0_F]^T$ and which satisfies the condition that for all $1 \leq j \leq n$ there exists a scalar $c_j \in F$ such that the jth column of A is just $c_j v^T$. Thus $A = v^T w$, where $w = [c_1, \ldots, c_n] \in F^n$. If A has rank 0, then it is just $[0_F, \ldots, 0_F]^T [0_F, \ldots 0_F]$.

A matrix of the form $v^T w$ is called the *outer product* of the vectors v and w; we will return to these products in more detail later.

Now let us rephrase Proposition 8.1.

(8.2) PROPOSITION. *Let F be a field, let $A \in \mathcal{M}_{k \times n}(F)$, and let $w \in F^k$. Then the system of linear equations $AX = w^T$ has a solution if and only if the rank of A equals the rank of the augmented matrix $[A \ w^T]$.*

PROOF. We have seen that the system of linear equations $AX = w^T$ has a solution if and only if w^T belongs to the space generated by the columns of A, i.e. if and only if the column space of A equals the column space of $[A \ w^T]$. Since the column space of A is always a subspace of the column space of $[A \ w^T]$, these two spaces are equal if and only if they have the same dimension. $\square$

Now let us return to systems of linear equations.

(8.3) PROPOSITION. *Let F be a field and let $A \in \mathcal{M}_{k \times n}(F)$. Assume that the columns $y_1, \ldots, y_n$ of this matrix are arranged such that $\{y_1, \ldots, y_r\}$ is a basis for the column space of A. Moreover, for all $r < h \leq n$ let us choose scalars $b_{h1}, \ldots, b_{hn}$ in F such that:*

(1) $y_h = \sum_{j=1}^r b_{hj} y_j$;
(2) $b_{hh} = -1_F$;
(3) $b_{hj} = 0_F$ otherwise.

For all $r < h \leq n$ let $v_h = [b_{h1}, \ldots, b_{hn}] \in F^n$. Then $\{v_{r+1}, \ldots, v_n\}$ is a basis for the space of solutions of the homogeneous system of linear equations $AX = [0_F, \ldots, 0_F]^T$.

(COMMENT: Since $\{y_1, \ldots, y_n\}$ is a generating set for the column space of A, we know that it has a subset which is a basis for this space. The hypothesis that this subset consists of the first r columns, which may involve renumbering the indeterminates if necessary, is there solely to simplify the notation.)

PROOF. If $r < h \leq n$ then $Av_h^T = b_{h1}y_1 + \cdots + b_{hr}y_r - y_h = [0_F, \ldots, 0_F]^T$ and so v_h indeed belongs to the solution space of $AX = [0_F, \ldots, 0_F]^T$. Moreover, the set $\{v_{r+1}, \ldots, v_n\}$ is linearly independent since if $\sum_{h=r+1}^n c_h v_h = [0_F, \ldots 0_F]$ then for each $r < h \leq n$ the hth component of the lefthand sum is $-c_h$, from which we see that $c_h = 0_F$.

We are left to show that $\{v_{r+1}, \ldots, v_n\}$ is a generating set for the solution space of $AX = [0_F, \ldots, 0_F]^T$. Indeed, assume that $w = [d_1, \ldots d_n]$ belongs to that space. Then

$$w + \sum_{h=r+1}^n d_h v_h = [e_1, \ldots, e_r, 0_F, \ldots, 0_F],$$

where the $e_1, \ldots, e_r$ are scalars. This vector belongs to the solution space of the system, and so $\sum_{h=1}^r e_h y_h = [0_F, \ldots, 0_F]^T$. But the set $\{y_1, \ldots, y_r\}$ is linearly independent, and so $e_1 = \cdots = e_r = 0_F$. Thus $w = -\sum_{h=r+1}^n d_h v_h$, proving that $\{v_{r+1}, \ldots, v_n\}$ is a generating set for the solution space, and hence a basis for it. $\square$

As an immediate consequence of this result we have the following corollary.

COROLLARY. *Let F be a field and let $A \in M_{k \times n}(F)$. Then the solution space of the homogeneous system of linear equations $AX = [0_F, \ldots, 0_F]^T$ has dimension $n - r$, where r is the rank of A.*

We now return to the concept of rank.

(8.4) PROPOSITION. *If F is a field and if $A \in M_{k \times n}(F)$ then the rank of the matrix A equals the dimension of its row space.*

PROOF. Let $v_1, \ldots, v_k$ be the rows of A, which belong to F^n. Again, we can change the order of the rows if necessary so that $\{v_1, \ldots, v_t\}$ is a basis for the row space of A, this without altering the solution space of the homogeneous system of linear equations $AX = [0_F, \ldots, 0_F]^T$ and hence without changing the rank of A. Let $B \in M_{t \times n}(F)$ be the matrix the rows of which are $\{v_1, \ldots, v_t\}$. The columns of B belong to $M_{t \times 1}(F)$ and so the rank r_B of B satisfies $r_B \leq t$. But we have already noted that the systems $AX = [0_F, \ldots, 0_F]^T$ and $BX = [0_F, \ldots, 0_F]^T$ have the same solution space and so, by the Corollary to Proposition 8.3, we have $r_B = r_A$, where r_A is the rank of A. Thus $r_A \leq t$.

Thus we have proven that the rank of any matrix is less than or equal to the dimension of its row space. In particular, this is true also for the matrix A^T. But the rank of A^T is t and the dimension of the row space of A^T is the rank of A, i.e. r_A. Thus $t \leq r_A$ and so we have equality. $\square$

EXAMPLE

Consider the system of linear equations over $\mathbb{R}$

$$X_1 + 2X_2 - 3X_4 + X_4 = 0$$
$$X_1 + X_2 + X_3 + X_4 = 0$$

the coefficient matrix of which is $\begin{bmatrix} 1 & 2 & -3 & 1 \\ 1 & 1 & 1 & 1 \end{bmatrix}$. This matrix reduces to the matrix in echelon form $\begin{bmatrix} 1 & 2 & -3 & 1 \\ 0 & -1 & 4 & 0 \end{bmatrix}$ which has rank 2. Therefore the solution space to the given system of linear equations has dimension $4 - 2 = 2$ and, indeed, a basis for this space is given $\{[-5, 4, 1, 0], [-1, 0, 0, 1]\}$.

If A is a nonempty set then a relation $\equiv$ between the elements of A is called an *equivalence relation* if and only if the following conditions are satisfied:
   (1) *(reflexivity)* $a \equiv a$ for all $a \in A$;
   (2) *(symmetry)* If $a \equiv a'$ then $a' \equiv a$;
   (3) *(transitivity)* If $a \equiv a'$ and $a' \equiv a''$ then $a \equiv a''$.

Equivalence relations in general play an important role in mathematics. For example, we have already seen that isomorphism of vector spaces satisfies the above three conditions. Here, however, we want to consider equivalence relations of a special type. Let V be a vector space over a field F. A nonempty subset G of $Aut(V)$ is called a *group of automorphisms* if and only if the follow conditions are satisfied:

   (1) $\sigma_1 \in G$;
   (2) If $\alpha \in G$ then $\alpha^{-1} \in G$;
   (3) If $\alpha, \beta \in G$ then $\beta\alpha \in G$.

EXAMPLE

Let V be a vector space over a field F and let Ω be a nonempty set. If τ is a permutation of Ω then we have an endomorphism α_τ of the vector space V^Ω defined by $\alpha_\tau(f): i \mapsto f(\tau(i))$. This is in fact an automorphism, for which $(\alpha_\tau)^{-1} = \alpha_{\tau^{-1}}$. Moreover, the set of all such endomorphisms is a group of automorphisms of V^Ω.

Let V be a vector space over a field F. Any group G of automorphisms of V defines an equivalence relation $\sim_G$ between the elements of V as follows: $v \sim_G v'$ if and only if there exists an element $\alpha \in G$ such that $\alpha(v) = v'$. Indeed, since $\sigma_1 \in G$ we note that $v \sim_G v$ for all $v \in G$. If $\alpha(v) = v'$ then $\alpha^{-1}(v') = v$ so $v \sim_G v'$ implies $v' \sim_G v$. Finally, if $\alpha(v) = v'$ and $\beta(v') = v''$ then $\beta\alpha(v) = v''$ so $v \sim_G v'$ and $v' \sim_G v''$ imply $v \sim_G v''$.

Let F be a field and let k, n be positive integers. If $P \in \mathcal{M}_{k \times k}(F)$ and $Q \in \mathcal{M}_{n \times n}(F)$ are nonsingular matrices then the function ω_{PQ} from $\mathcal{M}_{k \times n}(F)$ to itself defined by $C \mapsto PCQ$ is an endomorphism of $\mathcal{M}_{k \times n}(F)$ which is in fact an automorphism, the inverse of which is $\omega_{P^{-1}Q^{-1}}$. The set G of all automorphisms of $\mathcal{M}_{k \times n}(F)$ of this form is a group of automorphisms and so defines an equivalence relation $\sim_G$ between the elements of $\mathcal{M}_{k \times n}(F)$. Matrices C and C' satisfying $C \sim_G C'$ are simply said to be *equivalent*.

As we have already noted, multiplying a matrix C on the left by a nonsingular matrix has the same effect as performing a series of elementary operations on the rows of C. In a similar manner, one sees that multiplying C on the right by a nonsingular matrix has the same effect as performing a series of elementary operations on the columns of C. In fact, making use of both types of operations, any matrix can be reduced to a matrix of the form $\begin{bmatrix} I & O \\ O & O \end{bmatrix}$, where $I \in \mathcal{M}_{r \times r}(F)$. The way to do this is the following:

(I) Multiply C on the left by a nonsingular matrix in order to obtain a matrix C' in echelon form, the first r rows of which are nonzero.

(II) If $1 \leq h \leq r$, let c'_{hj} be the leftmost nonzero entry in the hth row. We can multiply C' on the left by a suitable elementary matrix in order to assume that this entry equals 1. If $h > 1$ then, again multiplying C' on the left by suitable elementary matrices, we can subtract suitable scalar products of this row from the rows above it in order to insure that $c'_{ij} = 0_F$ for all $1 \leq i < h$. At the end of this, we obtain a matrix C'' in echelon form such that in each of the first r rows the leftmost entry is 1 and the entries above it are all equal to 0_F.

(III) By multiplying C'' on the right by suitable elementary matrices, we can obtain a matrix of the block form $\begin{bmatrix} I & B \\ O & O \end{bmatrix}$, where I is the $r \times r$ identity matrix. By multiplying on the right by additional elementary matrices, we can subtract scalar multiples of the first r columns from the others and so obtain a matrix of the desired form.

EXAMPLE

Consider the matrix $\begin{bmatrix} 1 & 2 & 3 & 4 \\ -1 & -2 & 1 & 1 \\ 1 & 2 & 7 & 9 \\ -2 & -4 & 6 & 7 \end{bmatrix} \in \mathcal{M}_{4 \times 4}(\mathbb{R})$. Reducing this matrix to echelon form, we obtain $\begin{bmatrix} 1 & 2 & 3 & 4 \\ 0 & 0 & 4 & 5 \\ 0 & 0 & 0 & 0 \\ 0 & 0 & 0 & 0 \end{bmatrix}$. Multiplying the second row by $\frac{1}{4}$ we obtain a matrix the (2,3) entry of which is 1 and then, subtracting the second row of this matrix from the first, we obtain the matrix $\begin{bmatrix} 1 & 2 & 0 & 1/4 \\ 0 & 0 & 1 & 5/4 \\ 0 & 0 & 0 & 0 \\ 0 & 0 & 0 & 0 \end{bmatrix}$.

EXAMPLE (Continued)

Multiplying on the right by a suitable elementary matrix, we interchange the second

and third columns to obtain $\begin{bmatrix} 1 & 0 & 2 & 1/4 \\ 0 & 1 & 0 & 5/4 \\ 0 & 0 & 0 & 0 \\ 0 & 0 & 0 & 0 \end{bmatrix}$ and then, by subtracting suitable

multiples of the first and second columns from the third and fourth columns, we

obtain $\begin{bmatrix} 1 & 0 & 0 & 0 \\ 0 & 1 & 0 & 0 \\ 0 & 0 & 0 & 0 \\ 0 & 0 & 0 & 0 \end{bmatrix}$.

And finally, a warning! No matter which method we use to calculate the solution of a system of linear equations on a computer, we must be wary of the possibility that the system can be very delicate: a small change in one of the entries of the coefficient matrix can lead to a large change in the solutions.

EXAMPLE

The system of linear equations

$$\begin{bmatrix} 1 & -10 & 0 & 0 & 0 & 0 \\ 0 & 1 & -10 & 0 & 0 & 0 \\ 0 & 0 & 1 & -10 & 0 & 0 \\ 0 & 0 & 0 & 1 & -10 & 0 \\ 0 & 0 & 0 & 0 & 1 & -10 \\ 0 & 0 & 0 & 0 & 0 & 1 \end{bmatrix} \begin{bmatrix} X_1 \\ X_2 \\ X_3 \\ X_4 \\ X_5 \\ X_6 \end{bmatrix} = \begin{bmatrix} -9 \\ -9 \\ -9 \\ -9 \\ -9 \\ 1 \end{bmatrix}$$

over $\mathbb{R}$ has a solution $w = [1, 1, 1, 1, 1, 1]$. If we change the coefficient matrix so that the $(6, 6)$ entry is $1/1.001$ instead 1 (in other words, a change of about $1/10$ of 1 percent) then the new system also has a solution $w' = [101, 11, 2, 1.1, 1.01, 1.001]$.

Since, in real-life applications of linear algebra, the coefficient data comes from external observations often outside our control, this sensitivity can have devastating effects on the accuracy of computation. Numerical analysts study this matter in great detail. Here we will bring a rule of thumb which can give you an estimate of what loss of accuracy to expect. Given a positive integer n, define the function $\theta: \mathcal{M}_{n \times n}(\mathbb{R}) \to \mathbb{R}$ by setting

$$\theta: [a_{ij}] \mapsto max\{|a_{1j}| + \cdots + |a_{nj}| \mid 1 \le j \le n\}.$$

If A is a nonsingular matrix satisfying the condition that $\theta(A)\theta(A^{-1}) = g \times 10^t$, where $0.1 \le g < 1$ then it is reasonable to expect that the components of the solution of a system of linear equations $AX = w^T$ will have t digits accuracy *less* than the accuracy of the entries of A.

EXAMPLE

If A is the matrix in the previous example then $\theta(A) = 11$. Moreover,

$$A^{-1} = \begin{bmatrix} 1 & 10 & 100 & 1000 & 10000 & 100000 \\ 0 & 1 & 10 & 100 & 1000 & 10000 \\ 0 & 0 & 1 & 10 & 100 & 1000 \\ 0 & 0 & 0 & 1 & 10 & 100 \\ 0 & 0 & 0 & 0 & 1 & 10 \\ 0 & 0 & 0 & 0 & 0 & 1 \end{bmatrix}$$

and so $\theta(A^{-1}) = 111111$. Therefore $\theta(A)\theta(A^{-1}) = 0.12 \times 10^7$ and so we if our original data was good to only four significant digits then we should expect no significant digits in the answer, which is in fact what we saw happened.

Problems

1. Bring the matrix $\begin{bmatrix} 1 & 2 & 3 & 4 \\ 1 & 2 & 4 & 3 \\ 2 & 3 & 1 & 4 \end{bmatrix} \in \mathcal{M}_{3\times4}(\mathbb{R})$ to echelon form.

2. Bring the matrix $\begin{bmatrix} 1 & 2 & 1 & 0 \\ 2 & 3 & 1 & 1 \\ 1 & 2 & 4 & 0 \end{bmatrix} \in \mathcal{M}_{3\times4}(\mathbb{Z}/(5))$ to echelon form.

3. Solve the system of linear equations

$$(3 - i)X_1 + (4 + 2i)X_2 = 2 + 6i$$
$$(4 + 2i)X_1 - (2 + 3i)X_2 = 5 + 4i$$

over the field $\mathbb{C}$.

4. Solve the system of linear equations

$$X_1 + 2X_2 + 4X_3 = 31$$
$$5X_1 + X_2 + 2X_3 = 29$$
$$3X_1 - X_2 + X_3 = 10$$

over the field $\mathbb{R}$.

5. Solve the system of linear equations

$$X_1 + 2X_2 + X_3 = 1$$
$$X_1 + X_2 + X_3 = 0$$

over the field $\mathbb{Z}/(3)$.

6. Solve the system of linear equations

$$X_1 + (\sqrt{2})X_2 + (\sqrt{2})X_3 = 3$$
$$X_1 + (1 + \sqrt{2})X_2 + X_3 = 3 + \sqrt{2}$$
$$X_1 + X_2 - (\sqrt{2})X_3 = 4 + \sqrt{2}$$

over the field $\mathbb{Q}(\sqrt{2})$.

7. Solve the system of linear equations

$$4X_1 - 3X_2 = 3$$
$$2X_1 - X_2 + 2X_3 = 1$$
$$3X_1 + 2X_3 = 4$$

over the field $\mathbb{Z}/(5)$.

8. Find a real number a such that the system of linear equations

$$2X_1 - X_2 + X_3 + X_4 = 1$$
$$X_1 + 2X_2 - X_3 + 4X_4 = 2$$
$$X_1 + 7X_2 - 4X_3 + 11X_4 = a$$

over $\mathbb{R}$ is consistent.

9. Find all solutions of the following systems of linear equations over $\mathbb{R}$:

(i)
$$\begin{bmatrix} 1 & 2 & 3 & 4 \\ 2 & 1 & 2 & 3 \\ 3 & 2 & 1 & 2 \\ 4 & 3 & 2 & 1 \end{bmatrix} X = \begin{bmatrix} 5 \\ 1 \\ 1 \\ -5 \end{bmatrix}.$$

(ii)
$$\begin{bmatrix} 1 & 2 & 3 & 4 & 5 \\ 2 & 1 & 2 & 3 & 4 \\ 2 & 2 & 1 & 2 & 3 \\ 2 & 2 & 2 & 1 & 2 \\ 2 & 2 & 2 & 2 & 1 \end{bmatrix} X = \begin{bmatrix} 13 \\ 10 \\ 11 \\ 6 \\ 3 \end{bmatrix}.$$

(iii)
$$\begin{bmatrix} 1 & 3 & 2 \\ 2 & -1 & 3 \\ 3 & -5 & 4 \\ 1 & 17 & 4 \end{bmatrix} X = \begin{bmatrix} 0 \\ 0 \\ 0 \\ 0 \end{bmatrix}.$$

(iv)
$$\begin{bmatrix} 4 & 6 & 2 \\ 1 & -a & -2 \\ 7 & 3 & a-5 \end{bmatrix} X = \begin{bmatrix} 8 \\ -5 \\ 7 \end{bmatrix} \text{ for various values of } a.$$

(v)
$$\begin{bmatrix} a & 1 & 1 \\ 1 & a & 1 \\ 1 & 1 & a \end{bmatrix} X = \begin{bmatrix} 1 \\ 1 \\ 1 \end{bmatrix} \text{ for various values of } a.$$

10. Find the sets of all real numbers c such that the system of linear equations

$$X_1 + X_2 - X_3 = 1$$
$$X_1 + cX_2 + 3X_3 = 2$$
$$2X_1 + 3X_2 + cX_3 = 3$$

has

(i) A unique solution;
(ii) Infinitely-many solutions;
(iii) No solutions.

11. Find bases for the row space and the column space of the following matrix over $\mathbb{R}$:

$$\begin{bmatrix} 1 & 2 & -3 & -7 & -2 \\ -1 & -2 & 1 & 1 & 0 \\ 1 & 2 & 0 & 2 & 1 \end{bmatrix}.$$

12. Express the rows of the matrix $A = \begin{bmatrix} i & 2 & 0 \\ i-1 & 2 & i \\ 0 & 2 & -i \end{bmatrix} \in \mathcal{M}_{3\times3}(\mathbb{C})$ as linear combinations of the rows of A^T.

13. Calculate the ranks of the following matrices over $\mathbb{R}$:

(i)
$$\begin{bmatrix} 1 & -1 & 2 & 3 & 4 \\ 2 & 1 & -1 & 2 & 0 \\ -1 & 2 & 1 & 1 & 3 \\ 1 & 5 & -8 & -5 & -12 \\ 3 & -7 & 8 & 9 & 13 \end{bmatrix}.$$

(ii)
$$\begin{bmatrix} 1 & 0 & 1 & 0 & 0 \\ 1 & 1 & 0 & 0 & 0 \\ 0 & 1 & 1 & 0 & 0 \\ 0 & 0 & 1 & 1 & 0 \\ 0 & 1 & 0 & 1 & 1 \end{bmatrix}.$$

(iii)
$$\begin{bmatrix} a & -1 & 2 & 1 \\ -1 & a & 5 & 2 \\ 10 & -6 & 1 & 1 \end{bmatrix} \text{ for various values of } a.$$

(iv)
$$\begin{bmatrix} 3 & 1 & 1 & 4 \\ a & 4 & 10 & 1 \\ 1 & 7 & 17 & 3 \\ 2 & 2 & 4 & 3 \end{bmatrix} \text{ for various values of } a.$$

14. What is the rank of the matrix $\begin{bmatrix} 1 & 1 & 0 \\ 0 & 1 & 1 \\ 1 & 0 & 1 \end{bmatrix} \in \mathcal{M}_{3\times3}(\mathbb{Z}/(2))$?

15. Let $F = \mathbb{Z}/(5)$. Find the rank of the matrix $\begin{bmatrix} 1 & 2 & 3 & 4 & a \\ 4 & 3 & a & 1 & 2 \\ a & 1 & 2 & 3 & 4 \\ 2 & 3a & 2 & 4a & 1 \end{bmatrix}$ for various elements $a \in F$.

16. Let F be a field and let $A = \begin{bmatrix} B & C \\ D & E \end{bmatrix}$ be a matrix in $\mathcal{M}_{k \times n}(F)$, where k and n are positive integers and B is a nonsingular $r \times r$ matrix. Show that the rank of A equals r if and only if $DB^{-1}C = E$.

17. Let $\{a, b, c\}$ and $\{d, e, f\}$ be two sets of distinct elements of a field F. Calculate the rank of the matrix $\begin{bmatrix} 1 & a & d & ad \\ 1 & b & e & be \\ 1 & c & f & cf \end{bmatrix}$.

18. Find a rational number a such that the rank of the matrix

$$\begin{bmatrix} 1 & 7 & 17 & 3 \\ 4 & 4 & 8 & 6 \\ 3 & 1 & 1 & 4 \\ 2a & 8 & 20 & 2 \end{bmatrix} \in \mathcal{M}_{4 \times 4}(\mathbb{Q})$$

is minimal.

19. Let n be a positive integer, let F be a field, and let $O \neq A \in \mathcal{M}_{n \times n}(F)$. Show that there exists a positive integer k such that the rank of A^k equals the rank of A^h for all $h > k$.

20. Let F be a field and let $A \in \mathcal{M}_{k \times n}(F)$ be a matrix having rank h. Show that we can write $A = BC$, where $B \in \mathcal{M}_{k \times h}(F)$ and $C \in \mathcal{M}_{h \times n}(F)$.

21. Let V be a vector space over a field F. Decide which of the following relations $\equiv$ between the elements of V is an equivalence relation:
   (i) $v \equiv w$ if and only if $Rv = Rw$;
   (ii) $v \equiv w$ if and only if $v - w$ belongs to some fixed preselected subspace Y of V;
   (iii) $v \equiv w$ if and only if there exists a vector space V' and a linear transformation $\alpha: V \rightarrow V'$ satisfying $\alpha(v) = \alpha(w)$;
   (iv) $v \equiv w$ if and only if $Rv \cap Rw = \{0_V\}$.

22. Let V and W be vector spaces over a field F and let G be a nonempty subset of $Hom(V, W)$. Define a relation $\equiv$ between elements of V by setting $v \equiv v'$ if and only if $\alpha(v) = \alpha(v')$ for all $\alpha \in G$. Is this an equivalence relation between elements of V?

23. Find matrices P and Q in $\mathcal{M}_{3 \times 3}(\mathbb{R})$ which satisfy

$$P \begin{bmatrix} 1 & 2 & 3 \\ 2 & -2 & 1 \\ 3 & 0 & 4 \end{bmatrix} Q = \begin{bmatrix} 1 & 0 & 0 \\ 0 & 1 & 0 \\ 0 & 0 & 0 \end{bmatrix}.$$

24. Let F be a field and let $A \in \mathcal{M}_{2\times3}(F)$ be a matrix satisfying the condition that there exists an element $b \in F$ such that the sum of the elements in each row and each column of A is b. Show that $b = 0_F$.

25. Let F be a field. A matrix $A \in \mathcal{M}_{3\times3}(F)$ is a *magic matrix* if and only if there exists an element $b \in F$ such that the sum of the elements of each row and each column of A equals b. Show that the set of all magic matrices in $\mathcal{M}_{3\times3}(F)$ is a subspace of that vector space over F and find its dimension.

26. Let n be a positive integer and let F be a field. A matrix $A = [a_{ij}] \in \mathcal{M}_{n\times n}(F)$ is *tridiagonal* if and only if $a_{ij} \neq 0$ implies $i = j - 1$, $i = j$, or $i = j + 1$. If $A \in \mathcal{M}_{n\times n}(F)$ is a tridiagonal matrix and if $w \in F^n$, calculate the number of operations needed to solve the system of linear equations $AX = w^T$ using Gaussian elimination.

27. Use matrix methods to find a polynomial $p(X) = aX^2 + bX + c \in \mathbb{R}[X]$ satisfying $p(1) = -1$, $p(-1) = 9$, and $p(2) = -3$.

28. Calculate the inverse of $\begin{bmatrix} 1 & 2 & 3 & 4 & 5 \\ 0 & 1 & 2 & 3 & 4 \\ 0 & 0 & 1 & 2 & 3 \\ 0 & 0 & 0 & 1 & 2 \\ 0 & 0 & 0 & 0 & 1 \end{bmatrix} \in \mathcal{M}_{5\times5}(\mathbb{R}).$

29. Let n be a positive integer, let F be a field, and let $A \in \mathcal{M}_{n\times n}(F)$. Suppose that $w \in F^n$ and that the system of linear equations $AX = w^T$ has a nonempty set of solutions satisfying the property that the hth component of each of them equals some fixed scalar c. What can we deduce from this about the columns of the matrix A?

CHAPTER IX

DETERMINANTS

Let F be a field and let n be a positive integer. We want to find a function from $\delta_n: \mathcal{M}_{n \times n}(F) \rightarrow F$ which will act as an indicator of invertibility: a matrix A is nonsingular if $\delta_n(A) \neq 0_F$ and is singular if $\delta_n(A) = 0_F$. In quest of such a function, we begin with a definition. A function $\delta_n: \mathcal{M}_{n \times n}(F) \rightarrow F$ is a *determinant function* if and only if it satisfies the following conditions:

(1) $\delta_n(I) = 1_F$;
(2) $\delta_n(A) = 0_F$ if A has a row equal to $[0_F, \ldots, 0_F]$;
(3) $\delta_n(E_{ij}A) = -\delta_n(A)$ for all $A \in \mathcal{M}_{n \times n}(F)$ and all $1 \leq i \neq j \leq n$;
(4) $\delta_n(E_{ij;c}A) = \delta_n(A)$ for all $A \in \mathcal{M}_{n \times n}(F)$ and all $1 \leq i \neq j \leq n$;
(5) $\delta_n(E_{i;c}A) = c\delta_n(A)$ for all $A \in \mathcal{M}_{n \times n}(F)$ and all $1 \leq i \leq n$.

Note the fact that we have defined such functions does not imply that they exist. It is immediate, however, that we can define such functions for the smallest values of n.

EXAMPLE

If $n = 1$ then the function $\delta_1: [a] \mapsto a$ is a determinant function.

If $n = 2$ then the function $\delta_2: \begin{bmatrix} a & b \\ c & d \end{bmatrix} \mapsto ad - bc$ is a determinant function.

We want to show that a determinant function δ_n exists for every positive integer n and, moreover, that such a function is unique. As it turns out, it is simpler to consider the matter of uniqueness first.

(9.1) PROPOSITION. *Let F be a field. For every positive integer n there exists at most one determinant function $\delta_n: \mathcal{M}_{n \times n}(F) \rightarrow F$.*

PROOF. Assume that δ_n and δ_n' are determinant functions from $\mathcal{M}_{n \times n}(F)$ to F and let $\beta = \delta_n - \delta_n'$. Then the function β satisfies the following conditions:

(1) $\beta(I) = 0_F$;
(2) $\beta(A) = 0_F$ if A has a row equal to $[0_F, \ldots, 0_F]$;
(3) $\beta(E_{ij}A) = -\beta(A)$ for all $A \in \mathcal{M}_{n \times n}(F)$ and all $1 \leq i \neq j \leq n$;
(4) $\beta(E_{ij;c}A) = \beta(A)$ for all $A \in \mathcal{M}_{n \times n}(F)$ and all $1 \leq i \neq j \leq n$;
(5) $\beta(E_{i;c}A) = c\beta(A)$ for all $A \in \mathcal{M}_{n \times n}(F)$ and all $1 \leq i \leq n$.

In particular, if $A \in \mathcal{M}_{n \times n}(F)$ and if E is a elementary matrix then $\beta(A)$ and $\beta(EA)$ are either both equal to 0_F or both not equal to 0_F, and so, by induction, we see that the same is true if E is a product of elementary matrices. But for any matrix A there exists a product of elementary matrices E such that $EA = I$ or EA is a matrix having a row equal to $[0_F, \ldots, 0_F]$. Therefore, by (1) and (2) above, $\beta(A) = 0_F$ for any matrix A, proving that $\delta_n(A) = \delta'_n(A)$ for all A. $\square$

(9.2) PROPOSITION. *If $\delta_n: \mathcal{M}_{n \times n}(F) \to F$ is a determinant function then:*

(1) *A matrix $A \in \mathcal{M}_{n \times n}(F)$ is nonsingular if and only if $\delta_n(A) \neq 0_F$.*

(2) *If $A \in \mathcal{M}_{n \times n}(F)$ has two identical rows then $\delta_n(A) = 0_F$.*

(3) *If $A, B \in \mathcal{M}_{n \times n}(F)$ then $\delta_n(AB) = \delta_n(A)\delta_n(B)$.*

(4) *If $A \in \mathcal{M}_{n \times n}(F)$ is nonsingular then $\delta_n(A^{-1}) = \delta_n(A)^{-1}$.*

PROOF. (1) If A is a nonsingular matrix then there exist elementary matrices $E_1, \ldots, E_t \in \mathcal{M}_{n \times n}(F)$ such that $A = E_1 \cdot \ldots \cdot E_t I$ and so, by definition of a determinant function, $\delta_n(A)$ equals a nonzero scalar times $\delta_n(I)$ and so differs from 0_F. Conversely, if A is singular then there exist elementary matrices $F_1, \ldots, F_t \in \mathcal{M}_{n \times n}(F)$ such that $F_1 \cdot \ldots \cdot F_t A$ equals a matrix B having one row equal to $[0_F, \ldots, 0_F]$. If $\delta_n(A) \neq 0_F$ then also $\delta_n(B) \neq 0_F$, which contradicts the definition of a determinant function. Thus we must have $\delta_n(A) = 0_F$.

(2) Assume that the ith and hth rows of A are identical. If the characteristic of F is not equal to 2 then $E_{ih}A = A$ which implies that $-\delta_n(A) = \delta_n(A)$ and hence $\delta_n(A) = 0_F$. If the characteristic of F equals 2 then $\delta_n(A) = \delta_n(E_{ih;1}A)$ and this equals 0_F since since the matrix $E_{ih;1}A$ has a row equal to $[0_F, \ldots, 0_F]$.

(3) By Proposition 7.1 we know that A and B are both nonsingular if and only if AB is nonsingular. Therefore $\delta_n(A) = 0_F$ or $\delta_n(B) = 0_F$ if and only if $\delta_n(AB) = 0_F$. If $\delta_n(A)$ and $\delta_n(B)$ are both nonzero then, by (1), both of these matrices are nonsingular and so there exist elementary matrices $E_1, \ldots, E_t$ and $F_1, \ldots, F_s$ satisfying $A = E_1 \cdot \ldots \cdot E_t$ and $B = F_1 \cdot \ldots \cdot F_s$. Hence $AB = E_1 \cdot \ldots \cdot E_t F_1 \cdot \ldots \cdot F_s$ and so $\delta_n(AB) = \delta_n(A)\delta_n(B)$ by the definition of a determinant function.

(4) By (3) we see that $\delta_n(A^{-1})\delta_n(A) = \delta_n(A^{-1}A) = \delta_n(I) = 1_F$ and so the result follows immediately. $\square$

(9.3) PROPOSITION. *If $\delta_n: \mathcal{M}_{n \times n}(F) \to F$ is a determinant function and if $A \in \mathcal{M}_{n \times n}(F)$ then:*

(1) $\delta_n(AE_{ij}) = -\delta_n(A)$;

(2) $\delta_n(AE_{ij;c}) = \delta_n(A)$;

(3) $\delta_n(AE_{i;c}) = c\delta_n(A)$.

PROOF. This follows immediately from the definition of a determinant function and the fact that $\delta_n(AB) = \delta_n(A)\delta_n(B) = \delta_n(B)\delta_n(A) = \delta_n(BA)$ for all $A, B \in \mathcal{M}_{n \times n}(F)$. $\square$

COROLLARY. *If $\delta_n: \mathcal{M}_{n \times n}(F) \to F$ is a determinant function and if A belongs to $\mathcal{M}_{n \times n}(F)$ then $\delta_n(A^T) = \delta_n(A)$.*

Of course, we have not yet shown that a determinant function $\delta_n: \mathcal{M}_{n \times n}(F) \to F$ exists when $n > 2$, and this we set about to do now. Recall that a bijective function from a set A to itself is called a *permutation* of A. If n is a positive integer, let us denote the set of all permutations of the set $\{1, \ldots, n\}$ by S_n. This set is always nonempty since it includes the identity map $h \mapsto h$, which we will denote by ι. Since every permutation $\tau \in S_n$ is bijective, we know that for each such τ there exists a permutation $\tau^{-1} \in S_n$ defined by the condition that $\tau^{-1}(i) = j$ if and only if $\tau(j) = i$. Then $\tau\tau^{-1} = \iota = \tau^{-1}\tau$ for all $\tau \in S_n$. We also note that S_n is closed under composition of functions: if $\sigma, \tau \in S_n$ then $\sigma\tau \in S_n$.

(9.4) PROPOSITION. *The number of elements in S_n is $n!$.*

PROOF. Suppose we wanted to construct a permutation $\tau \in S_n$. There are n possibilities to choose the element $\tau(1)$. Once we have decided on this element, there are $n-1$ possibilities left for $\tau(2)$, etc. Continuing in this manner, we see that the number of different ways of constructing τ is $n! = n(n-1)(n-2)\cdot\ldots\cdot 2\cdot 1$. □

Let $\tau \in S_n$. A pair (i, j) of integers satisfying $1 \le i < j \le n$ is *inverted* by τ if and only if $\tau(i) > \tau(j)$. That is to say, (i, j) is inverted if and only if the rational number $\frac{i-j}{\tau(i)-\tau(j)}$ is negative. A permutation $\tau \in S_n$ is *even* or *odd* if and only if the number of pairs of integers (i, j) satisfying $1 \le i < j \le n$ inverted by τ is even or odd. We will define the *sign* of τ, denoted $sgn(\tau)$, to be 1 if τ is even and -1 if τ is odd. Since ι is clearly an even permutation it follows that τ is even if and only if τ^{-1} is even and so $sgn(\tau) = sgn(\tau^{-1})$ for all $\tau \in S_n$.

EXAMPLE

Let $\tau \in S_4$ be defined by

$$1 \mapsto 1$$
$$2 \mapsto 4$$
$$3 \mapsto 3$$
$$4 \mapsto 2.$$

Then the pairs inverted by τ are $(1, 4)$, $(2, 3)$, and $(3, 4)$. Thus τ is odd and $sgn(\tau) = -1$.

EXAMPLE

Let $\tau \in S_4$ be defined by

$$1 \mapsto 3$$
$$2 \mapsto 4$$
$$3 \mapsto 1$$
$$4 \mapsto 2.$$

Then the pairs inverted by τ are $(1, 4)$, $(2, 3)$, $(2, 4)$, and $(1, 3)$. Thus τ is even and $sgn(\tau) = 1$.

Let F be a field, let n be a positive integer, and let $A = [a_{ij}] \in \mathcal{M}_{n \times n}(F)$. Define the scalar $|A| \in F$ by

$$|A| = \sum_{\tau \in S_n} sgn(\tau)a_{1,\tau(1)}a_{2,\tau(2)}\cdot\ldots\cdot a_{n,\tau(n)}.$$

For each $\tau \in S_n$ we have $a_{1,\tau(1)} a_{2,\tau(2)} \cdot \cdots \cdot a_{n,\tau(n)} = a_{\mu(1),1} a_{\mu(2),2} \cdot \cdots \cdot a_{\mu(n),n}$, where $\mu = \tau^{-1}$ in S_n, and so we have

$$|A| = \sum_{\mu \in S_n} sgn(\mu) a_{\mu(1),1} a_{\mu(2),2} \cdot \cdots \cdot a_{\mu(n),n}.$$

and so we see immediately that $|A| = |A^T|$ for all matrices $A \in \mathcal{M}_{n \times n}(F)$.

EXAMPLE

If $A = [a_{ij}] \in \mathcal{M}_{3 \times 3}(F)$ then

$$|A| = a_{11}a_{22}a_{33} + a_{12}a_{23}a_{31} + a_{13}a_{21}a_{32} - a_{32}a_{23}a_{11} - a_{13}a_{31}a_{22} - a_{12}a_{21}a_{33}.$$

(9.5) PROPOSITION. *Let F be a field, let n be a positive integer, and let $A = [a_{ij}] \in \mathcal{M}_{n \times n}(F)$. Suppose that $1 \le h \le n$ and suppose that for each $1 \le j \le n$ we write $a_{hj} = b_{hj} + c_{hj}$, where $b_{hj}, c_{hj} \in F$. Let $B = [b_{ij}]$ and $C = [c_{ij}]$ be the matrices in $\mathcal{M}_{n \times n}(F)$ defined as follows:*

(1) *If $i \ne h$ then $b_{ij} = c_{ij} = a_{ij}$ for all $1 \le j \le n$;*
(2) *The elements b_{hj} and c_{hj} for all $1 \le j \le n$ are given by the above decomposition.*

Then $|A| = |B| + |C|$.

PROOF. By definition of $|A|$ we have:

$$A = \sum_{\tau \in S_n} sgn(\tau) a_{1,\tau(1)} a_{2,\tau(2)} \cdot \cdots \cdot a_{n,\tau(n)}$$

$$= \sum_{\tau \in S_n} sgn(\tau) a_{1,\tau(1)} a_{2,\tau(2)} \cdot \cdots \cdot [b_{h,\tau(h)} + c_{h,\tau(h)}] \cdot \cdots \cdot a_{n,\tau(n)}$$

$$= \sum_{\tau \in S_n} sgn(\tau) a_{1,\tau(1)} a_{2,\tau(2)} \cdot \cdots \cdot b_{h,\tau(h)} \cdot \cdots \cdot a_{n,\tau(n)}$$

$$+ \sum_{\tau \in S_n} sgn(\tau) a_{1,\tau(1)} a_{2,\tau(2)} \cdot \cdots \cdot c_{h,\tau(h)} \cdot \cdots \cdot a_{n,\tau(n)}$$

$$= |B| + |C|,$$

as desired. $\square$

Similarly, let F be a field, let n be a positive integer, and let $A = [a_{ij}] \in \mathcal{M}_{n \times n}(F)$. If B is the matrix formed from A by multiplying one of the rows of A by a scalar $c \in F$ then, as an immediate consequence of the definition, we note that $|B| = c|A|$. Putting this observation together with Proposition 9.5, we see the following: If $1 \le h \le n$ and if $u_1, \ldots, u_{h-1}, u_{h+1}, \ldots, u_n$ are vectors in F^n then the function from F^n to F given by

$$v \longmapsto \begin{vmatrix} u_1 \\ \cdots \\ u_{h-1} \\ v \\ u_{h+1} \\ \cdots \\ u_n \end{vmatrix}$$

is a linear transformation.

(9.6) PROPOSITION. *If F is a field and if n is a positive integer then the function $A \mapsto |A|$ from $\mathcal{M}_{n \times n}(F)$ to F is a determinant function.*

PROOF. If $n = 1$ then this function is given by $[a] \mapsto a$ and that is clearly a determinant function. Hence we can assume that $n > 1$. In order to simplify notation, for each $\tau \in S_n$ and each $A \in \mathcal{M}_{n \times n}(F)$ we define $u(\tau, A) = sgn(\tau)a_{1,\tau(1)}a_{2,\tau(2)} \cdot \cdots \cdot a_{n,\tau(n)}$. We now check the properties required of a determinant function.

(1) It is clear that $u(\tau, I) = 0_F$ for all $\iota \neq \tau \in S_n$, while $u(\iota, I) = 1_F$. Therefore $|I| = 1_F$.

(2) If A is has one row equal to $[0_F, \ldots, 0_F]$ then we must have $|A| = 0_F$, since $u(\tau, A)$ contains a factor taken from each row of A for any $\tau \in S_n$ we must have $|A| = 0_F$.

(3) Let $1 \leq i \neq j \leq n$ and let $B = E_{ij}A$ and let $\pi \in S_n$ be the permutation defined by $\pi(i) = j$, $\pi(j) = i$, and $\pi(h) = h$ for $h \neq i, j$. If $\tau \in S_n$ then $sgn(\tau\pi) \neq sgn(\tau)$ and so for each $\tau \in S_n$ we have $u(\tau, B) = -u(\tau, A)$. From this we conclude that $|E_{ij}A| = -|A|$.

(4) Let $1 \leq i \neq j \leq n$, let $c \in F$, and let $B = E_{ij;c}A$. Then $B = [b_{ht}]$, where $b_{ht} = a_{ht}$ for all $1 \leq t \leq n$ and all $h \neq j$ and $b_{jt} = a_{jt} + ca_{it}$ for all $1 \leq t \leq n$. Therefore by Proposition 9.5 we have $|B| = |A| + |C|$, where $C = [c_{ht}]$ satisfies $c_{ht} = a_{ht}$ for all $h \neq j$ and all $1 \leq t \leq n$ while $c_{jt} = ca_{it}$ for all $1 \leq t \leq n$. Thus $|C| = c|C'|$, where C' is a matrix having two identical rows (namely the ith and the jth). If the characteristic of F differs from 2, then $C' = E_{ij}C'$ implies that $|C'| = -|C'|$ and so $|C'| = 0_F$ and so, in this case $|B| = |A|$, which is what we want to prove. If the characteristic of F equals 2, the proof is a bit more complicated. Let us assume therefore that we are in that situation and let $\pi \in S_n$ be the permutation defined in (3). Let H be the set of all even permutations in S_n and let H' be the set of all odd permutations in S_n. These sets are disjoint and their union is all of S_n. Moreover, the function from H to H' given by $\tau \mapsto \pi\tau$ is monic, for if $\pi\tau = \pi\mu$ then $\tau = \pi^{-1}\pi\tau = \pi^{-1}\pi\mu = \mu$. Indeed, this function is clearly epic since any element τ of H' is of the form $\pi(\pi^{-1}\tau)$. Since the characteristic of F equals 2, we have $u(\tau, C') = u(\pi\tau, C')$ for all $\tau \in H$ and so $u(\tau, C') + u(\pi\tau, C') = 0_F$. But then

$$|C'| = \sum_{\tau \in H} u(\tau, C') + u(\pi\tau, C') = 0_F$$

from which we deduce, again, that $|C| = 0_F$ and so $|B| = |A|$, as desired.

(5) It is clear from the definition of $|A|$ that if $B = E_{i;c}A$ then $|B| = c|A|$.

Therefore the function $A \mapsto |A|$ satisfies the conditions needed for it to be a determinant function. □

Thus we have shown that determinant functions on $\mathcal{M}_{n \times n}(F)$ exist, and are unique, for all positive integers n. The value $|A|$ defined by a square matrix A is simply referred to as the *determinant* of A. The first systematic treatment of the theory of determinants was given by the French mathematician Augustin-Louis Cauchy, based on work done earlier for special cases by many mathematicians, including Leibnitz and the Japanese Seki Kova. His work was carried on extensively by the British mathematicians Arthur Cayley and James J. Sylvester.

Unfortunately, there remains the problem of actually calculating $|A|$, especially when A is relatively large. After all, our definition involves adding up $n!$ summands, each of which requires $n - 1$ multiplications to compute, so that the total number of operations involved is $(n - 1)n! + (n! - 1) = nn! - 1$. Thus, if $A \in \mathcal{M}_{25 \times 25}(F)$ and given a computer capable of a billion calculations per second, it would take us more than $12,200,000,000$ years to calculate $|A|$ from the definition. This is obviously not viable, and we want to look for better methods, at least for matrices which are relatively small.

Let F be a field, let n be a positive integer, and let $A = [a_{ij}]$ be a matrix in $\mathcal{M}_{n \times n}(F)$. For each pair (i, j), where $1 \leq i, j \leq n$, we define the *minor* of the entry a_{ij} of A to be $|A_{ij}|$, where A_{ij} is the matrix in $\mathcal{M}_{(n-1) \times (n-1)}(F)$ obtained from A by deleting the ith row and jth column.

EXAMPLE

If $A = \begin{bmatrix} 4 & 3 & 1 \\ 8 & 9 & 0 \\ 5 & 6 & 2 \end{bmatrix} \in \mathcal{M}_{3 \times 3}(\mathbb{Q})$ then $A_{13} = \begin{bmatrix} 8 & 9 \\ 5 & 6 \end{bmatrix}$ and $A_{22} = \begin{bmatrix} 4 & 1 \\ 5 & 2 \end{bmatrix}$.

(9.7) PROPOSITION. *Let F be a field and let n be a positive integer. If $A \in \mathcal{M}_{n \times n}(F)$ and if $1 \leq h \leq n$ then $|A| = \sum_{j=1}^{n}(-1)^{h+j}a_{hj}|A_{hj}|$.*

PROOF. Let $v_1, \ldots, v_n \in F^n$. In order to simplify notation, we will denote by $D(v_1, \ldots, v_n)$ the determinant of the matrix the ith row of which is equal to v_i.

We will first prove the proposition for the special case of $h = 1$. That is to say, we will show that $|A| = \sum_{j=1}^{n}(-1)^{j+1}a_{1j}|A_{1j}|$. Let $v_1, \ldots, v_n$ be the canonical basis for F^n. The ith row of A is then equal to $w_i = \sum_{j=1}^{n} a_{ij}v_j$. Moreover,

$$|A| = D(w_1, \ldots, w_n)$$

$$= D\left(\sum_{j=1}^{n} a_{1j}v_j, w_2, \ldots, w_n\right)$$

$$= \sum_{j=1}^{n} a_{1j} D(v_j, w_2, \ldots, w_n)$$

and so we will be done if we can show that $D(v_j, w_2, \ldots, w_n) = (-1)^{j+1}|A_{1j}|$. But

$$D(v_j, w_2, \ldots, w_n) = \sum_{\tau \in S_n} sgn(\tau) a_{1,\tau(1)} \cdot \ldots \cdot a_{n,\tau(n)}$$

$$= \sum_{j=1}^{n} \left(\sum \{ sgn(\tau) a_{2,\tau(2)} \cdot \ldots \cdot a_{n,\tau(n)} \mid \tau \in S_n; \tau(1) = j \} \right)$$

If $j = 1$ this is just $\sum \{ sgn(\tau) a_{2,\tau(2)} \cdot \ldots \cdot a_{n,\tau(n)} \mid \tau \in S \}$, where S is the set of all permutations of $\{2, \ldots, n\}$ (the number 1 does not appear in any inversion and so the value of sgn does not change) and that is precisely equal to $|A_{11}|$. Thus we have shown that $D(v_1, w_2, \ldots, w_n) = |A_{11}| = (-1)^{1+1}|A_{11}|$. If $j > 1$, we can bring the jth column of A to the leftmost position by $j - 1$ transpositions with neighboring columns (and in doing so we have not changed any of the other columns). Then $D(v_j, w_2, \ldots, w_n) = (-1)^{j-1}|A_{1j}| = (-1)^{j+1}|A_{1j}|$. This finishes the special case of $h = 1$.

Now assume that $h > 1$. We will bring the hth row to the top by $h - 1$ transpositions with the rows above it, in order to get $|A| = (-1)^{h-1}|C|$, where C is a matrix satisfying the condition $|C_{1j}| = |A_{hj}|$ for all $1 \leq j \leq n$. Then $|A|$ equals $(-1)^{h-1}|C| = \sum_{j=1}^{n}(-1)^{j+1}c_{1j}|C_{1j}| = \sum_{j=1}^{n}(-1)^{h+j} a_{hj}|A_{hj}|$, as desired. $\square$

Since $|A| = |A^T|$ we see that we also have $|A| = \sum_{h=1}^{n}(-1)^{h+j} a_{hj}|A_{hj}|$ for all $1 \leq j \leq n$.

EXAMPLE

Let $A = \begin{bmatrix} 1 & 2 & 3 & 0 \\ 4 & 0 & 1 & 3 \\ 0 & 2 & 4 & 0 \\ 3 & 1 & 5 & 1 \end{bmatrix} \in \mathcal{M}_{4 \times 4}(\mathbb{Q})$. Then

$$|A| = 1 \begin{vmatrix} 0 & 1 & 3 \\ 2 & 4 & 0 \\ 1 & 5 & 1 \end{vmatrix} - 2 \begin{vmatrix} 4 & 1 & 3 \\ 0 & 4 & 0 \\ 3 & 5 & 1 \end{vmatrix} + 3 \begin{vmatrix} 4 & 0 & 3 \\ 0 & 2 & 0 \\ 3 & 1 & 1 \end{vmatrix} - 0 \begin{vmatrix} 4 & 0 & 1 \\ 0 & 2 & 4 \\ 3 & 1 & 5 \end{vmatrix}$$

$$= 16 + 40 - 30 + 0$$

$$= 26.$$

Another way of calculating the same value is

$$|A| = 0 \begin{vmatrix} 2 & 3 & 0 \\ 0 & 1 & 3 \\ 1 & 5 & 1 \end{vmatrix} - 2 \begin{vmatrix} 1 & 3 & 0 \\ 4 & 1 & 3 \\ 3 & 5 & 1 \end{vmatrix} + 4 \begin{vmatrix} 1 & 2 & 0 \\ 4 & 0 & 3 \\ 3 & 1 & 1 \end{vmatrix} - 0 \begin{vmatrix} 1 & 2 & 3 \\ 4 & 0 & 1 \\ 3 & 1 & 5 \end{vmatrix}$$

$$= 0 - 2 + 28 - 0$$

$$= 26.$$

It is important to emphasize that the calculation of determinants using minors, while it may save considerable effort if there are several rows and/or columns having entries mostly equal to 0_F, is also, in general, considerably tedious. Let us denote by p_n the number of arithmetic operations needed in computing the determinant of a matrix in $\mathcal{M}_{n \times n}(F)$ using minors. Clearly $p_2 = 3$. Now suppose we have already found p_{n-1}. In order to calculate the determinant of a matrix of size $n \times n$ we must calculate the determinants of n matrices of size $(n-1) \times (n-1)$ and then perform n multiplications and $n-1$ additions or subtractions. That is to say,

$$p_n = np_{n-1} + n + (n-1) = np_{n-1} + 2n - 1$$

for $n > 2$. If we divide by $n!$ we obtain

$$\frac{p_n}{n!} - \frac{p_{n-1}}{(n-1)!} = \frac{2}{(n-1)!} - \frac{1}{n!}.$$

Setting $t_n = p_n/n!$, we thus obtain

$$
\begin{aligned}
t_n &= [t_n - t_{n-1}] + [t_{n-1} - t_{n-2}] + \cdots + [t_3 - t_2] + t_2 \\
&= \left[\frac{2}{(n-1)!} - \frac{1}{n!} \right] + \cdots + \left[\frac{2}{2!} - \frac{1}{3!} \right] + \frac{3}{2} \\
&= 2\left[\frac{1}{(n-1)!} + \cdots + \frac{1}{1!} \right] - \left[\frac{1}{n!} + \frac{1}{(n-1)!} + \cdots + \frac{1}{2!} + \frac{1}{1!} \right] + 1 \\
&= \frac{1}{(n-1)!} + \cdots + \frac{1}{2!} + \frac{1}{1!} + 1 - \frac{1}{n!} \\
&= \left[\frac{1}{n!} + \frac{1}{(n-1)!} + \cdots + \frac{1}{2!} + \frac{1}{1!} + 1 \right] - \frac{2}{n!}
\end{aligned}
$$

and hence

$$p_n = n!\left[1 + \frac{1}{1!} + \cdots \frac{1}{(n-1)!} + \frac{1}{n!} \right] - 2.$$

We know that e, the base of the natural logarithms, can be written as

$$1 + \frac{1}{1!} + \cdots + \frac{1}{n!} + \frac{e^c}{(n+1)!},$$

where $0 < c < 1$. Therefore $p_n = n![e - e^c(n+1)!]$. But for $n \geq 2$ we have $0 < e^c/(n+1) < e/(n+1) \leq e/3 < 1$ and so we see that $en! - 3 < p_n < en! - 2$. Since p_n is a positive integer, we get $p_n = \lfloor en! \rfloor - 2$ where, for a real number r, the symbol $\lfloor r \rfloor$ stands for the greatest integer less than or equal to r. In particular, we see that p_n also increases very rapidly as a function of n.

We conclude, therefore, that while the notion of determinant is important theoretically, and is necessary for the definition of the characteristic polynomial in the next chapter, calculating determinants is not a useful tool for solving numerical problems in linear algebra. Except for special cases in which we are dealing with matrices of very specific types, it is rarely necessary or desirable to actually compute determinants.

Let F be a field and let n be a positive integer. If $A \in \mathcal{M}_{n \times n}(F)$ then its *adjoint* is the matrix $adj(A) = [(-1)^{i+j}|A_{ji}|]$. The relation between nonsingular matrices and their adjoints is given by the following result.

(9.8) PROPOSITION. *Let F be a field and let n be a positive integer. If $A \in \mathcal{M}_{n \times n}(F)$ is nonsingular then $A^{-1} = |A|^{-1} adj(A)$.*

PROOF. Let $A = [a_{ij}]$ and $adj(A) = [b_{ij}]$. Then $adj(A)A = [c_{ij}]$, where

$$c_{ij} = \sum_{k=1}^{n} b_{ik} a_{kj} = \sum_{k=1}^{n} (-1)^{i+k} |A_{ki}| a_{kj}.$$

If $i = j$ this is the development of $|A|$ by minors and so equals $|A|$. Otherwise, this is the development by minors of the determinant of a matrix which equals A in all of its columns except for the ith column, which is equal to the jth column of A. But the determinant of such a matrix equals 0_F. Therefore $adj(A)A = |A|I$, where I is the identity matrix of $\mathcal{M}_{n \times n}(F)$, and this proves that $A^{-1} = |A|^{-1} adj(A)$. $\square$

(9.9) PROPOSITION. *Let F be a field and let n be a positive integer. If $A \in \mathcal{UT}_{n \times n}(F)$ then $|A|$ equals the product of the diagonal elements of A.*

PROOF. We will prove this result by induction on n. The case of $n = 1$ is immediate, so assume that $n > 1$ and that the result has already been established for upper triangular matrices of size $(n-1) \times (n-1)$. Then

$$|A| = \sum_{j=1}^{n} (-1)^{j+1} a_{j1} |A_{j1}| = a_{11} |A_{11}|.$$

By the induction hypothesis, $|A_{11}| = a_{22} \cdot \ldots \cdot a_{nn}$ and so we are done. $\square$

As a consequence of Proposition 9.9 we see that computations of the determinant can be simplified if we first perform elementary operations on the matrix to bring it into echelon form.

(9.10) PROPOSITION. **(Cramer's Theorem)** *Let F be a field and let n be a positive integer. If $A = [a_{ij}] \in \mathcal{M}_{n \times n}(F)$ is nonsingular and if $w = [b_1, \ldots, b_n] \in F^n$ then the system of linear equations $AX = w^T$ has a unique solution $v = A|^{-1}[|A_{(1)}|, \ldots, |A_{(n)}|]$, where, for each $1 \le i \le n$, the matrix $A_{(i)}$ is obtained from A by replacing the ith column with w^T.*

PROOF. We already know that the given system of linear equations has a unique solution $v = [d_1, \ldots, d_n]$, where $v^T = A^{-1} w^T$. Therefore $|A| v^T = (|A| A^{-1}) A v^T = adj(A) A v^T = adj(A) w^T$ and so for all $1 \le i \le n$ we have

$$|A| d_i = \sum_{j=1}^{n} (-1)^{i+j} b_i |A_{ij}|.$$

But the expression on the righthand side of this equation is just the development by minors of $|A_{(i)}|$, and that gives the desired result. $\square$

EXAMPLE

Consider the system of linear equations over $\mathbb{R}$:

$$\begin{bmatrix} 1 & -1 & 1 \\ 1 & 2 & 0 \\ 1 & 0 & -1 \end{bmatrix} X = \begin{bmatrix} 2 \\ 1 \\ 4 \end{bmatrix}.$$

Then $|A| = -5$ and

$$|A_{(1)}| = \begin{vmatrix} 2 & -1 & 1 \\ 1 & 2 & 0 \\ 4 & 0 & -1 \end{vmatrix} = -13$$

$$|A_{(2)}| = \begin{vmatrix} 1 & 2 & 1 \\ 1 & 1 & 0 \\ 1 & 4 & -1 \end{vmatrix} = 4$$

$$|A_{(3)}| = \begin{vmatrix} 1 & -1 & 2 \\ 1 & 2 & 1 \\ 1 & 0 & 4 \end{vmatrix} = 7$$

so that the unique solution is $[\frac{13}{5}, -\frac{4}{5}, -\frac{7}{5}]$.

Finally, we note that the definition of $|A|$, where A is a matrix of size $n \times n$ over a field F, has nothing to do with the fact that the entries in the matrix are elements of a field. The formal definition would work perfectly well if the entries of A were elements of an arbitrary ring and would yield an element of that ring. In particular, if F is a field, if n is a positive integer, and if $p_{ij}(X) \in F[X]$ for all $1 \le i, j \le n$ then it is quite possible to talk about $|[p_{ij}(X)]|$ and, indeed, this is equal to the polynomial $\sum_{\tau \in S_n} sgn(\tau) p_{1,\tau(1)}(X) p_{2,\tau(2)}(X) \cdot \ldots \cdot p_{n,\tau(n)}(X) \in F[X]$.

We will return to determinants of matrices of polynomials in the next chapter.

Problems

1. For $a, b \in \mathbb{R}$, calculate $\begin{vmatrix} sin(a) & cos(a) \\ sin(b) & cos(b) \end{vmatrix}$ and $\begin{vmatrix} cos(a) & sin(a) \\ sin(b) & cos(b) \end{vmatrix}$.

2. Calculate the determinant of $\begin{bmatrix} 1 & i & 1+i \\ -i & 1 & 0 \\ 1-i & 0 & 1 \end{bmatrix} \in \mathcal{M}_{3\times3}(\mathbb{C})$.

3. Let c be a real number and let $A = [a_{ij}] \in \mathcal{M}_{n \times n}(\mathbb{R})$ be the matrix defined as follows:

   (i) $a_{ij} = c$ if $i < j$;
   (ii) $a_{ij} = i$ if $i = j$;
   (iii) $a_{ij} = 0$ if $i > j$.

Calculate $|A|$.

4. Let n be an odd positive integer and let $A = [a_{ij}] \in \mathcal{M}_{n \times n}(\mathbb{R})$ be a matrix satisfying the condition that $a_{ij} + a_{ji} = 0$ for all $1 \leq i, j \leq n$. Prove that $|A| = 0$.

5. For real numbers a and b, calculate $\begin{vmatrix} a & b & a+b \\ b & a+b & a \\ a+b & a & b \end{vmatrix}$.

6. Let c and d be real numbers and let $A = [a_{ij}] \in \mathcal{M}_{n \times n}(\mathbb{R})$ be the matrix defined by the condition that $a_{ij} = c$ if $1 \leq i \neq j \leq n$ and $a_{ii} = d$ if $1 \leq i \leq n$. Show that $|A| = [d + (n-1)c](d-c)^{n-1}$.

7. Let $n > 1$ be an integer and let d be a real number. Let $A = [a_{ij}] \in \mathcal{M}_{n \times n}(\mathbb{R})$ be the matrix defined as follows:

(i) $a_{ij} = 0$ if $i = j$;
(ii) $a_{k1} = a_{1k} = 1$ if $1 < k \leq n$;
(iii) $a_{ij} = d$ otherwise.

Prove that $|A| = (-1)^{n-1}(n-1)d^{n-2}$.

8. Let $b_1, \ldots, b_n$ be nonzero real numbers (not necessarily distinct) and let $A = [a_{ij}] \in \mathcal{M}_{n \times n}(\mathbb{R})$ be the matrix defined by

$$a_{ij} = \begin{cases} 1 + b_i, & \text{if } i = j \\ 1, & \text{otherwise.} \end{cases}$$

Calculate $|A|$.

9. For each integer $n = 2^m$ we define the matrix $A(m) \in \mathcal{M}_{n \times n}(\mathbb{Q})$ inductively as follows:

(i) $A(1) = \begin{bmatrix} 1 & 1 \\ 1 & -1 \end{bmatrix}$;
(ii) If we have defined $A(m)$ for all then $A(m+1)$ is the matrix constructed from $A(m)$ by replacing each "1" in $A(m)$ by $A(1)$ and each "-1" by $-A(1)$.

Thus, for example,

$$A(2) = \begin{bmatrix} 1 & 1 & 1 & 1 \\ 1 & -1 & 1 & -1 \\ 1 & 1 & -1 & -1 \\ 1 & -1 & -1 & 1 \end{bmatrix}.$$

Calculate the determinant of $A(m)$ for all positive integers m.

10. Find all triples (a, b, c) of real numbers satisfying

$$\begin{vmatrix} 1 & a & a^3 \\ 1 & b & b^3 \\ 1 & c & c^3 \end{vmatrix} = (b-c)(c-a)(a-b)(a+b+c).$$

11. Let n be a positive integer and let $A = [a_{ij}] \in \mathcal{M}_{n \times n}(\mathbb{C})$. Let $B = [b_{ij}]$ be the matrix defined by $b_{ij} = \bar{a}_{ji}$ for all $1 \leq i, j \leq n$. Show that $|AB|$ is a nonnegative real number.

12. For positive real numbers a and b, calculate $\begin{vmatrix} 1 & log_b a \\ log_a b & 1 \end{vmatrix}$.

13. Let $n > 1$ be an integer and let $A = [a_{ij}] \in \mathcal{M}_{n \times n}(\mathbb{R})$ be the matrix defined by $a_{ii} = 0$ for all $1 \le i \le n$ and $a_{ij} = 1$ for all $1 \le i \ne j \le n$. Calculate $|A|$.

14. Let n be a positive integer, let F be a field, and let A be a matrix in $\mathcal{M}_{n \times n}(F)$. Let $B = [b_{ij}]$ be the matrix defined by $b_{ij} = (-1)^{i+j} a_{ij}$ for all $1 \le i, j \le n$. Show that $|A| = |B|$.

15. Let n be a positive integer, let F be a field, and let A be a matrix in $\mathcal{M}_{n \times n}(F)$. Let $B = [b_{ij}]$ be the matrix defined by $b_{ij} = (-1)^{i+j+1} a_{ij}$ for all $1 \le i, j \le n$. Show that $(-1)^n |A| = |B|$.

16. Let $A = [a_{ij}]$ be a matrix in $\mathcal{M}_{3 \times 3}(\mathbb{R})$ satisfying $a_{i2} \ne 0$ for all $1 \le i \le 3$. Show that

$$|A| = \frac{1}{a_{12}} \begin{vmatrix} |A_{21}| & |A_{23}| \\ |A_{31}| & |A_{32}| \end{vmatrix} + \frac{1}{a_{22}} \begin{vmatrix} |A_{11}| & |A_{13}| \\ |A_{31}| & |A_{33}| \end{vmatrix} + \frac{1}{a_{32}} \begin{vmatrix} |A_{11}| & |A_{13}| \\ |A_{21}| & |A_{23}| \end{vmatrix}.$$

17. For each real number a, calculate $\begin{vmatrix} a-6 & 0 & 0 & -8 \\ 5 & a-4 & 0 & 12 \\ -1 & 3 & a-2 & -6 \\ 0 & -1/2 & 1 & 1 \end{vmatrix}$.

18. If $\{a_i\}$ is the Fibonacci sequence, show that

$$\begin{vmatrix} a_1 & a_2 \\ a_3 & a_4 \end{vmatrix} = \begin{vmatrix} a_3 & a_4 \\ a_5 & a_6 \end{vmatrix} = \begin{vmatrix} a_5 & a_6 \\ a_7 & a_8 \end{vmatrix} = \cdots = 1$$

and

$$\begin{vmatrix} a_2 & a_3 \\ a_4 & a_5 \end{vmatrix} = \begin{vmatrix} a_4 & a_5 \\ a_6 & a_7 \end{vmatrix} = \begin{vmatrix} a_6 & a_7 \\ a_8 & a_9 \end{vmatrix} = \cdots = -1.$$

19. Let a and b be real numbers and let $n > 2$ be an integer. Let $D = [d_{ij}]$ be the matrix in $\mathcal{M}_{n \times n}(\mathbb{R})$ defined by $d_{ij} = sin(ia + jb)$ for all $1 \le i, j \le n$. Show that $|D| = 0$.

20. If a, b, c, d, e, f, g are elements of a field F, show that

$$\begin{vmatrix} a & b & b \\ c & d & e \\ f & g & g \end{vmatrix} + \begin{vmatrix} a & b & b \\ e & c & d \\ f & g & g \end{vmatrix} + \begin{vmatrix} a & b & b \\ d & e & c \\ f & g & g \end{vmatrix} = 0_F.$$

21. Let n be a positive integer, let F be a field, and let A be a matrix in $\mathcal{M}_{n \times n}(F)$. For all $1 \le i \le n$ and all $1 \le j < n$, let $b_{ij} = a_{ij} + a_{i,j+1}$ and let $b_{in} = a_{in}$ for all $1 \le i \le n$. Show that $|B| = |A|$.

22. Let F be a field. For elements $a, b, c \in F$, calculate

$$\begin{vmatrix} -2a & a+b & a+c \\ a+b & -2b & b+c \\ a+c & b+c & -2c \end{vmatrix}.$$

23. Show that the value of the determinant $\begin{vmatrix} 1 & a & a^2 & a^3 \\ a^3 & 1 & a & a^2 \\ b & a^3 & 1 & a \\ c & d & a^3 & 1 \end{vmatrix}$ is independent of $b, c,$ and d.

24. Let n be a positive integer, let F be a field, and let $A \in \mathcal{M}_{n \times n}(F)$ be a matrix satisfying the condition that the sum of the odd-numbered rows of A (as vectors in F^n) equals the sum of the even-numbered rows. What is $|A|$?

25. Let n be a positive integer and let $A \in \mathcal{M}_{n \times n}(\mathbb{Q})$ be a matrix all of the entries of which are ± 1. Show that $|A|$ is a multiple of 2^{n-1}.

26. Let a, b, c, d, e, f be nonzero elements of a field F. Show that

$$\begin{vmatrix} 0 & a^2 & b^2 & c^2 \\ a^2 & 0 & f^2 & e^2 \\ b^2 & f^2 & 0 & d^2 \\ c^2 & e^2 & d^2 & 0 \end{vmatrix} = \begin{vmatrix} 0 & ad & be & cf \\ ad & 0 & cf & be \\ be & cf & 0 & ad \\ cf & be & ad & 0 \end{vmatrix}.$$

27. Let a, b, c, d be elements of a field F. Find all elements y of F satisfying

$$\begin{vmatrix} a & b & c & d \\ b & c & d & a \\ c & d & a & b \\ d & a & b & c \end{vmatrix} = y \begin{vmatrix} 0 & 1 & -1 & 1 \\ 1 & c & d & a \\ 1 & d & a & b \\ 1 & a & b & c \end{vmatrix}.$$

28. If $a, b,$ and c are elements of a field F, is it necessarily true that

$$\begin{vmatrix} a & b & c & 0 \\ b & a & 0 & c \\ c & 0 & a & b \\ 0 & c & b & a \end{vmatrix} = \begin{vmatrix} -a & b & c & 0 \\ b & -a & 0 & c \\ c & 0 & -a & b \\ 0 & c & b & -a \end{vmatrix}?$$

29. Let $f: \mathbb{R}^4 \to \mathbb{R}$ be a function satisfying the following conditions:
   (1) $f([1, 0, 0, 1]) = 1$;
   (2) If $b, d \in \mathbb{R}$ then the function from $\mathbb{R}^2$ to $\mathbb{R}$ defined by $[a, c] \mapsto f([a, b, c, d])$ is a linear transformation.
   (3) If $a, b, c, d \in \mathbb{R}$ then $f(a, b, c, d) = -f(b, a, d, c)$.

Show that $f(a, b, c, d) = \begin{vmatrix} a & b \\ c & d \end{vmatrix}$.

30. Let n be an even positive integer and let $c, d \in \mathbb{Q}$ be given. Define the matrix $A = [a_{ij}] \in \mathcal{M}_{n \times n}(\mathbb{Q})$ by setting

$$a_{ij} = \begin{cases} c, & \text{if } i = j \\ d, & \text{if } i + j = n + 1 \\ 0, & \text{otherwise.} \end{cases}$$

Calculate $|A|$.

31. For real numbers a, b, and c, show that

$$\begin{vmatrix} 0 & (a-b)^2 & (a-c)^2 \\ (b-a)^2 & 0 & (b-c)^2 \\ (c-a)^2 & (c-b)^2 & 0 \end{vmatrix} \geq 0.$$

32. Let $n > 2$ be an integer and let F be a field. Let $A = [a_{ij}] \in M_{n \times n}(\mathbb{R})$ be a matrix satisfying $a_{11} \neq 0$. For all $2 \leq i, j \leq n$, set $b_{ij} = \begin{vmatrix} a_{11} & a_{1j} \\ a_{i1} & a_{ij} \end{vmatrix}$ and let $B = [b_{ij}] \in M_{(n-1) \times (n-1)}(\mathbb{R})$. Show that $|A| = a_{11}^{2-n}|B|$.

33. Let $A \in M_{4 \times 4}(\mathbb{Q})$ be a matrix each entry of which is -2 or 3. Show that $|A|$ is divisible by 125.

34. Let n be a positive integer and let h and k be fixed integers between 1 and n. Are the functions $f, g \in \mathbb{R}^\mathbb{R}$ defined by $f: c \mapsto |E_{h;c}|$ and $g: c \mapsto |E_{hk;c}|$ continuous on all of $\mathbb{R}$?

35. Let F be a field and let $A = [a_{ij}] \in M_{2 \times 2}(F)$. Show that the matrix $A^2 + |A|I$ belongs to the subspace of $M_{2 \times 2}(F)$ generated by $\{A\}$.

36. Calculate the adjoint of the matrix $\begin{bmatrix} 3 & -1 & 1 \\ 0 & 2 & 4 \\ 1 & -1 & 1 \end{bmatrix}$ in $M_{3 \times 3}(\mathbb{R})$.

37. Calculate the adjoint of the matrix $\begin{bmatrix} 1 & 1 & 0 \\ 0 & 1 & 1 \\ 1 & 1 & 1 \end{bmatrix}$ in $M_{3 \times 3}(\mathbb{Z}/(2))$.

38. Let n be a positive integer, let F be a field, and let A and B be nonsingular matrices in $M_{n \times n}(F)$. Show that $adj(AB) = adj(B) \cdot adj(A)$.

39. Use the adjoint matrix to calculate A^{-1}, where

$$A = \begin{bmatrix} 1 & 2 & 3 \\ 1 & 3 & 4 \\ 1 & 4 & 3 \end{bmatrix} \in M_{3 \times 3}(\mathbb{R}).$$

40. Let n be a positive integer, let F be a field, and let A be a nonsingular matrix in $M_{n \times n}(F)$. Show that $adj(adj(A)) = |A|^{n-2}A$.

41. Is the function $adj: M_{2 \times 2}(\mathbb{R}) \to M_{2 \times 2}(\mathbb{R})$ epic?

42. Does there exist a matrix $A \in M_{3 \times 3}(\mathbb{R})$ satisfying $adj(A) = \begin{bmatrix} 1 & 0 & 0 \\ 0 & 1 & 0 \\ 0 & 0 & 0 \end{bmatrix}$?

43. Let a, b, c, and d be real numbers, not all of which are equal to 0. Show that the matrix $\begin{bmatrix} a & b & c & d \\ b & -a & d & -c \\ c & -d & -a & b \\ d & c & -b & -a \end{bmatrix}$ is nonsingular.

44. For real numbers a, b, c, and d, show that the matrix

$$\begin{bmatrix} a^2 & (a+1)^2 & (a+2)^2 & (a+3)^2 \\ b^2 & (b+1)^2 & (b+2)^2 & (b+3)^2 \\ c^2 & (c+1)^2 & (c+2)^2 & (c+3)^2 \\ d^2 & (d+1)^2 & (d+2)^2 & (d+3)^2 \end{bmatrix}$$

is singular.

45. Let A be a nonsingular matrix in $\mathcal{M}_{n \times n}(\mathbb{C})$, which we will write in the form $B + iC$ for matrices $B, C \in \mathcal{M}_{n \times n}(\mathbb{R})$. Show that there exists a real number d such that $B + dC$ is nonsingular.

46. Let n be a positive integer and let F be a field. Let $D \in \mathcal{M}_{n \times n}(F)$ be the matrix all of the entries of which equal 1_F. Show that for a matrix $A \in \mathcal{M}_{n \times n}(F)$ precisely one of the following conditions holds:

(1) There exists a unique scalar c such that the matrix $A + cD$ is singular;
(2) The matrix $A + cD$ is nonsingular for all $c \in F$;
(3) The matrix $A + cD$ is singular for all $c \in F$.

47. Let A, B, C, and D be matrices in $\mathcal{M}_{2 \times 2}(\mathbb{R})$ and let M be the matrix $\begin{bmatrix} A & B \\ C & D \end{bmatrix} \in \mathcal{M}_{4 \times 4}(\mathbb{R})$. If M is nonsingular, does it follow that at least one of the "formal derivatives" $AD - BC$, $AD - CB$, $DA - BC$, or $DA - CB$ has to be a nonsingular matrix?

48. Let A, B, C, and D be matrices in $\mathcal{M}_{2 \times 2}(\mathbb{R})$ and let M be the matrix $\begin{bmatrix} A & B \\ C & D \end{bmatrix} \in \mathcal{M}_{4 \times 4}(\mathbb{R})$. If all of the "formal derivatives" $AD - BC$, $AD - CB$, $DA - BC$, and $DA - CB$ are nonsingular, is M necessarily nonsingular?

49. Let n be a positive integer and let $A = [a_{ij}] \in \mathcal{M}_{n \times n}(\mathbb{R})$ satisfying the condition $|a_{ii}| > \sum_{j \neq i} |a_{ij}|$ for all $1 \leq i \leq n$. Show that A is nonsingular.

50. Let n be a positive integer and let $A \in \mathcal{M}_{n \times n}(\mathbb{Q})$ be a nonsingular matrix all of the entries of which are integers and having determinant equal to ± 1. Show that all of the entries of A^{-1} are integers.

51. Find all matrices $A \in \mathcal{M}_{2 \times 2}(\mathbb{R})$ satisfying $A \neq I$ and $A^3 = I$.

52. Let F be a field and let k and n be integers greater than 1. Let A be a matrix in $\mathcal{M}_{k \times n}(F)$ the top row of which is not $[0_F, \ldots, 0_F]$. For all $2 \leq i \leq k$ and all $2 \leq j \leq n$, let $d_{ij} = \begin{vmatrix} a_{11} & a_{1j} \\ a_{1i} & a_{ij} \end{vmatrix}$. Show that the rank of the matrix

$$D = \begin{bmatrix} d_{22} & \cdots & d_{2n} \\ & \cdots & \\ d_{k2} & \cdots & d_{kn} \end{bmatrix} \in \mathcal{M}_{(k-1) \times (n-1)}(F)$$

is 1 less than the rank of A.

53. Let $V = \mathbb{R}^2$ and the function $f: V^3 \to \mathbb{R}$ as follows: if $v_i = [a_i, b_i]$ for $1 \leq i \leq 3$ then

$$f(v_1, v_2, v_3) = \begin{vmatrix} a_1 & b_1 & 1 \\ a_2 & b_2 & 1 \\ a_3 & b_3 & 1 \end{vmatrix}.$$

Prove that for all $v_1, v_2, v_3, v_4 \in V$ we have

$$f(v_1, v_2, v_3) = f(v_4, v_2, v_3) + f(v_1, v_4, v_3) + f(v_1, v_2, v_4).$$

CHAPTER X

EIGENVECTORS AND EIGENVALUES

We have already referred to one of the major problems of linear algebra: given a vector space V finitely generated over a field F and given an endomorphism α of V, find a basis D for V over F such that the matrix $\Phi_{DD}(\alpha)$ which represents α will be as "nice" and as "handy" as possible. The time has come to address this problem in more detail and, in particular, to specify what we mean by "nice" and "handy".

Indeed, let F, V, and α be as stated. A scalar $c \in F$ is an *eigenvalue* of α if and only if there exists a vector $0_V \neq v \in V$ satisfying $\alpha(v) = cv$. Such a vector v is said to be an *eigenvector* of α *associated* with the eigenvalue c. Note that a vector $0_V \neq v \in V$ is an eigenvector of α if and only if the subspace Fv is stable under α. Every eigenvector of α is associated to precisely one eigenvalue of α but, on the contrary, a given eigenvalue may have many eigenvectors associated with it. The set of all eigenvalues of an endomorphism α of V is called the *spectrum* of α and is denoted by $spec(\alpha)$.

EXAMPLE

An endomorphism of a vector space need have no eigenvalues or eigenvectors. For example, let $F = \mathbb{R}$, let $V = \mathbb{R}^2$, and let α be the endomorphism of V defined by $\alpha: [a, b] \mapsto [-b, a]$. Then it is easy to see that α has no eigenvalues and hence no eigenvectors. Thus $spec(\alpha) = \varnothing$.

EXAMPLE

Let $F = \mathbb{R}$, let $V = \mathbb{R}^2$, and let α be the endomorphism of V defined by $\alpha: [a, b] \mapsto [b, a]$. Then $spec(\alpha) = \{-1, 1\}$, where:

(1) The set of eigenvectors associated with the eigenvalue 1 is

$$\{[a, a] \mid 0_F \neq a \in F\}.$$

(2) The set of eigenvectors associated with the eigenvalue -1 is

$$\{[a, -a] \mid 0_F \neq a \in F\}.$$

EXAMPLE

Let $F = \mathbb{R}$, let $V = \{f \in \mathbb{R}^{\mathbb{R}} \mid f$ is infinitely differentiable everywhere on $\mathbb{R}\}$, and let δ be the endomorphism of V which assigns to each $f \in V$ its derivative f'. A scalar $c \in F$ is an eigenvalue of δ if and only if there exists a nonzero function $f \in V$ satisfying $f' = cf$. But for any $c \in F$ such a function exists, namely the function $x \mapsto e^{cx}$. Thus $spec(\delta) = \mathbb{R}$ and the set of eigenvectors associated with $c \in \mathbb{R}$ is $\{ae^{cx} \mid a \neq 0\}$.

EXAMPLE

Let V be a vector space over a field F and let α be an endomorphism of V. A vector $v \in V$ is a *fixed point* of α if and only if $\alpha(v) = v$. Thus we see that an endomorphism α of V has nonzero fixed points if and only if $1_F \in spec(\alpha)$, and in this case the set of fixed points is precisely the set of all eigenvectors associated with that eigenvalue.

(10.1) PROPOSITION. *Let V be a vector space over a field F and let α be an endomorphism of V having an eigenvalue c. Then the subset W of V consisting of 0_V and all eigenvectors of α associated with c is a subspace of V.*

PROOF. Note that $\alpha(0_V) = 0_V = c0_V$. Thus, If $w, w' \in W$ then $\alpha(w + w') = \alpha(w) + \alpha(w') = cw + cw' = c(w + w')$ and so $w + w' \in W$. Similarly, if $a \in F$ then $\alpha(aw) = a\alpha(w) = a(cw) = c(aw)$ and so $aw \in W$. Thus W is a subspace of V. $\square$

If V is a vector space over a field F and α is an endomorphism of V having an eigenvalue c then the subspace W of V defined in Proposition 10.1 is called the *eigenspace* of α associated with the eigenvalue c. Eigenspaces will play a very important role in our discussion in this and subsequent chapters.

EXAMPLE

Let V be the vector space $\mathbb{R}^3$ over the field $\mathbb{R}$ and let α be the endomorphism of V defined by $\alpha: [a, b, c] \mapsto [a, 0, c]$. Then 1 is an eigenvalue of α and its associated eigenspace is the subspace of V generated by $\{[1, 0, 0], [0, 0, 1]\}$.

(10.2) PROPOSITION. *Let V be a vector space finitely generated over a field F and let α be an endomorphism of V. Then the following conditions on a scalar $c \in F$ are equivalent:*

(1) *c is an eigenvalue of α.*
(2) *The endomorphism $\alpha - c\sigma_1$ of V is not an automorphism.*
(3) *If A is the matrix representing α with respect to some basis of V then $|cI - A| = 0_F$.*

PROOF. Condition (1) holds if and only if there exists a vector $0_V \neq v$ satisfying $\alpha(v) = cv$ and hence $(\alpha - c\sigma_1)(v) = 0_V$. This is so if and only if $\alpha - c\sigma_1$ is not monic and hence, by Proposition 5.1, if and only if $\alpha - c\sigma_1$ is not an automorphism of V.

Thus (1) and (2) are equivalent. But an endomorphism is an automorphism if and only if it is represented by a nonsingular matrix and a matrix is nonsingular if and only if its determinant is nonzero. Thus (2) and (3) are equivalent as well. $\square$

Proposition 10.2 shows how we may define eigenvalues of square matrices over a field F as well as of endomorphisms: if $A \in \mathcal{M}_{n \times n}(F)$ then a scalar $c \in F$ is an *eigenvalue* of A if and only if $|cI - A| = 0_F$. The set of eigenvalues of A is the *spectrum* of A and is denoted by $spec(A)$. Since $|cI - A| = |(cI - A)^T| = |cI - A^T|$, we see that $spec(A) = spec(A^T)$. A vector $[0_F, \ldots, 0_F] \neq v \in F^n$ is an *eigenvector* of A associated with c if and only if $Av^T = cv^T$. We note from Proposition 10.2 that a matrix A is nonsingular if and only if 0_F is not an eigenvalue of A.

EXAMPLE

If F is a field and $A \in \mathcal{M}_{n \times n}(F)$ then, as we have noted, both A and A^T have the same eigenvalues. However, these eigenvalues may have different associated eigenvectors. For example, let

$$A = \begin{bmatrix} 1 & 1 & -2 \\ -1 & 2 & 1 \\ 0 & 1 & -1 \end{bmatrix} \in \mathcal{M}_{3 \times 3}(\mathbb{R}).$$

The eigenvalues of A are -1, 1, and 2, and so these are also the eigenvalues of A^T. However:

(1) The eigenspace of A associated with -1 is $\mathbb{R}[1, 0, 1]$ while the eigenspace of A^T associated with -1 is $\mathbb{R}[1, 2, -1]$;
(2) The eigenspace of A associated with 1 is $\mathbb{R}[3, 2, 1]$ while the eigenspace of A^T associated with 1 is $\mathbb{R}[1, 0, -1]$;
(3) The eigenspace of A associated with 2 is $\mathbb{R}[1, 3, 1]$ while the eigenspace of A^T associated with 2 is $\mathbb{R}[1, -1, -1]$;

(10.3) PROPOSITION. *Let V be a vector space finitely generated over a field F and let α and β be endomorphisms of V. Then $spec(\alpha\beta) = spec(\beta\alpha)$.*

PROOF. Let c be an eigenvalue of $\alpha\beta$. If $c = 0_F$ then $\alpha\beta$ is not an automorphism of V and so, by the corollary to Proposition 5.2, $\beta\alpha$ is not an automorphism of V. Therefore c is also an eigenvalue of $\beta\alpha$. Now assume that $c \neq 0_F$ and that v is an eigenvector of $\alpha\beta$ associated with c. Set $w = \beta(v)$. Then $\alpha(w) = \alpha\beta(v) = cv \neq 0_V$ and so $w \neq 0_V$. Moreover, $\beta\alpha(w) = \beta\alpha\beta(v) = \beta(cv) = c\beta(v) = cw$ so c is an eigenvalue of $\beta\alpha$ with associated eigenvector w.

Thus every eigenvalue of $\alpha\beta$ is an eigenvalue of $\beta\alpha$. A similar argument shows that every eigenvalue of $\beta\alpha$ is an eigenvalue of $\alpha\beta$ and thus we have equality. $\square$

The relation between eigenvectors and the problem we presented at the beginning of the chapter is given by the following theorem.

(10.4) PROPOSITION. *Let V be a vector space finitely generated over a field F and let α be an endomorphism of V. The following conditions on a basis $D =$*

$\{v_1, \ldots, v_n\}$ *for V over F are equivalent:*

(1) *Each v_i is an eigenvector of α.*

(2) *$\Phi_{DD}(\alpha)$ is a diagonal matrix.*

PROOF. Assume (1). Then for each $1 \leq i \leq n$ there exists a scalar $c_i \in F$ satisfying $\alpha(v_i) = cv_i$ and so $\Phi_{DD}(\alpha) = \begin{bmatrix} c_1 & 0_F & \cdots & 0_F \\ 0_F & c_2 & \cdots & 0_F \\ & & \cdots & \\ 0_F & 0_F & \cdots & c_n \end{bmatrix}$. Conversely, if $\Phi_{DD}(\alpha)$ is a diagonal matrix $[a_{ij}]$ then $\alpha(v_i) = a_{ii}v_i$ for each $1 \leq i \leq n$ and so each v_i is an eigenvector of α. $\square$

We have previously seen an example of a vector space, not finitely-generated over its field of scalars, having an endomorphism for which every scalar is an eigenvalue. This cannot happen for finitely-generated vector spaces over infinite fields, as the following result shows.

(10.5) PROPOSITION. *Let V be a vector space finitely generated over a field F and let α be an endomorphism of V having distinct eigenvalues $c_1, \ldots, c_k$. For each $1 \leq i \leq k$, let v_i be an eigenvector of α associated with c_i. Then the set $\{v_1, \ldots, v_k\}$ is linearly independent.*

PROOF. Assume that the set $\{v_1, \ldots, v_k\}$ is linearly dependent. Then there exists an index t satisfying $1 \leq t < k$, such that the set $\{v_1, \ldots, v_t\}$ is linearly independent while the set $\{v_1, \ldots, v_{t+1}\}$ is linearly dependent. That is to say, there exist scalars $a_1, \ldots, a_{t+1}$, not all of which equal 0_F, satisfying

$$(*) \qquad\qquad a_1 v_1 + \cdots + a_{t+1} v_{t+1} = 0_V$$

and so

$$(**) \qquad\qquad c_{t+1} a_1 v_1 + \cdots + c_{t+1} a_{t+1} v_{t+1} = 0_V.$$

If we apply α to both sides of (*) we get

$$a_1 \alpha(v_1) + \cdots + a_{t+1} \alpha(v_{t+1}) = 0_V,$$

namely

$$(***) \qquad\qquad c_1 a_1 v_1 + \cdots + c_{t+1} a_{t+1} v_{t+1} = 0_V.$$

If we subtract each side of (**) from the corresponding side of (***) we obtain

$$(c_1 - c_{t+1}) a_1 v_1 + \cdots + (c_t - c_{t+1}) a_t v_t = 0_V.$$

However, we know that the set $\{v_1, \ldots, v_t\}$ is linearly independent and $c_i - c_{t+1} \neq 0_F$ for all $1 \leq i \leq t$. Therefore we must have $a_1 = \cdots = a_t = 0_F$, which, by (*), implies that $a_{t+1} = 0_F$ as well. Thus we have arrived at a contradiction, proving that $\{v_1, \ldots, v_k\}$ is linearly independent. $\square$

The following corollaries of Proposition 10.5 are immediate.

COROLLARY. *Let V be a vector space finitely generated over a field F and let α be an endomorphism of V having distinct eigenvalues $c_1, \ldots, c_k$. For each $1 \leq i \leq k$, let W_i be the eigenspace of α associated with c_i. Then the set $\{W_1, \ldots, W_k\}$ is independent.*

COROLLARY. *If V is a vector space finite-dimensional over a field F and if α is an endomorphism of V then the number of distinct eigenvalues of α is at most $dim(V)$.*

Moreover, the following result is an immediate consequence of Proposition 10.4 and Proposition 10.5.

COROLLARY. *If V is a vector space of finite dimension n over a field F and if α is an endomorphism of V having n distinct eigenvalues then there exists a basis D of V over F satisfying the condition that $\Phi_{DD}(\alpha)$ is a diagonal matrix.*

EXAMPLE

Let $F = \mathbb{R}$ and $V = \mathbb{R}^2$. Let α be the endomorphism of V defined by $\alpha: [a, b] \mapsto [3a - b, 3b - a]$. Then we see that:

(1) $\alpha([1, 1]) = [2, 2]$ and so $[1, 1]$ is an eigenvector of α associated with the eigenvalue 2;

(2) $\alpha([1, -1]) = [4, -4]$ and so $[1, -1]$ is an eigenvector of α associated with the eigenvalue 4.

Therefore $D = \{[1, 1], [1, -1]\}$ is a basis of V satisfying $\Phi_{DD}(\alpha) = \begin{bmatrix} 2 & 0 \\ 0 & 4 \end{bmatrix}$.

EXAMPLE

Let $F = \mathbb{R}$ and $V = \mathbb{R}^2$. Let α be the endomorphism of V defined by $\alpha: [a, b] \mapsto [a + b, b]$. If $[a, b] \neq [0, 0]$ and $\alpha([a, b]) = c[a, b]$ then $cb = b$ and $ca = a + b$ and that can happen only when $c = 1$ and $b = 0$. Therefore α has precisely one eigenvalue, namely 1, with which is associated the eigenspace $\mathbb{R}[1, 0]$. This space is one-dimensional and so is properly contained in V. We conclude that there is no basis of V relative to which α can be represented by a diagonal matrix.

Thus we see that if $A \in \mathcal{M}_{n \times n}(F)$ we know that A can have at most n eigenvalues but may in fact have less – in fact it may have none at all. Suppose we have a matrix in $\mathcal{M}_{n \times n}(\mathbb{R})$ the entries of which are chosen independently and randomly from a standard normal distribution. How many distinct eigenvalues in $\mathbb{R}$ should we expect? This probabilistic question has been recently solved by Edelman, Kostlan, and Shub: if E_n is the expected number of real eigenvalues of such a matrix then

$$\lim_{n \to \infty} \frac{E_n}{\sqrt{n}} = \sqrt{\frac{2}{\pi}}.$$

Let $F[X]$ be a polynomial ring over a field F and let n be a positive integer. If $A \in \mathcal{M}_{n \times n}(F)$ then the polynomial $|XI - A| \in F[X]$ is called the *characteristic polynomial* of the matrix A. Note that this polynomial is monic and has degree n.

EXAMPLE

The characteristic polynomial of the matrix $\begin{bmatrix} 3 & -1 \\ -1 & 3 \end{bmatrix} \in \mathcal{M}_{2\times2}(\mathbb{Q})$ equals

$$\begin{vmatrix} X-3 & 1 \\ 1 & X-3 \end{vmatrix} = X^2 - 6X + 8 = (X-2)(X-4).$$

EXAMPLE

The characteristic polynomial of the matrix $\begin{bmatrix} 1 & 2 & 0 \\ 0 & 1 & 2 \\ 0 & 0 & 1 \end{bmatrix} \in \mathcal{M}_{3\times3}(\mathbb{Q})$ equals

$$\begin{vmatrix} X-1 & -2 & 0 \\ 0 & X-1 & -2 \\ 0 & 0 & X-1 \end{vmatrix} = (X-1)^3.$$

We see from this definition that the eigenvalues of a matrix A (or of the endomorphism which the matrix represents) are precisely the zeros of the characteristic polynomial of A.

EXAMPLE

The characteristic polynomial of $\begin{bmatrix} 0 & -1 \\ 1 & 0 \end{bmatrix} \in \mathcal{M}_{2\times2}(\mathbb{R})$ equals X^2+1. If we think of this matrix as representing an endomorphism of $\mathbb{R}^2$ then that endomorphism has no eigenvalues. If we think of this matrix as representing an endomorphism of $\mathbb{C}^2$ then that endomorphism has two eigenvalues: i and $-i$.

EXAMPLE

The characteristic polynomial of $A = \begin{bmatrix} 1 & 1 & -2 \\ -1 & 2 & 1 \\ 0 & 1 & -1 \end{bmatrix} \in \mathcal{M}_{3\times3}(\mathbb{R})$ is

$$X^3 - 2X^2 - X + 2 = (X+1)(X-1)(X-2)$$

and so A has eigenvalues -1, 1, and 2. The eigenspace associated with -1 is $\mathbb{R}[1,0,1]$; the eigenspace associated with 1 is $\mathbb{R}[3,2,1]$; the eigenspace associated with 2 is $\mathbb{R}[1,3,1]$.

EXAMPLE

The characteristic polynomial of the matrix $A = \begin{bmatrix} 5 & 4 & 2 \\ 4 & 5 & 2 \\ 2 & 2 & 2 \end{bmatrix} \in \mathcal{M}_{3 \times 3}(\mathbb{R})$ is $(X - 10)(X - 1)^2$ and so A has eigenvalues 1 and 10. The eigenspace associated with 1 is $\mathbb{R}\{[-1, 1, 0], [-\frac{1}{2}, 0, 1]\}$; the eigenspace associated with 10 is $\mathbb{R}[2, 2, 1]$.

Let F be a field and let n be a positive integer. Each nonsingular matrix $P \in \mathcal{M}_{n \times n}(F)$ defines a function $\omega_P : \mathcal{M}_{n \times n}(F) \to \mathcal{M}_{n \times n}(F)$ by $\omega_P : A \mapsto P^{-1}AP$. Indeed, this is an automorphism of $\mathcal{M}_{n \times n}(F)$, the inverse of which is $\omega_{P^{-1}}$. The set of all automorphisms of $\mathcal{M}_{n \times n}(F)$ of this form is a group of endomorphisms and so defines an equivalence relation $\sim$ between elements of $\mathcal{M}_{n \times n}(F)$ given by $A \sim B$ if and only if there exists a nonsingular matrix $P \in \mathcal{M}_{n \times n}(F)$ satisfying $B = P^{-1}AP$. In such a case we say that the matrices A and B are *similar*. Note that similar matrices are surely equivalent, in the sense which we defined previously.

We have already seen a very important example of similarity of matrices.

EXAMPLE

Let V be a vector space finite dimensional over a field F and let B and D be bases for V over F. Let P be the change of basis matrix from B to D. If $\alpha \in End(V)$ then $\Phi_{DD}(\alpha) = P^{-1}\Phi_{BB}(\alpha)P$ and so the matrices $\Phi_{BB}(\alpha)$ and $\Phi_{DD}(\alpha)$ are similar.

The following proposition shows why similarity is important for us.

(10.6) PROPOSITION. *Let F be a field and let n be a positive integer. If matrices A and B in $\mathcal{M}_{n \times n}(F)$ are similar they have the same characteristic polynomial.*

PROOF. If A and B are similar there exists a nonsingular matrix $P \in \mathcal{M}_{n \times n}(F)$ satisfying $B = P^{-1}AP$. Then $|XI - B| = |XI - P^{-1}AP| = |P^{-1}(XI - A)P| = |P^{-1}||XI - A||P| = |P|^{-1}|XI - A||P| = |XI - A|$. $\square$

Combining this result with the previous proposition, we see that if V is a vector space finite dimensional over a field F and if α is an endomorphism of V then we can unambiguously define the *characteristic polynomial* of α to be the characteristic polynomial of the matrix representing α with respect to any basis of V.

The converse of Proposition 10.6 is false, as the following example shows.

EXAMPLE

The matrices $\begin{bmatrix} 1 & 1 \\ 0 & 1 \end{bmatrix}$ and $\begin{bmatrix} 1 & 0 \\ 0 & 1 \end{bmatrix}$ in $\mathcal{M}_{2 \times 2}(\mathbb{R})$ are not similar but both of them have the same characteristic polynomial, namely $(X - 1)^2$.

Let α be an endomorphism of a vector space V which is finitely generated over a field F and let $c \in F$ be an eigenvalue of α. The *algebraic multiplicity* of c is defined to be the largest integer k such that $(X - c)^k$ is a factor of the characteristic polynomial of α; the *geometric multiplicity* of c is defined to be the dimension of the eigenspace of α associated with c. The geometric multiplicity of c is always no greater than the algebraic multiplicity. However, these numbers need not be equal, as the following example shows.

EXAMPLE

Let α be the endomorphism of $\mathbb{R}^2$ defined by $\alpha: [a, b] \mapsto [a + b, b]$. Then $c = 1$ is an eigenvalue of α with associated eigenspace $\mathbb{R}[1, 0]$ and so c has geometric multiplicity 1. On the other hand, α is represented with respect to the canonical basis by the matrix $\begin{bmatrix} 1 & 1 \\ 0 & 1 \end{bmatrix}$ and so the characteristic polynomial of α is

$$\begin{vmatrix} X - 1 & -1 \\ 0 & X - 1 \end{vmatrix} = (X - 1)^2.$$

Therefore c has algebraic multiplicity 2.

(10.7) PROPOSITION. *Let V be a vector space finitely generated over a field F and let α be an endomorphism of V having distinct eigenvalues $c_1, \ldots c_k$. Then the following conditions are equivalent:*

(1) *There exists a basis D of V such that $\Phi_{DD}(\alpha)$ is a diagonal matrix.*
(2) *The algebraic multiplicity of c_j equals its geometric multiplicity for each $1 \leq j \leq k$.*

PROOF. (1) $\Rightarrow$ (2): Let $D = \{v_1, \ldots, v_n\}$ and, for each $1 \leq j \leq k$, let $m(j)$ be the number of times that c_j appears on the diagonal of $\Phi_{DD}(\alpha)$. Then $\sum_{j=1}^{k} m(j) = n$ and, by Proposition 10.4, for each $1 \leq j \leq k$ there exists a subset of D having $m(j)$ elements which forms a basis for the eigenspace of α associated with c_j. Moreover, the characteristic polynomial of α is $(X - c_1)^{m(1)} \cdot \ldots \cdot (X - c_k)^{m(k)}$ and so $m(j)$ is both the algebraic multiplicity and the geometric multiplicity of c_j.

(2) $\Rightarrow$ (1): For each $1 \leq j \leq k$, let $m(j)$ be the algebraic (and geometric) multiplicity of c_j and let D_j be a basis for the eigenspace of α associated with c_j. Set $D = \bigcup_{j=1}^{k} D_j$. This is a set of n vectors in V which is linearly independent by Proposition 3.12 and the first corollary to Proposition 10.5. Thus, (1) follows by Proposition 10.4. $\square$

Let F be a field, let n be a positive integer, and let $A \in M_{n \times n}(F)$. Each polynomial $p(X) \in F[X]$ defines a function $M_{n \times n}(F) \to M_{n \times n}(F)$ given by $A \mapsto p(A)$ defined as follows: if $p(X) = \sum_{i \geq 0} c_i X^i \in F[X]$ then $p(A) \in M_{n \times n}(F) = \sum_{i \geq 0} c_i A^i$, where $A^0 = I$ and where this sum is in fact finite – and hence well-defined – since only finitely-many of the coefficients c_i are nonzero. If $p(A) = O$ then A is said to *annihilate* the polynomial $p(X)$. Note that the set of all polynomials in $F[X]$ annihilated by a fixed matrix A is a subspace of $F[X]$: if $p(A) = O = q(A)$ then surely $(p + q)(A) = p(A) + q(A) = O$ and $cp(A) = O$ for every scalar $c \in F$.

Furthermore, if $A \sim B$ in $\mathcal{M}_{n \times n}(F)$ and if A annihilates a polynomial $p(X) = \sum_{i \geq 0} c_i X^i \in F[X]$ then there exists a nonsingular matrix $P \in \mathcal{M}_{n \times n}(F)$ satisfying $B = P^{-1}AP$ and so $p(B) = \sum_{i \geq 0} c_i (P^{-1}AP)^i = P^{-1}(\sum_{i \geq 0} c_i A^i)P = \mathbf{O}$ and so B annihilates $p(X)$ as well. Thus we see that similar matrices annihilate the same set of polynomials.

EXAMPLE

Let $A = \begin{bmatrix} 2 & 1 \\ 1 & 0 \end{bmatrix}$ in $\mathcal{M}_{2 \times 2}(\mathbb{R})$.

(1) If $p(X) = X^2 - X + 2 \in \mathbb{R}[X]$ then $p(A) = A^2 - A + 2I = \begin{bmatrix} 5 & 1 \\ 1 & 3 \end{bmatrix}$.

(2) If $q(X) = X^2 - 2X - 1 \in \mathbb{R}[X]$ then $q(A) = A^2 - 2A - I = \mathbf{O}$ so A annihilates $q(X)$.

(10.8) PROPOSITION. *Let F be a field and let n be a positive integer. If $A \in \mathcal{M}_{n \times n}(F)$ then there exists a polynomial $p(X) \in F[X]$ of positive degree satisfying $p(A) = \mathbf{O}$.*

PROOF. If $A = \mathbf{O}$ we can choose $p(X) = X^h$ for any $h > 0$. Thus assume $A \neq \mathbf{O}$. The dimension of $\mathcal{M}_{n \times n}(F)$ as a vector space over F is n^2 and so the subset $\{I, A, \ldots, A^{n^2}\}$ of this space is linearly dependent. That is to say, there exist elements $c_0, \ldots, c_{n^2}$ of F, not all of them equal to 0_F, such that

$$c_0 I + c_1 A + \ldots c_{n^2} A^{n^2} = \mathbf{O}.$$

Indeed, there must be some $i > 0$ such that $c_i \neq 0_F$. Thus the polynomial $p(X) = c_0 + c_1 X + \ldots c_{n^2} X^{n^2}$ has positive degree and is annihilated by A. □

As a consequence of Proposition 10.8, we see that, given a matrix $A \in \mathcal{M}_{n \times n}(F)$, there exists a smallest positive integer k such that A annihilates a polynomial of degree k in $F[X]$. Among all such polynomials, only one of them can be monic. Indeed, if $p(X)$ and $q(X)$ are distinct monic polynomials of degree k annihilated by A then $p(X) - q(X)$ is a nonzero polynomial of degree less than k annihilated by A, which is a contradiction. This monic polynomial of least degree which annihilates A is called the *minimal polynomial* of A over the field F. We will denote it by $m_A(X)$. By what we have seen before, if A and B are similar matrices then $m_A(X) = m_B(X)$. Therefore, if V is a vector space finite dimensional over a field F and if α is an endomorphism of V then we can define the *minimal polynomial* of α to be the minimal polynomial of any matrix representing α with respect to a basis of V. Thus, by definition, if $m(X)$ is the minimal polynomial of α then $m(\alpha) = \sigma_0$ and $p(\alpha) \neq \sigma_0$ for any polynomial $p(X) \in F[X]$ satisfying $0 \leq deg(p) < deg(m)$.

(10.9) PROPOSITION. *Let F be a field and let n be a positive integer. If $A \in \mathcal{M}_{n \times n}(F)$ annihilates a polynomial $p(X) \in F[X]$ then there exists a polynomial $u(X) \in \mathcal{M}_{n \times n}(F)$ satisfying $p(X) = u(X)m_A(X)$.*

PROOF. By Proposition 1.3 we note that there exist unique polynomials $u(X)$ and $v(X)$ in $F[X]$ satisfying $p(X) = u(X)m_A(X) + v(X)$, where $deg(v) < deg(m_A)$.

If $deg(v) = -\infty$ we are done. Otherwise, we have $O = p(A) = u(A)m_A(A)+v(A) = v(A)$ and this contradicts the minimality of $m_A(X)$. Thus $v(X)$ must be the 0-polynomial. $\square$

EXAMPLE

The characteristic polynomial of $I \in \mathcal{M}_{n \times n}(F)$ is $(X - 1_F)^n$ and its minimal polynomial is $X - 1_F$. Thus these two polynomials need not be equal.

EXAMPLE

If $A = \begin{bmatrix} 1 & 0 \\ 0 & 0 \end{bmatrix} \in \mathcal{M}_{2 \times 2}(\mathbb{Q})$ then $A^2 = A$ so A annihilates the polynomial $X^2 - X = X(X - 1)$. Since A does not annihilate each of the factors of this polynomial, we see that $m_A(X) = X^2 - X$. Thus we see that the minimal polynomial of a matrix need not be irreducible.

EXAMPLE

The matrix $A = \begin{bmatrix} 1 & 0 & 0 \\ 0 & 0 & 0 \\ 0 & 0 & 0 \end{bmatrix} \in \mathcal{M}_{3 \times 3}(\mathbb{Q})$ also annihilates $X^2 - X$ and this is indeed its minimal polynomial. The characteristic polynomial of A is $(X - 1)X^2 = X^3 - X^2$.

EXAMPLE

One can check that the matrices $\begin{bmatrix} 1 & 0 & 0 \\ 0 & 3 & 0 \\ 0 & 0 & 3 \end{bmatrix}$ and $\begin{bmatrix} 1 & 0 & 0 \\ 0 & 1 & 0 \\ 0 & 0 & 3 \end{bmatrix}$ in $\mathcal{M}_{3 \times 3}(\mathbb{Q})$ are not similar, but they have the same minimal polynomial, namely $(X - 1)(X - 3)$.

(10.10) PROPOSITION. (Cayley-Hamilton Theorem) *If F is a field and n is a positive integer then any matrix $A \in \mathcal{M}_{n \times n}(F)$ annihilates its characteristic polynomial.*

PROOF. Let $A \in \mathcal{M}_{n \times n}(F)$ and let $p(X) = X^n + a_{n-1}X^{n-1} + \cdots + a_1 X + a_0$ be its characteristic polynomial. Consider the matrix $adj(XI - A) \in \mathcal{M}_{n \times n}(F[X])$. This matrix is of the form $[p_{ij}(X)]$, where each $p_{ij}(X)$ belongs to the subspace W of $F[X]$ consisting of those polynomials of degree at most $n-1$. Since $\{1, X, X^2, \ldots, X^{n-1}\}$ is a basis for W, we can, in $\mathcal{M}_{n \times n}(W)$, write $adj(XI - A) = B_1 M_{n-1} + \cdots + B_{n-1}M_1 + B_n$ where each B_i belongs to $\mathcal{M}_{n \times n}(F)$ and M_i is the scalar matrix in $\mathcal{M}_{n \times n}(F[X])$ having X^i on the diagonal for each $i > 0$. Then, in $\mathcal{M}_{n \times n}(F[X])$, we have

$$IM_n + a_{n-1}IM_{n-1} + \cdots + a_1IM_1 + a_0I = p(X)$$
$$= |XI - A|I$$
$$= (XI - A)adj(XI - A)$$
$$= (XI - A)(B_1M_{n-1} + \cdots + B_{n-1}M_1 + B_n).$$

Equating coefficients of the matrices M_i, we see that

$$I = B_1$$
$$a_{n-1}I = B_2 - AB_1$$
$$a_{n-2}I = B_3 - AB_2$$
$$\cdots$$
$$a_1I = B_n - AB_{n-1}$$
$$a_0I = -AB_n$$

and if we multiply both sides of the jth equation $(1 \leq j \leq n-1)$ on the left by A^{n+1-j} and then add up the respective sides of the equations, we obtain $p(A) = 0$. $\square$

Let V be a vector space over a field F and let $\alpha \in End(V)$. In Proposition 10.4 we saw that α can be represented by a diagonal matrix with respect to some basis for V over F if and only if there exists such a basis composed of eigenvectors of α. Assume that we are in that situation and that $c_1, \ldots, c_k$ are the distinct eigenvalues of α. For each $1 \leq i \leq k$, let W_i be the eigenspace associated with c_i. Then the set $\{W_1, \ldots, W_k\}$ is independent and, indeed, $V = W_1 \oplus \cdots \oplus W_k$. For each $1 \leq i \leq k$, let $\pi_i: V \to W_i$ be a projection of V onto W_i associated with the above decomposition. Then:

(1) $\pi_1 + \cdots + \pi_k = \sigma_1$;
(2) $\pi_i\pi_j = \sigma_0$ when $i \neq j$;
(3) $\alpha = c_1\pi_1 + \cdots + c_k\pi_k$.

For each $1 \leq h \leq k$ let $p_h(X)$ be the Lagrange interpolation polynomial defined by $\{c_1, \ldots, c_k\}$. Then one can check that $\pi_h = p_h(\alpha)$ for all h since $p_h(\alpha)(X - c_h)$ is a scalar multiple of the minimal polynomial of α.

What is the situation when we are given two endomorphisms of a vector space simultaneously? Is it possible to choose a basis of V relative to which both of them are represented by diagonal matrices?

Let V be a vector space finitely generated over a field F and let $\alpha, \beta \in End(V)$. In order to choose a basis of V relative to which both of them are represented by diagonal matrices we have to find a basis of V composed of vectors which are eigenvectors of both endomorphisms. Indeed, suppose $D = \{u_1, \ldots, u_n\}$ is such a basis, where $\alpha(u_i) = a_iu_i$ and $\beta(u_i) = b_iu_i$ for all $1 \leq i \leq n$. Let $A = \Phi_{DD}(\alpha)$ and $B = \Phi_{DD}(\beta)$. Since A and B are both diagonal matrices, we have $AB = BA$ and so $\alpha\beta = \beta\alpha$. Thus we see that a necessary condition for such a basis to exist is that α and β commute.

Given a basis D of V, let $Y(D)$ be the set of all endomorphisms of V which can be represented by a diagonal matrix with respect to D. That is to say, $\alpha \in Y(D)$ if and only if $\alpha = \sigma_0$ or every element of D is an eigenvector of α. Since $Y(D)$ is clearly closed under sums and scalar multiplication, we see that it is a subspace of the vector space $End(V)$ over F.

(10.11) PROPOSITION. *Let V be a vector space over a field F and let $\alpha, \beta \in End(V)$ satisfy $\alpha\beta = \beta\alpha$. If $p(X) \in F[X]$ then $p(\alpha)\beta = \beta p(\alpha)$*

PROOF. We note that $\alpha^2\beta = \alpha(\alpha\beta) = \alpha(\beta\alpha) = (\alpha\beta)\alpha = \beta\alpha^2$ and by an easy induction argument we in fact have $\alpha^k\beta = \beta\alpha^k$ for each positive integer k. Therefore, if $p(X) = \sum_{i \geq 0} c_i X^i$ then $\beta p(\alpha) = \beta(\sum_{i \geq 0} c_i \alpha^i) = \sum_{i \geq 0} c_i \beta\alpha^i = \sum_{i \geq 0} c_i \alpha^i \beta = (\sum_{i \geq 0} c_i \alpha^i)\beta = p(\alpha)\beta$ and we are done. $\square$

Applying Proposition 10.11 twice, we obtain the following corollary.

COROLLARY. *Let V be a vector space over a field F and let $\alpha, \beta \in End(V)$ satisfy $\alpha\beta = \beta\alpha$. If $p(X), q(X) \in F[X]$ then $p(\alpha)q(\beta) = q(\beta)p(\alpha)$.*

Moreover, if α and β are idempotent endomorphisms of V satisfying $\alpha\beta = \beta\alpha$ then $(\alpha\beta)^2 = \alpha\beta\alpha\beta = \alpha^2\beta^2 = \alpha\beta$ and so $\alpha\beta$ is idempotent as well.

(10.12) PROPOSITION. *Let V be a vector space finitely generated over a field F and let $\alpha, \beta \in End(V)$ be endomorphisms each of which can be represented by a diagonal matrix with respect to some basis of V. Then there exists a basis of V over F relative to which both α and β can be represented by diagonal matrices if and only if $\alpha\beta = \beta\alpha$.*

PROOF. We have already noted that if α and β can be respectively represented by diagonal matrices A and B with respect by some basis of V over F then $AB = BA$ and so $\alpha\beta = \beta\alpha$.

Conversely, assume that $\alpha\beta = \beta\alpha$. From what we have already seen, there exists scalars $c_1, \ldots, c_k$ in F and idempotent endomorphisms $\pi_1, \ldots, \pi_k$ of V such that:

(1) $\sigma_1 = \pi_1 + \cdots + \pi_k$;
(2) $\pi_i \pi_j = \sigma_0$ if $i \neq j$;
(3) $\alpha = c_1 \pi_1 + \cdots + c_k \pi_k$;
(4) For each i there exists a polynomial $p_i(X) \in F[X]$ such that $\pi_i = p_i(\alpha)$.

and there exist scalars $d_1, \ldots, d_t$ in F and idempotent endomorphisms $\mu_1, \ldots, \mu_t$ of V such that:

(1') $\sigma_1 = \mu_1 + \cdots + \mu_t$;
(2') $\mu_i \mu_j = \sigma_0$ if $i \neq j$;
(3') $\beta = d_1 \mu_1 + \cdots + d_t \mu_t$;
(4') For each j there exists a polynomial $q_j(X) \in F[X]$ such that $\mu_j = q_j(\beta)$.

Therefore

$$\alpha = \alpha\sigma_1 = \left(\sum_{i=1}^{k} c_i \pi_i\right)\left(\sum_{j=1}^{t} \mu_j\right) = c_1 \pi_1 \mu_1 + \cdots + c_k \pi_k \mu_t$$

and

$$\beta = \beta\sigma_1 = \left(\sum_{j=1}^{t} d_j \mu_j\right)\left(\sum_{i=1}^{k} \pi_i\right) = d_1 \mu_1 \pi_1 + \cdots + d_t \mu_t \pi_k.$$

By conditions (4) and (4'), we have $\mu_j \pi_i = \pi_i \mu_j$ for all $1 \le i \le k$ and all $1 \le j \le t$. Therefore $\theta_{ij} = \pi_i \mu_j$ is an idempotent endomorphism of V for all such i and j. Moreover, $\theta_{ij} \theta_{st} = \sigma_0$ if $(i,j) \ne (s,t)$ and

$$\sigma_1 = \left(\sum_{i=1}^{k} \pi_i \right) \left(\sum_{j=1}^{t} \mu_j \right) = \sum_{i=1}^{k} \sum_{j=1}^{t} \theta_{ij}.$$

This suffices to show that α and β can be simultaneously represented by diagonal matrices with respect to some basis of V over F. $\square$

Problems

1. Let α be the endomorphism of $\mathbb{R}^3$ defined by

$$\alpha: [a,b,c] \mapsto [a - b, a + 2b + c, -2a + b - c].$$

Find the eigenvalues of α and the eigenspaces associated with them.

2. Let $F = \mathbb{Z}/(2)$ and let A be a nonempty set having a given subset B. Let α be the endomorphism of the vector space F^A over F given by $\alpha: f \mapsto \chi_B f$, where χ_B is the characteristic function on B. Find the eigenvectors of α.

3. Let α be an endomorphism of a a vector space V finitely generated over a field F.

  (i) Show that every eigenvector of α is also an eigenvector of $p(\alpha)$ for any $p(X) \in F[X]$;
  (ii) Show that the converse of (i) is false.

4. Find the eigenvalues of the following matrices and the eigenspaces associated with them.

(i) $\begin{bmatrix} 5 & 6 & -3 \\ -1 & 0 & 1 \\ 2 & 2 & -1 \end{bmatrix} \in \mathcal{M}_{3\times3}(\mathbb{R})$;

(ii) $\begin{bmatrix} 0 & 2 & 1 \\ -2 & 0 & 3 \\ -1 & -3 & 0 \end{bmatrix} \in \mathcal{M}_{3\times3}(\mathbb{C})$;

(iii) $\begin{bmatrix} 3 & 1 & 0 \\ -4 & -1 & 0 \\ 4 & -8 & -2 \end{bmatrix} \in \mathcal{M}_{3\times3}(\mathbb{C})$.

5. Let n be a positive integer, let F be a field and let $A \in \mathcal{M}_{n\times n}(F)$ be nonsingular. Given the eigenvalues of A, find the eigenvalues of A^{-1}.

6. Let n be a positive integer, let F be a field and let $A = [a_{ij}] \in \mathcal{M}_{n\times n}(F)$ be a matrix having eigenvalue c. If $b, d \in F$, show that $bc + d$ is an eigenvalue of the matrix $bA + dI$.

7. Let $A = \begin{bmatrix} a & b \\ c & d \end{bmatrix} \in M_{2\times2}(\mathbb{R})$ If $t \in \mathbb{R}$ is a zero of the polynomial $bX^2 + (a-d)X - c \in \mathbb{R}[X]$, show that $[1, t]$ is an eigenvector of A associated with the eigenvalue $a + bt$.

8. Let $A \in M_{2\times2}(\mathbb{C})$ have two distinct nonzero eigenvalues. Show that there are precisely four matrices $B \in M_{2\times2}(\mathbb{C})$ satisfying $B^2 = A$.

9. Find a matrix $A = \begin{bmatrix} 1 & a_{12} & a_{13} \\ 1 & a_{22} & a_{23} \\ 1 & a_{32} & a_{33} \end{bmatrix} \in M_{3\times3}(\mathbb{R})$ having eigenvectors $[1, 1, 1]$, $[1, 0, -1]$, and $[1, -1, 0]$.

10. Find the eigenvalues of the matrix $\begin{bmatrix} 1 & -1 & 1 \\ 1 & 0 & 0 \\ 0 & 1 & 0 \end{bmatrix} \in M_{3\times3}(\mathbb{C})$ and for each eigenvalue find the associated eigenspace.

11. Let c be a nonzero complex number and let m and n be positive integers. Let $A = [a_{ij}] \in M_{n\times n}(\mathbb{C})$ and let $B = [b_{ij}] \in M_{n\times n}(\mathbb{C})$ be the matrix defined by $b_{ij} = c^{m+i-j}a_{ij}$ for all $1 \leq i, j \leq n$. Show that if $d \in \mathbb{C}$ is an eigenvalue of A then $r^m d$ is an eigenvalue of B.

12. Show that every matrix in $S_{2\times2}(\mathbb{R})$ has a real eigenvalue.

13. Characterize magic matrices in terms of their eigenvalues.

14. Let $A \in M_{2\times2}(\mathbb{C})$ have distinct eigenvalues a and b. Show that

$$A^n = \frac{a^n}{a-b}(A - bI) + \frac{b^n}{b-a}(A - aI)$$

for all integers $n > 1$.

15. Let $A \in M_{2\times2}(\mathbb{C})$ have a unique eigenvalue c. Show that $A^n = c^{n-1}(cA - (n-1)cI)$ for all integers $n > 1$.

16. Let n be a positive integer. Show that every eigenvector of a matrix $A \in M_{n\times n}(\mathbb{C})$ is also an eigenvector of $adj(A)$.

17. Let n be a positive integer and let $\mathcal{A}$ be the set of all matrices in $M_{n\times n}(\mathbb{C})$ be the set of all matrices having the property that their eigenvectors generate all of $\mathbb{C}^n$. Is $\mathcal{A}$ closed under addition? Is it closed under multiplication?

18. Let n be a positive integer and let $A \in M_{n\times n}(\mathbb{C})$. If $p(X) \in \mathbb{C}[X]$, calculate $|p(A)|$ using the eigenvalues of A.

19. Let n be a positive integer and let c be a nonzero real number. Let $A \in M_{n\times n}(\mathbb{R})$ be the matrix every entry of which equals c. Find the eigenvalues of A and, for each eigenvalue, find the associated eigenspace.

20. Let n be a positive integer and let $A \in M_{n\times n}(\mathbb{Q})$. Find infinitely-many different matrices in $M_{n\times n}(\mathbb{Q})$ having the same eigenvectors as A.

21. Find the spectra of the following matrices.

(i) $\begin{bmatrix} a & b & c \\ a-d & b+d & c \\ a-e & b & c+e \end{bmatrix} \in M_{3\times3}(\mathbb{R})$;

(ii) $\begin{bmatrix} 0 & 1 \\ 1 & 1 \end{bmatrix} \in M_{2\times2}(\mathbb{Z}/(2))$.

22. Find the characteristic polynomials of the following matrices.

(i) $\begin{bmatrix} 3 & 2 & 2 \\ 1 & 4 & 1 \\ 2 & -4 & -1 \end{bmatrix} \in M_{3\times3}(\mathbb{R})$;

(ii) $\begin{bmatrix} 7 & 4 & -4 \\ 4 & -8 & -1 \\ -4 & -1 & -8 \end{bmatrix} \in M_{3\times3}(\mathbb{R})$;

(iii) $\begin{bmatrix} 1 & 3 & 0 & 0 & 0 \\ 0 & 1 & 0 & 0 & 0 \\ 0 & 0 & 1 & 0 & 0 \\ 0 & 0 & 2 & 2 & 0 \\ 0 & 0 & 0 & 0 & 2 \end{bmatrix} \in M_{5\times5}(\mathbb{R})$;

(iv) $\begin{bmatrix} 0 & 0 & 0 & a \\ 1 & 0 & 0 & b \\ 0 & 1 & 0 & c \\ 0 & 0 & 1 & d \end{bmatrix} \in M_{4\times4}(\mathbb{R})$.

23. Let n be a positive integer and let $\alpha: M_{n\times n}(\mathbb{C}) \to \mathbb{C}^n$ be the function defined by $\alpha: A \mapsto [b_0, \dots, b_{n-1}]$, where $X^n + b_{n-1}X^{n-1} + \cdots + b_1 X + b_0$ is the characteristic polynomial of A. Is α a linear transformation?

24. Let n be a positive integer, let F be a field, and let $A, B \in M_{n\times n}(F)$. Show that AB and BA have the same characteristic polynomial.

25. Find six different matrices in $M_{2\times2}(\mathbb{R})$ which annihilate the polynomial $X^2 - 1$.

26. Are the matrices $\begin{bmatrix} 4 & 2 \\ 3 & 1 \end{bmatrix}$ and $\begin{bmatrix} 1 & 2 \\ 3 & 4 \end{bmatrix}$ in $M_{2\times2}(\mathbb{R})$ similar?

27. Find diagonal matrices in $M_{3\times3}(\mathbb{R})$ which are similar to each of the following matrices:

(i) $\begin{bmatrix} 1 & 0 & 1 \\ 0 & 1 & 0 \\ 1 & 0 & 1 \end{bmatrix}$;

(ii) $\begin{bmatrix} 0 & 0 & 1 \\ 0 & 0 & 0 \\ 1 & 0 & 0 \end{bmatrix}$;

(iii) $\begin{bmatrix} 8 & 3 & -3 \\ -6 & -1 & 3 \\ 12 & 6 & -4 \end{bmatrix}$;

(iv) $\begin{bmatrix} 1 & -1 & 1 \\ -2 & 1 & 2 \\ -2 & -1 & 4 \end{bmatrix}$;

(v) $\begin{bmatrix} 1 & 0 & 0 \\ 0 & 1 & 1 \\ 0 & 1 & 1 \end{bmatrix}$.

28. Show that

$$\begin{bmatrix} 0 & 1 & a \\ 0 & 0 & 1 \\ 0 & 0 & 0 \end{bmatrix} \sim \begin{bmatrix} 0 & 1 & b \\ 0 & 0 & 1 \\ 0 & 0 & 0 \end{bmatrix}$$

for all $a, b \in \mathbb{R}$.

29. Let $F = \mathbb{Z}/(5)$. Show that the matrices $\begin{bmatrix} 0 & 0 & 0 \\ 1 & 0 & 0 \\ 0 & 1 & 0 \end{bmatrix}$ and $\begin{bmatrix} 0 & 1 & 0 \\ 0 & 0 & 1 \\ 0 & 0 & 0 \end{bmatrix}$ in $\mathcal{M}_{3 \times 3}(F)$ are similar.

30. Show that the matrix $A = \begin{bmatrix} 1 & i \\ i & -1 \end{bmatrix} \in \mathcal{M}_{2 \times 2}(\mathbb{C})$ is not similar to a diagonal matrix.

31. Let a be an element of a field F and let $A = \begin{bmatrix} a & 1 & 0 \\ 0 & a & 1 \\ 0 & 0 & a \end{bmatrix} \in \mathcal{M}_{3 \times 3}(F)$. For any polynomial $p(X) \in F[X]$, show that

$$p(A) = \begin{bmatrix} p(a) & p'(a) & \frac{1}{2}p''(a) \\ 0 & p(a) & p'(a) \\ 0 & 0 & p(a) \end{bmatrix},$$

where $f'(X)$ denotes the formal derivative of a polynomial $f(X) \in F[X]$.

32. Let n be a positive integer and let V be the vector space over $\mathbb{R}$ consisting of all polynomial functions from $\mathbb{R}$ to itself having degree no more than n. Find the minimal polynomial of the differentiation endomorphism $\delta: f \mapsto f'$ of V.

33. Let n be a positive integer and let $A \in \mathcal{M}_{n \times n}(\mathbb{R})$ be a matrix of rank h. Show that the degree of the minimal polynomial of A is at most $h + 1$.

34. Let n be a positive integer and let $A \in \mathcal{M}_{n \times n}(\mathbb{R})$. Show that there exist nonsingular matrices B and C in $\mathcal{M}_{n \times n}(\mathbb{R})$ satisfying $A = B + C$.

35. Let α and β be the endomorphisms of $\mathbb{Q}^4$ represented respectively by the matrices $\begin{bmatrix} 1 & 0 & -1 & 0 \\ 0 & 1 & 0 & -1 \\ 1 & 0 & -1 & 0 \\ 0 & 1 & 0 & -1 \end{bmatrix}$ and $\begin{bmatrix} 4 & -1 & -1 & 0 \\ -1 & 4 & 0 & -1 \\ 1 & 0 & 2 & -1 \\ 0 & 1 & -1 & 2 \end{bmatrix}$. Does there exist a basis of $\mathbb{Q}^4$ relative which both endomorphisms can be represented by diagonal matrices?

36. Let $a \neq -1$ be a real number and let $A = \begin{bmatrix} 1 - a + a^2 & 1 - a \\ a - a^2 & a \end{bmatrix} \in \mathcal{M}_{2 \times 2}(\mathbb{R})$. Calculate A^n for each $n > 1$.

37. Let n be a positive integer and let F be a field. Show that a matrix $A \in \mathcal{M}_{n \times n}(F)$ is nonsingular if and only if $m_A(0_F) \neq 0_F$.

38. Let n be a positive integer and let F be a field. Given an element c of F, find a matrix $A \in \mathcal{M}_{n \times n}(F)$ having minimal polynomial $(X - c)^n$.

39. Use the Cayley-Hamilton theorem to calculate the inverse of the matrix

$$\begin{bmatrix} 5 & 1 & -1 \\ -6 & 0 & 2 \\ 0 & 0 & 2 \end{bmatrix} \in \mathcal{M}_{3 \times 3}(\mathbb{R}).$$

CHAPTER XI

THE JORDAN CANONICAL FORM

We are still in the midst of considering the following problem: given a vector space V finitely generated over a field F and given an endomorphism α of V, we want to find a basis for V relative to which α can be represented in a "nice" manner. In Chapter 10 we saw that if V has a basis composed of eigenvectors of α then, relative to that basis, α is represented by a diagonal matrix. However, what happens if such a basis is not available?

Consider a vector space V over a field F let α be an endomorphism of V. If $v \in V$ consider the subspace of V generated by $\{v, \alpha(v), \alpha^2(v), \dots\}$. This space is called the α-*cyclic subspace* of V determined by v. We know that for each polynomial $g(X) = \sum_{i \geq 0} a_i X^i \in F[X]$ we have an endomorphism $g(\alpha)$ of V given by $g(\alpha): v \mapsto \sum_{i \geq 0} a_i \alpha^i(v)$ and it is easy to see that this space is precisely the set of all vectors in V of the form $g(\alpha)(v)$ for some $g(X) \in F[X]$; we will denote it by $F[\alpha]v$.

(11.1) PROPOSITION. *Let V be a vector space over a field F and let α be an endomorphism of V. If $0_V \neq v \in V$ then:*

(1) *$dim(F[\alpha]v) = 1$ if and only if v is an eigenvector of α;*
(2) *$F[\alpha]v$ is the intersection of all those subspaces of V containing v and stable under α.*

PROOF. (1) If v is an eigenvector of α associated with an eigenvalue c then for all $h > 0$ we have $\alpha^h(v) = c^h v \in Fv$ and so $F[\alpha]v \subseteq Fv$. The reverse containment is surely true and so we have equality. Thus $dim(F[\alpha]v) = 1$. Conversely, suppose that $dim(F[\alpha]v) = 1$. Since $v \in F[\alpha]v$, this implies that $F[\alpha]v = Fv$. In particular, $\alpha(v) \in Fv$ and so there exists a scalar c satisfying $\alpha(v) = cv$, which proves that v is an eigenvector of α.

(2) Since $F[\alpha]v$ itself contains v and is stable under α, it certainly contains this intersection. Conversely, let W be a subspace of V containing v and stable under α. Then $\alpha^h(v) \in W$ for all $h \geq 0$ and so $F[\alpha]v \subseteq W$. Thus $F[\alpha]v$ is also contained in this intersection, and so we have equality. $\square$

Proposition 11.1(1) suggests that the number $dim(F[\alpha]v)$ can be used as a measure of how "far" the vector $v \in V$ is from being an eigenvector of α.

If the space V is finitely generated over F then $F[\alpha]v$ is also finitely generated over F and so there must exist a positive integer k such that the set $\{v, \alpha(v), \dots, \alpha^{k-1}(v)\}$ is linearly independent while $\{v, \alpha(v), \dots, \alpha^k(v)\}$ is linearly dependent. Indeed, in this case $\{v, \alpha(v), \dots, \alpha^{k-1}(v)\}$ is a basis for $F[\alpha]v$ over F, which is called the *canonical basis* for $F[\alpha]v$. With respect to this basis, the restriction of α to $F[\alpha]v$ (which, by stability, is an endomorphism of $F[\alpha]v$) is represented by the matrix

$$\begin{bmatrix} 0_F & 0_F & \cdots & 0_F & c_1 \\ 1_F & 0_F & \cdots & 0_F & c_2 \\ 0_F & 1_F & \cdots & 0_F & c_3 \\ & & \cdots & & \\ 0_F & 0_F & \cdots & 0_F & c_{k-1} \\ 0_F & 0_F & \cdots & 1_F & c_k \end{bmatrix},$$

where the scalars c_i satisfy $a^k(v) = c_1 v + c_2 \alpha(v) + \cdots + c_k \alpha^k(v)$.

We begin by considering a special class of endomorphisms of V. Let V be a vector space finitely generated over a field F. An endomorphism α of V is *nilpotent* if and only if there exists a positive integer k satisfying $\alpha^k = \sigma_0$. The least such positive integer is the *index of nilpotency* of α.

EXAMPLE

Let F be a field. The endomorphism α of the vector space F^3 defined by $\alpha: [a, b, c] \mapsto [0_F, a, b]$ is nilpotent and has index of nilpotency 3.

If α is a nilpotent endomorphism of a vector space V and if $v \in V$ does not belong to $ker(\alpha)$ then the restriction of α to $F[\alpha]v$ is represented with respect to the canonical basis of $F[\alpha]v$ by a matrix of the form

$$\begin{bmatrix} 0_F & 0_F & \cdots & 0_F & 0_F \\ 1_F & 0_F & \cdots & 0_F & 0_F \\ 0_F & 1_F & \cdots & 0_F & 0_F \\ & & \cdots & & \\ 0_F & 0_F & \cdots & 0_F & 0_F \\ 0_F & 0_F & \cdots & 1_F & 0_F \end{bmatrix}.$$

(11.2) PROPOSITION. *Let V be a vector space over a field F and let α be a nilpotent endomorphism of V having index of nilpotency k. Then there exists an element v of V satisfying $dim(F[\alpha]v) = k$.*

PROOF. We know that $\alpha^k = \sigma_0$ but that there exists a vector $v \in V$ satisfying $\alpha^{k-1}(v) \neq 0_V$. We will prove the theorem if we succeed in showing that the set $\{v, \alpha(v), \ldots, \alpha^{k-1}(v)\}$ is linearly independent. And, indeed, assume that there exist scalars $a_0, \ldots, a_{k-1}$ not all of which are equal to 0_F, such that $\sum_{i=0}^{k-1} a_i \alpha^i(v) = 0_V$. Let t be the smallest index such that $a_t \neq 0_F$. Then $\sum_{i=t}^{k-1} a_i \alpha^i(v) = 0_V$. If we let the endomorphism α^{k-t-1} act on both sides of this equation we get $a_t \alpha^{k-1}(v) + a_{t+1} \alpha^k(v) + \cdots = 0_V$. Since $\alpha^i = \sigma_0$ for all $i \geq k$, we conclude that $a_t \alpha^{k-1}(v) = 0_V$ and so $a_t = 0_F$, which is a contradiction. Hence we see that all of the a_i must equal 0_F, and so the set is in fact linearly independent, as we wanted. $\square$

(11.3) PROPOSITION. *Let V be a vector space finitely generated over a field F and let α be a nilpotent endomorphism of V having index of nilpotency k. If the* *vector $0_V \neq v \in V$ satisfies the condition $dim(F[\alpha]v) = k$ then the subspace $F[\alpha]v$ has a complement W in V which is stable under α.*

PROOF. We will prove the theorem by induction on k. If $k = 1$ then $\alpha = \sigma_0$ and since any subspace of V is stable with respect to this endomorphism, we are done. Therefore, assume that $k > 1$ and that the result is true for all finitely-generated vector spaces having a nilpotent endomorphism having index of nilpotency less than k.

In order to simplify our notation, set $U = F[\alpha]v$ and $V' = im(\alpha)$. We know that V' is stable under α and so the restriction α' of α to V' is an endomorphism of V'. This endomorphism is also nilpotent and has index of nilpotency $k - 1$. Let $\{v, \alpha(v), \ldots, \alpha^{k-1}(v)\}$ be a basis for U. Then $\{\alpha(v), \alpha^2(v), \ldots, \alpha^{k-1}(v)\}$ is a basis for the image U' of the restriction of α to U. In particular, U' is a subspace of V' of the form $F[\alpha] \cdot \alpha(v)$, and by the induction hypothesis it has a complement W' in V' which is stable under α.

Let $W_0 = \{w \in V \mid \alpha(w) \in W'\}$. This is a subspace of V which contains W' since W' is stable under α. Moreover, $\alpha(w_0) \in W' \subseteq W_0$ for all $w_0 \in W_0$ and so W_0 is also stable under α. We claim that $V = U + W_0$. Indeed, if $x \in V$ is an arbitrary vector then $\alpha(x) \in V' = U' \oplus W'$ and so we can write $\alpha(x) = u' + w'$, where $u' \in U'$ and $w' \in W'$. But u' is of the form $\alpha(u)$ for some element u of U and, moreover, $x = u + (x - u)$. The first summand in this representation belongs to U, whereas the second satisfies $\alpha(x - u) = \alpha(x) - \alpha(u) = \alpha(x) - u' = w' \in W'$. Thus $x - u \in W_0$ and so $x \in U + W_0$. Thus our first claim is established.

We next claim that $U \cap W_0 \subseteq U'$. Indeed, if $x \in U \cap W_0$ then we see that $\alpha(x) \in U' \cap W' = \{0_V\}$ and so $\alpha(x) = 0_V$. Since $x \in U$ we know that there exists scalars $a_0, \ldots, a_{k-1}$ satisfying $x = a_0 v + a_1 \alpha(v) + \cdots + a_{k-1} \alpha^{k-1}(v)$ and hence $0_V = \alpha(x) = a_0 \alpha(v) + \cdots + a_{k-2} \alpha^{k-1}(v)$, which implies that $a_0 = \cdots = a_{k-2} = 0_F$. Therefore $x = a_{k-1} \alpha^{k-1}(v) \in U'$, proving the second claim.

In particular, this implies that $W' \cap (U \cap W_0) = \{0_V\}$ and so $W' \oplus (U \cap W_0)$ is a subspace of W_0. This space then has a complement W'' in W_0, which satisfies $W_0 = W'' \oplus [W' \oplus (U \cap W_0)]$. Our third claim is that $W = W'' \oplus W'$ is a complement of U in V which is stable under α, namely exactly what we are seek. Indeed, we first note that if $w \in W$ then $\alpha(w) \in W' \subseteq W$ and so W is indeed stable under α. Furthermore, $U \cap W = \{0_V\}$ since this subspace is contained in $(U \cap W_0) \cap W$, which equals $\{0_V\}$ by the choice of W. Finally, $V = U + W_0 = U + [W'' + W' + (U \cap W_0)] = U + W'' + W' = U + W$ and so $V = U \oplus W$, as required. $\square$

(11.4) PROPOSITION. *Let V be a vector space over a field F and let α be a nilpotent endomorphism of V having index of nilpotency k. Then:*

(1) *There exist natural numbers $k = k_1 \geq k_2 \geq \cdots \geq k_t$ satisfying $k_1 + \cdots + k_t = n$ and there exist vectors $v_1, \ldots, v_t$ in V such that the set*

$$D = \{v_1, \alpha(v_1), \ldots, \alpha^{k_1-1}(v_1), \ldots, v_t, \ldots, \alpha^{k_t-1}(v_t)\}$$

is a basis for V.

(2) *The matrix $\Phi_{DD}(\alpha)$ is of the form* $\begin{bmatrix} A_1 & O & \cdots & O \\ O & A_2 & \cdots & O \\ & & \cdots & \\ O & O & \cdots & A_t \end{bmatrix}$, *where each A_i is*

a matrix of size $k_i \times k_i$ of the form

$$\begin{bmatrix} 0_F & 0_F & 0_F & \ldots & 0_F \\ 1_F & 0_F & 0_F & \ldots & 0_F \\ 0_F & 1_F & 0_F & \ldots & 0_F \\ & & \ldots & & \\ 0_F & 0_F & 0_F & \ldots & 0_F \end{bmatrix}.$$

PROOF. Choose $k_1 = k$ and let v_1 be any vector not in $ker(\alpha^{k_1-1})$. Let $U_1 = F[\alpha]v_1$, which is a subspace of V having canonical basis

$$\{v_1, \alpha(v_1), \ldots, \alpha^{k_1-1}(v_1)\}.$$

Then U_1 is stable under α and $dim(U_1) = k_1$. Moreover, by Proposition 11.3 we know that V can be decomposed into a direct sum $V = U_1 \oplus W_1$, where W_1 is a subspace of V which is also stable under α. The restriction of α to W_1 is a nilpotent endomorphism of W_1 having index of nilpotency k_2, where $k_2 \leq k_1$. If W_1 is nontrivial we can now repeat the previous process: we select an element $v_2 \in W_1 \setminus ker(\alpha^{k_2-1})$ and define U_2 to be the subspace $F[\alpha]v_2$ of W_1 having canonical basis $\{v_2, \alpha(v_2), \ldots, \alpha^{k_2-1}(v_2)\}$. Then there exists a subspace W_2 of W_1 which is stable under α and which is a complement for U_2 in W_1.

Continue in this manner until we end up with a decomposition $V = U_1 \oplus \cdots \oplus U_t$ in which each U_i is a subspace of V stable under α and having a basis of the form $\{v_i, \alpha(v_i), \ldots, \alpha^{k_i-1}(v_i)\}$, where $\alpha^{k_i}(v_i) = 0_V$ for all i. This proves (1).

As for (2), since $U_i = F[\alpha]v_i$ for each i, the matrix A which represents α with respect to the basis of V given by the union of the canonical bases of the U_i is in the desired form. $\square$

A matrix of the form defined in Proposition 11.4(2) is said to be in *Jordan canonical form*.

Let V be a vector space over a field F and let α be an endomorphism of V having an eigenvalue c. A nonzero vector $v \in V$ is a *generalized eigenvector* of α *associated with c of degree k* if and only if it belongs to the kernel of $(c\sigma_1 - \alpha)^k$ but not to the kernel of $(c\sigma_1 - \alpha)^{k-1}$.

EXAMPLE

Let α be the endomorphism of $\mathbb{R}^3$ represented with respect to the canonical basis by the matrix $\begin{bmatrix} 2 & 1 & 4 \\ 0 & 2 & -1 \\ 0 & 0 & 3 \end{bmatrix}$. This endomorphism has an eigenvector $[3, -1, 1]$ which is associated with the eigenvalue 3 and an eigenvector $[1, 0, 0]$ which is associated with the eigenvalue 2. It also has a generalized eigenvector $[0, 1, 0]$ of degree 2 associated with the eigenvalue 2. Note that the set $\{[1, 0, 0], [0, 1, 0]\}$ is linearly independent.

We now extend Proposition 10.1.

(11.5) PROPOSITION. *Let V be a vector space over a field F and let α be an endomorphism of V having an eigenvalue c. Then the subset W of V consisting of 0_V and all generalized eigenvectors of α associated with c is a subspace of V.*

PROOF. Let v and w be generalized eigenvectors of α associated with c. Then there exist positive integers h and k satisfying $(c\sigma_1 - \alpha)^h(w) = (c\sigma_1 - \alpha)^k(v) = 0_V$. Therefore v and w both belong to the kernel of $(c\sigma_1 - \alpha)^{h+k}$, whence so does their sum and so does av for all $a \in F$. $\square$

If V is a vector space over a field F and if α is an endomorphism of V having an eigenvalue c. Then the subspace W of V consisting of 0_V and all generalized eigenvectors of α associated with c is called the *generalized eigenspace* of α associated with the eigenvalue c.

(11.6) PROPOSITION. *Let V be a vector space over a field F and let α be an endomorphism of V having an eigenvalue c. Let v be a generalized eigenvector of α of degree k associated with c. Then the set*

$$\{v, (c\sigma_1 - \alpha)(v), \ldots, (c\sigma_1 - \alpha)^{k-1}(v)\}$$

is linearly independent.

PROOF. Set $v_j = (c\sigma_1 - \alpha)^{k-j}(v)$ for each $1 \leq j \leq k$ and assume that there exist scalars $c_0, \ldots, c_{k-1}$ in F satisfying $\sum_{j=0}^{k-1} c_j v_j = 0_V$. Then

$$0_V = (c\sigma_1 - \alpha)^{k-1} \left[\sum_{j=0}^{k-1} c_j v_j \right] = c_{k-1}(c\sigma_1 - \alpha)^{k-1}(v)$$

and from this we conclude that $c_{k-1} = 0_F$. A similar argument now implies that $c_{k-1} = 0_F$ and, indeed, that $c_j = 0_F$ for all $0 \leq j \leq k - 1$, proving the result we want. $\square$

Let F be a field. A polynomial $p(X) \in F[X]$ is *completely reducible* if and only if it can be written as a product of polynomials in $F[X]$ of degree 1.

EXAMPLE

The polynomial $X^2 + 1$ is not completely reducible when considered as an element of $\mathbb{Q}[X]$ or of $\mathbb{R}[X]$; it is completely reducible when considered as an element of $\mathbb{C}[X]$, since there $X^2 + 1 = (X - i)(X + i)$.

A field F is *algebraically closed* if every polynomial in $F[X]$ of positive degree is completely reducible. The previous example shows that the fields $\mathbb{Q}$ and $\mathbb{R}$ are not algebraically closed. The field $\mathbb{C}$ is algebraically closed. The theorem establishing this fact is called the Fundamental Theorem of Algebra, and a proof of it can be found in most books on advanced algebra. Despite the name, the theorem is essentially analytic rather than algebraic and its proof depends in one way or another on analytic properties of the complex plane, so that it is beyond the scope of this book.

(11.7) PROPOSITION. *Let V be a vector space over a field F and let α be an endomorphism of V satisfying the condition that the characteristic polynomial $p(X)$ of α is completely reducible:*

$$p(X) = (X - c_1)^{n_1}(X - c_2)^{n_2} \cdot \ldots \cdot (X - c_m)^{n_m},$$

where $c_1, \ldots, c_m$ are the distinct eigenvalues of α in F. Then V can be decomposed as $U_1 \oplus \cdots \oplus U_m$, where the U_j are submodules of V satisfying the following conditions:

(1) *For each $1 \leq j \leq m$, the space U_j is stable under α;*
(2) *$dim(U_j) = n_j$ for all $1 \leq j \leq m$;*
(3) *For each $1 \leq j \leq m$, the restriction of α to U_j is of the form $c_j \tau_j + \beta_j$, where β_j is a nilpotent endomorphism of U_j and τ_j is the identity endomorphism of U_j.*

PROOF. For each $1 \leq j \leq m$, consider the endomorphism $\beta_j = \alpha - c_j \sigma_1$ of V. Let U_j be the generalized eigenspace of α associated with c_j. This is a subspace of V stable under β_j. It is also stable under α since for each $v \in U_j$ we have $\beta_j^n \alpha(v) = \alpha \beta_j^n(v) = \alpha(0_V) = 0_V$. We claim that there exists a fixed positive integer k such that each element of U_j is a generalized eigenvector of degree at most k. Indeed, we see that $ker(\beta_j) \subseteq ker(\beta_j^2) \subseteq \ldots$. Since V has finite dimension over F, not all of these inclusions can be proper and so there exists a positive integer k satisfying $ker(\beta_j^k) = ker(\beta_j^{k+1}) = \ldots$. From here it is clear that $ker(\beta_j^k) = U_j$.

In particular, this claim shows that the restriction of β_j to U_j is a nilpotent endomorphism having index of nilpotency k. Moreover, $\alpha = c_j \sigma_1 + \beta_j$ and so we have proven (3). We now note that if $h \neq j$ then U_h is stable under β_j. We claim that the restriction of β_j to U_h is an automorphism. Since U_h is finitely generated, it suffices to show that this restriction is monic. Indeed, assume that $v \in U_h \cap ker(\beta_j)$. Then there exists a positive integer k satisfying $\beta_h^k(v) = 0_V$ and so $(\alpha - c_h \sigma_1)^k(v) = [(\alpha - c_j \sigma_1) + (c_j - c_h)\sigma_1]^k(v) = [(c_j - c_h)\sigma_1]^k(v)$ and this equals 0_V. Since $c_j - c_h \neq 0_F$, we must have $v = 0_V$, proving the claim.

The next step is to show that the set of subspaces $\{U_1, \ldots, U_m\}$ is independent. Indeed, let $1 \leq h \leq m$ and let U' be the subspace $U_h \cap \sum_{j \neq h} U_j$ of V. This subspace is stable under β_h. Moreover, the restriction of β_h to U' is an automorphism (since $U' \subseteq \sum_{j \neq h} U_j$) and nilpotent (since $U' \subseteq U_h$), something which is possible only when $U' = \{0_V\}$. Thus the given set of subspaces of V is independent.

Let $U'' = U_1 \oplus \cdots \oplus U_m$. We want to show that U'' is in fact equal to V. Assume that this is not the case and let $v \in V \setminus U''$. By the Cayley-Hamilton Theorem we know that α annihilates its characteristic polynomial $p(X)$ and so $[\beta_1^{n_1} \cdot \ldots \cdot \beta_m^{n_m}](v) = 0_V \in U''$. Let t be the smallest among the integers $1, \ldots, m$ satisfying $[\beta_1^{n_1} \cdot \ldots \cdot \beta_t^{n_t}](v) \in U''$ and set $v' = [\beta_1^{n_1} \cdot \ldots \cdot \beta_{t-1}^{n_{t-1}}](v)$. Then $v' \notin U''$ but $\beta_t^{n_t}(v') \in U''$. Thus we can write $\beta_t^{n_t}(v') = u_1 + \cdots + u_m$, where $u_j \in U_j$ for each j. If $j \neq t$ we have seen that the restriction of $\beta_t^{n_t}$ to U_j is an automorphism of U_j and so there exists an element w_j of U_j satisfying $u_j = \beta_t^{n_t}(w_j)$. Therefore $\beta_t^{n_t}(v' - \sum_{j \neq t} w_j) = u_t \in U_t$. By definition of U_t it follows that $v' - \sum_{j \neq t} w_j \in U_t$ and so this vector belongs to U'' as well. But $\sum_{j \neq t} w_j \in U''$ too and so $v' \in U''$, contradicting the choice of v'. From this contradiction we conclude, as desired, that $V = U''$.

We are left to show that $dim(U_j) = n_j$ for all $1 \leq j \leq m$. Choose a basis for V which is the union of bases chosen for each of the U_j. With respect to this basis, α is represented by a matrix of the form

$$\begin{bmatrix} A_1 & O & \cdots & O \\ O & A_2 & \cdots & O \\ & & \cdots & \\ O & O & \cdots & A_m \end{bmatrix},$$

where each A_j is a matrix represented the restriction of α to U_j. It is immediate that the characteristic polynomial of α decomposes in the form

$$|XI - A| = \prod_{j=1}^{m} |XI - A_j|.$$

From this decomposition and from the fact that the restriction of β_j to U_h is an automorphism whenever $h \neq j$, we see that the only possible eigenvalue of the restriction of α to U_j is c_j and that the algebraic multiplicity of this eigenvalue is at most n_j. But from the equality $\sum_{j=1}^{n} dim(U_j) = dim(V) = \sum_{j=1}^{m} n_j$ it follows that $dim(U_j) = n_j$ for each $1 \leq j \leq m$. $\square$

The previous proposition shows us that in nice situations, for example when the field of scalars is algebraically closed, when we are given an endomorphism α of a finitely-generated vector space we can choose a basis for V relative to which α is represented by a matrix of a very simple type:

(1) First, we write V as a direct sum $U_1 \oplus \cdots \oplus U_m$ as above. By choosing a basis of V which is the union of bases of the U_j, we obtain a matrix representing α made up of blocks strung along the diagonal, each of which represents the restriction of α to one of the U_j.

(2) For each $1 \leq j \leq m$ write the restriction of α to U_j as $c_j \tau_j + \beta_j$, where β_j is a nilpotent endomorphism of U_j and τ_j is the identity endomorphism of U_j.

Thus we have a representation of A in the form

$$\begin{bmatrix} A_1 & O & \cdots & O \\ O & A_2 & \cdots & O \\ & & \cdots & \\ O & O & \cdots & A_m \end{bmatrix},$$

where each A_j is a matrix of the form

$$\begin{bmatrix} c_j & 0_F & 0_F & \cdots & 0_F & 0_F \\ 1_F & c_j & 0_F & \cdots & 0_F & 0_F \\ 0_F & 1_F & c_j & \cdots & 0_F & 0_F \\ & & & \cdots & & \\ 0_F & 0_F & 0_F & \cdots & c_j & 0_F \\ 0_F & 0_F & 0_F & \cdots & 1_F & c_j \end{bmatrix}.$$

Such a matrix is said to be in *Jordan canonical form*. By Proposition 11.7 we know that if V is a vector space finitely generated over a field F and if α is an endomorphism of V having a completely reducible characteristic polynomial over F, then there exists a basis D of V such that $\Phi_{DD}(\alpha)$ is a matrix in Jordan canonical form. In other words, if A is a matrix representing α with respect to some basis then A is similar to a matrix in Jordan canonical form.

EXAMPLE

Consider the endomorphism α of $\mathbb{R}^4$ defined by

$$\alpha: [a, b, c, d] \mapsto [b, c, d, -a + 4b - 6c + 4d].$$

This endomorphism is represented with respect to the canonical basis by the matrix

$$A = \begin{bmatrix} 0 & 1 & 0 & 0 \\ 0 & 0 & 1 & 0 \\ 0 & 0 & 0 & 1 \\ -1 & 4 & -6 & 4 \end{bmatrix}.$$

The characteristic polynomial of α is $(X-1)^4 = X^4 - 4X^3 + 6X^2 - 4X + 1$ and so the only eigenvalue of α is 1. We therefore claim that there exists a basis of $\mathbb{R}^4$ relative to which α is represented by the matrix $B = \begin{bmatrix} 1 & 0 & 0 & 0 \\ 1 & 1 & 0 & 0 \\ 0 & 1 & 1 & 0 \\ 0 & 0 & 1 & 1 \end{bmatrix}$ or, in other words, that

A is similar to B. And, indeed, $B = PAP^{-1}$, where $P = \begin{bmatrix} -1 & 3 & -3 & 1 \\ 1 & -2 & 1 & 0 \\ -1 & 1 & 0 & 0 \\ 1 & 0 & 0 & 0 \end{bmatrix}.$

EXAMPLE

Let α be the endomorphism of $\mathbb{R}^4$ represented with respect to the canonical basis by the matrix $A = \begin{bmatrix} 4 & 0 & 1 & 0 \\ 2 & 2 & 3 & 0 \\ -1 & 0 & 2 & 0 \\ 4 & 0 & 1 & 2 \end{bmatrix}$. The characteristic polynomial of this

matrix is $(X-3)^2(X-2)^2$. If P is the nonsingular matrix $\begin{bmatrix} 1 & 0 & 1 & 0 \\ 0 & 0 & -1 & 0 \\ -3 & 1 & -4 & 0 \\ -1 & 0 & 2 & 1 \end{bmatrix}$ then

$PAP^{-1} = \begin{bmatrix} 3 & 0 & 0 & 0 \\ 1 & 3 & 0 & 0 \\ 0 & 0 & 2 & 0 \\ 0 & 0 & 0 & 2 \end{bmatrix}$, and so this is the Jordan canonical form of α.

EXAMPLE

Let α be the endomorphism of $\mathbb{C}^4$ represented with respect to the canonical basis

by the matrix $A = \begin{bmatrix} 0 & 0 & 2i & 0 \\ 1 & 0 & 0 & 2i \\ -2i & 0 & 0 & 0 \\ 0 & -2i & 1 & 0 \end{bmatrix}$. The characteristic polynomial of this

matrix is $(X^2-4)^2$. If P is the nonsingular matrix $\begin{bmatrix} 1 & 0 & i & 0 \\ 0 & 1 & 0 & i \\ 1 & 0 & -i & 0 \\ 1 & 1 & -i & i \end{bmatrix}$ then $PAP^{-1} =$

$\begin{bmatrix} 2 & 0 & 0 & 0 \\ 1 & 2 & 0 & 0 \\ 0 & 0 & -2 & 0 \\ 0 & 0 & 1 & -2 \end{bmatrix}$ and this is the representation of α in Jordan canonical form.

Problems

1. Let F be a field. Give an example of an endomorphism of the vector space F^5 having index of nilpotency 3.

2. Let α be the endomorphism of $\mathbb{R}^4$ represented with respect to a given basis

by the matrix $\begin{bmatrix} 2 & -8 & 12 & -60 \\ 2 & -5 & 9 & -48 \\ 6 & -17 & 29 & -152 \\ 1 & -3 & 5 & -26 \end{bmatrix}$. Show that α is nilpotent and find its index of nilpotency.

3. Let V be a vector space over a field F and let α be a nilpotent endomorphism of V having index of nilpotency k. Find the minimal polynomial of α.

4. Let α be the endomorphism of $\mathbb{R}^3$ represented with respect to the canonical

basis by the matrix $\begin{bmatrix} 1 & 2 & -2 \\ 3 & 0 & 3 \\ 1 & 1 & -2 \end{bmatrix}$. What is $\mathbb{R}[\alpha] \cdot [1,0,0]$?

5. Let V be the subspace of $\mathbb{R}^{\mathbb{R}}$ consisting of all those functions everywhere differentiable infinitely-many times and let $\delta: V \rightarrow V$ be the differentiation endomorphism $f \mapsto f'$. Calculate $\mathbb{R}[\delta] \cdot sin(x)$.

6. Let α be the endomorphism of $\mathbb{R}^3$ represented with respect to the canonical

basis by the matrix $\begin{bmatrix} 1 & 1 & 1 \\ 0 & 1 & 0 \\ 0 & 0 & 2 \end{bmatrix}$. Find the eigenvalues of α and the generalized eigenspace associated with each of them.

7. Let α and β be endomorphisms of a vector space V over a field F satisfying $\alpha\beta = \beta\alpha$ and let W be a generalized eigenspace of α associated with an eigenvalue c of α. Show that W is stable under β.

8. Find the characteristic polynomial of the matrix

$$\begin{bmatrix} 5 & 0 & 0 & 0 & 0 & 0 & 0 & 0 \\ 1 & 5 & 0 & 0 & 0 & 0 & 0 & 0 \\ 0 & 0 & 5 & 0 & 0 & 0 & 0 & 0 \\ 0 & 0 & 0 & 5 & 0 & 0 & 0 & 0 \\ 0 & 0 & 0 & 1 & 5 & 0 & 0 & 0 \\ 0 & 0 & 0 & 0 & 0 & 3 & 0 & 0 \\ 0 & 0 & 0 & 0 & 0 & 1 & 3 & 0 \\ 0 & 0 & 0 & 0 & 0 & 0 & 0 & 3 \end{bmatrix} \in \mathcal{M}_{8 \times 8}(\mathbb{R}).$$

9. Find the matrix in Jordan canonical form which represents the endomorphisms of $\mathbb{R}^3$ represented with respect to the canonical basis by the following matrices:

(i) $\begin{bmatrix} 1 & 2 & 0 \\ 0 & 2 & 0 \\ 2 & -2 & -1 \end{bmatrix}$;

(ii) $\begin{bmatrix} -2 & 8 & 6 \\ -4 & 10 & 6 \\ 4 & -8 & -4 \end{bmatrix}$;

(iii) $\begin{bmatrix} 1 & 0 & 0 \\ 5 & 1 & 0 \\ 0 & 0 & 3 \end{bmatrix}$.

10. Find the matrix in Jordan canonical form which represents the endomorphism of $\mathbb{R}^4$ represented with respect to the canonical basis by the matrix

$$\begin{bmatrix} 3 & 1 & 0 & 0 \\ -4 & -1 & 0 & 0 \\ 7 & 1 & 2 & 1 \\ -17 & -6 & -1 & 0 \end{bmatrix}.$$

11. Let n be a positive integer. If $A \in \mathcal{M}_{n \times n}(\mathbb{C})$, show that $A \sim A^T$.

12. Let A be a matrix in $\mathcal{M}_{5 \times 5}(\mathbb{Q})$ in Jordan canonical form having minimal polynomial $(X - 3)^2$. What does A look like?

13. Give an example of a matrix in $\mathcal{M}_{4 \times 4}(\mathbb{R})$ which is not similar to any matrix in Jordan canonical form.

14. Let a, b, and c be elements of a field F and let α be the endomorphism of F^2 defined by $\alpha: [d, e] \mapsto [(ab - ab^2)d, (a - c)e]$. Can we choose c in such a manner that this endomorphism will be nilpotent with index of nilpotency 2?

CHAPTER XII

THE DUAL SPACE

Let V be a vector space over a field F. A linear transformation from V to F, considered as a vector space of dimension 1 over itself, is called a *linear functional* on V. The space $Hom(V, F)$ of all linear functionals on V is called the *dual space* of V, and will be denoted by $D(V)$. Since $dim(F) = 1$, we note that every linear functional other than the 0-map is an epimorphism.

EXAMPLE

Let D be a basis for a vector space V over a field F. If $u \in D$ then there exists a linear functional $\delta_u \in D(V)$ uniquely defined by its action on the other elements of D:
$$\delta_u(u') = \begin{cases} 1_F, & \text{if } u = u' \\ 0_F, & \text{otherwise.} \end{cases}$$
In particular, if $V = F^n$ and if $D = \{v_1, \ldots, v_n\}$ is the canonical basis, then δ_{v_i} is given by $[a_1, \ldots, a_n] \mapsto a_i$.

EXAMPLE

Let F be a field and let n be a positive integer. Then we have a linear functional on the space $\mathcal{M}_{n \times n}(F)$, called the *trace*, defined by $tr: [a_{ij}] \mapsto \sum_{i=1}^{n} a_{ii}$. Note that if $A = [a_{ij}]$ and $B = [b_{ij}]$ are matrices in $\mathcal{M}_{n \times n}(F)$ then

$$tr(BA) = \sum_{i=1}^{n} \sum_{h=1}^{n} a_{ih} b_{hi} = tr(AB).$$

As a consequence of this, we see that if P is a nonsingular matrix then $tr(P^{-1}AP) = tr(AP^{-1}P) = tr(A)$ and so similar matrices have identical traces.

If $A \in \mathcal{M}_{n \times n}(\mathbb{C})$ is a matrix having characteristic polynomial $p(X) = \sum_{i=0}^{n} c_i X^i$ then from the fact that $\mathbb{C}$ is algebraically closed we note that $p(X)$ equals $(X - b_1) \cdot \ldots \cdot (X - b_n)$, where the b_i are eigenvalues (not necessarily distinct) of A. By multiplying these factors, we see that $c_{n-1} = -\sum_{i=1}^{n} b_i$. But from the definition of $p(X)$ it is easy to check that $c_{n-1} = -tr(A)$. Thus we see that, over $\mathbb{C}$, the trace of a square matrix is the sum of its eigenvalues (counted with the appropriate multiplicities).

EXAMPLE

More generally, let F be a field, let n be a positive integer and let $B \in \mathcal{M}_{n \times n}(F)$. Then the function $A \mapsto tr(BA)$ is a linear functional on $\mathcal{M}_{n \times n}(F)$.

EXAMPLE

Let F be a field, let A be a nonempty set, and let $V = F^A$. For each $a \in A$ we have a linear functional $\delta_a : V \to F$ defined by $\delta_a : f \mapsto f(a)$. In the special case $F = \mathbb{R}$ and $A = [0, 1]$, the functional δ_0 is known to physicists as the *Dirac functional*.

EXAMPLE

Let V be the set of all continuous functions from $[0, 1]$ to $\mathbb{R}$, which is a vector space over $\mathbb{R}$. Then we have a linear functional on V defined by $f \mapsto \int_0^1 f(t)dt$.

EXAMPLE

Let F be a field. Any expression of the form $a_1 X_1 + \cdots + a_n X_n$, where the a_i are elements of F and the X_i are indeterminates, is called a *linear form* over F. Each such linear form defines a linear functional $F^n \to F$ given by $[b_1, \ldots, b_n] \mapsto \sum_{i=1}^n a_i b_i$. Moreover, any linear functional $\alpha : F^n \to F$ can be constructed in this way. Indeed, if $\{v_1, \ldots, v_n\}$ is the canonical basis for F^n and if $a_i = \alpha(v_i)$ for all $1 \leq i \leq n$ then α is defined by the linear form $a_1 X_1 + \cdots + a_n X_n$.

(12.1) PROPOSITION. *Let V be a vector space over a field F. If $0_V \neq v \in V$ then there exists a linear functional $\alpha \in D(V)$ satisfying $\alpha(v) \neq 0_F$.*

PROOF. Since the set $\{v\}$ is linearly independent, it can be expanded to a basis D of V. Now define the function $f : D \to F$ by $f(v) = 1_F$ and $f(u) = 0_F$ for all $v \neq u \in D$. By Proposition 4.1 there exists a linear transformation $\alpha : V \to F$ satisfying $\alpha(u) = f(u)$ for all $u \in D$, and this is the linear functional we seek. $\square$

We can extend the idea used in the proof of Proposition 12.1 further. Let V be a vector space over a field F and let D be a basis for V over F. For each $u \in D$ we can define a function $f_u : D \to F$ by setting $f_u(u) = 1_F$ and $f_u(u') = 0_F$ if $u \neq u' \in D$. Then there exists a unique linear functional $\delta_u \in D(V)$ satisfying $\delta_u(u') = f_u(u')$ for all $u' \in D$. We claim that $E = \{\delta_u \mid u \in D\}$ is a linearly-independent subset of $D(V)$. Indeed, if $c_1, \ldots, c_n$ are scalars and if $\{u_1, \ldots, u_n\}$ is a subset of E satisfying the condition that $\sum_{i=1}^n c_i \delta_{u_i}$ is the 0-function then, for each $1 \leq h \leq n$, we have $0_F = (\sum_{i=1}^n c_i \delta_{u_i})(u_h) = \sum_{i=1}^n c_i \delta_{u_i}(u_h) = c_h$.

If V is finitely generated over F (and hence D is a finite set) then E is also a basis for $D(V)$ over F. Indeed, If $\delta \in D(V)$ then one easily sees that $\delta = \sum_{u \in D} \delta(u)\delta_u$ by showing that the linear functionals on both sides of the equal sign take the same value at each element u' of D. Thus E is a basis for $D(V)$ over F, called the *dual basis* of D.

If V is not finitely generated over F then FE is the subspace of $D(V)$ consisting of those linear functionals δ on V satisfying the condition that $\delta(u) \neq 0_F$ for at most finitely-many elements u on D. This space is called the *weak dual space* of V. As an immediate consequence of the above remarks, we have the following proposition.

(12.2) PROPOSITION. *If V is a vector space finitely generated over a field F then $D(V)$ is also finitely generated over F, and $dim(V) = dim(D(V))$.*

Let V and W be vector spaces over a field F and let $\alpha: V \to W$ be a linear transformation. If $\pi \in D(W)$ then $\pi\alpha \in D(V)$. Moreover, for all $\pi, \pi' \in D(W)$ and all $v \in V$ we have

$$[(\pi + \pi')\alpha](v) = (\pi + \pi')\alpha(v) = \pi\alpha(v) + \pi'\alpha(v) = (\pi\alpha + \pi'\alpha)(v)$$

and so $(\pi + \pi')\alpha = \pi\alpha + \pi'\alpha$. In the same manner, if $c \in F$ then

$$[c(\pi\alpha)](v) = (\pi\alpha)(cv) = c[\pi\alpha(v)] = (c\pi)\alpha(v) = [(c\pi)\alpha](v)$$

and so $c(\pi\alpha) = (c\pi)\alpha$. Thus we see that α defines a linear transformation $D(\alpha)$ from $D(W)$ to $D(V)$ given by $D(\alpha): \pi \mapsto \pi\alpha$.

If V, W, and Y are vector spaces over F and if we have linear transformations $\alpha: V \to W$ and $\beta: W \to Y$ then one easily sees that $D(\beta\alpha) = D(\alpha)D(\beta)$. If α is in fact an isomorphism then $D(\alpha^{-1}) = D(\alpha)^{-1}$.

(12.3) PROPOSITION. *Let V and W be vector spaces finitely generated over a field F. Let $\{v_1, \ldots, v_k\}$ be a basis for V over F and let $\{\delta_1, \ldots, \delta_k\}$ be the dual basis for $D(V)$ corresponding to it. Similarly, let $\{w_1, \ldots, w_n\}$ be a basis for W over F and let $\{\pi_1, \ldots, \pi_n\}$ be the dual basis for $D(V)$ corresponding to it. If $\alpha: V \to W$ is a linear transformation represented with respect to the given bases by a matrix $A = [a_{ij}]$ then $D(\alpha): D(W) \to D(V)$ is represented with respect to the given bases by A^T.*

PROOF. For all $1 \leq i \leq k$ we have $\alpha(v_i) = \sum_{h=1}^{n} a_{ih} w_h$ and so, for all $1 \leq i \leq k$ and all $1 \leq j \leq n$ we have

$$[D(\alpha)(\pi_j)](v_i) = \pi_j\alpha(v_i) = \sum_{h=1}^{n} \pi_j(a_{ih} w_h) = \sum_{h=1}^{n} a_{ih}(\pi_i(w_h)) = a_{ij}.$$

But for all $\delta \in D(V)$ we have $\delta = \sum_{i=1}^{k} \delta(v_i)\delta_i$ and so, in particular, we see that

$$D(\alpha)(\pi_j) = \sum_{i=1}^{k}[D(\alpha)(\pi_j)](v_i)\delta_i = \sum_{i=1}^{k} a_{ij}\delta_i = \sum_{i=1}^{k} b_{ji}\delta_i$$

where $[b_{ij}]$ is the matrix A^T. $\square$

Given a vector space V over a field F, we have seen how to build the vector space $D(V)$ over F. This process can be repeated to build the vector space $D^2(V) = D(D(V))$.

EXAMPLE

Let V be a vector space over a field F. Each vector $v \in V$ we defines the *evaluation functional* $\theta_v \in D^2(V)$ given by $\theta_v : \delta \mapsto \delta(v)$.

(12.4) PROPOSITION. *If V is a vector space over a field F then the function $v \mapsto \theta_v$ is a monomorphism from V to $D^2(V)$. It is an isomorphism when V is finitely generated over F.*

PROOF. First, we must show that the given function is a linear transformation. Indeed, if $v, w \in V$, if $a \in F$, and if $\delta \in D(V)$ then $\theta_{v+w}(\delta) = \delta(v + w) = \delta(v) + \delta(w) = \theta_v(\delta) + \theta_w(\delta) = [\theta_v + \theta_w](\delta)$ and so $\theta_{v+w} = \theta_v + \theta_w$. Similarly, $\theta_{av}(\delta) = \delta(av) = a\delta_v = a\theta_v(\delta)$ and so $\theta_{av} = a\theta_v$. Thus we have shown that the function $v \mapsto \theta_v$ belongs to $Hom(V, D^2(V))$. If $\theta_v(\delta) = 0_F$ for all $\delta \in D(V)$ then $\delta(v) = 0_F$ for all $\delta \in D(V)$ and so, by Proposition 12.1, we see that $v = 0_V$. This shows that we in fact have a monomorphism.

If V is finitely generated over F then, by Proposition 12.2, $dim(D^2(V)) = dim(D(V)) = dim(V)$ and so any monomorphism from $D^2(V)$ to V is in fact an isomorphism. $\square$

Note that what is important in Proposition 12.4 is not the mere fact that V and $D^2(V)$ are isomorphic in the case V is finitely generated – we could have deduced that from the fact that their dimensions are equal – but rather the fact that we have a specific "natural" description of this isomorphism.

A proper subspace W of a vector space V over a field F is *maximal* if and only if it is properly contained in no proper subspace of V. By the Hausdorff Maximum Principle, which we introduced in Chapter 3, it is clear that every nontrivial vector space over a field F contains a maximal subspace.

(12.5) PROPOSITION. *Let V be a vector space over a field F and let δ be a linear functional on V which is not the 0-map. Then $ker(\delta)$ is a maximal subspace of V.*

PROOF. Assume that there exists a proper subspace Y of V properly containing $ker(\delta)$. Pick elements $y \in Y \setminus ker(\delta)$ and $w \in V \setminus Y$. These elements are necessarily not equal to 0_V. The set $\{w, v\}$ is linearly independent by Proposition 3.3 since $Fy \subseteq Y$ so $w \notin Fy$. If $W = F\{w, v\}$ then $ker(\delta) \cap W = \{0_V\}$ and so the restriction δ' of δ to W is monic and so an isomorphism (since any linear functional not the 0-function is an epimorphism). But this is impossible since $dim(W) = 2$ and $dim(F) = 1$. Thus $ker(\delta)$ must be a maximal subspace of V. $\square$

(12.6) PROPOSITION. *Let V be a vector space over a field F and let $B = \{\delta_1, \ldots, \delta_n\}$ be a nonempty finite subset of $D(V)$. Then $FB = \{\delta \in D(V) \mid \bigcap_{i=1}^n ker(\delta_i) \subseteq ker(\delta)\}$.*

PROOF. Set $W = \bigcap_{i=1}^n ker(\delta_i)$. If $\delta \in FB$ then there exist scalars $a_1, \ldots, a_n$ such that $\delta = \sum_{i=1}^n a_i\delta_i$. If $w \in W$ then $\delta(w) = \sum_{i=1}^n a_i\delta_i(w) = 0_V$ so $W \subseteq ker(\delta)$. This proves one direction of the proposition.

Conversely, assume that $\delta \in D(V)$ satisfies the condition that $W \subseteq ker(\delta)$. If δ is the 0-function, then it clearly belongs to FB. Hence we can assume that this is not the case. In order to show that $\delta \in FB$, we will proceed by induction on n. If $n = 1$ then $ker(\delta_1) \subseteq ker(\delta)$ and so, by Proposition 12.5, we in fact have equality. Let $v \in V \setminus ker(\delta)$ and let $a = \delta_1(v)^{-1}\delta(v)$. Then $\delta(v) = a\delta_1(v) = (a\delta_1)(v)$ and so $v \in ker(\delta - a\delta_1)$. But $ker(\delta) \subseteq ker(\delta - a\delta_1)$ and so this containment is proper. By Proposition 12.5, this implies that $ker(\delta - a\delta_1) = V$ and so $\delta = a\delta_1$ so $\delta \in FB$.

Now assume that $n > 1$ and that the proposition has already been established for every subset B of $D(V)$ having fewer than n elements. Let $Y = ker(\delta_n)$ and for each $1 \le i \le n - 1$ let β_i be the restriction of δ_i to Y. Let β be the restriction of δ to Y. Then $\bigcap_{i=1}^{n-1} ker(\beta_i) \subseteq ker(\beta)$ and so by the induction hypothesis there exists scalars $a_1, \ldots, a_{n-1}$ in F such that $\beta = \sum_{i=1}^{n-1} a_i\beta_i$. From this we see that $ker(\delta_n) \subseteq ker(\delta - \sum_{i=1}^{n-1} a_i\delta_i)$. As in the case $n = 1$ it follows that there exists a scalar $a_n \in F$ such that $\delta - \sum_{i=1}^{n-1} a_i\delta_i = a_n\delta_n$, i.e., $\delta = \sum_{i=1}^{n} a_i\delta_i \in FB$. $\square$

Problems

1. Let V be the vector space of all continuous functions form $[0, 1]$ to $\mathbb{R}$. From calculus we know that for each $f \in V$ the set $\{f(t) \mid 0 \le t \le 1\}$ has a maximal element a_f. Is the function $f \mapsto a_f$ a linear functional on V?

2. Let $F = \mathbb{Z}/(2)$ and let $\delta: F^3 \to F$ be the function which assigns to each vector $v = [a, b, c]$ the value (0 or 1) which appears in most of the components of v. Does δ belong to $D(F^3)$?

3. Find a nonzero linear functional $\delta \in D(\mathbb{R}^3)$ satisfying $\delta([3, 2, -1]) = \delta([3, 2, 1]) = 0$.

4. Let F be a field and let n be a positive integer. For each matrix $A = [a_{ij}]$ in $\mathcal{M}_{n \times n}(F)$, define the *antitrace* of A by

$$antitr(A) = \sum_{i=1}^{n} a_{i,n+1-i}.$$

Is $antitr: \mathcal{M}_{n \times n}(F) \to F$ a linear functional?

5. Let $F = \mathbb{Z}/(2)$ and let $V = F^A$, where A is a nonempty set. Let D be a given nonempty subset of A and define the function $\gamma: V \to F$ by setting $\gamma(f) = 1$ if $f(d) = 1$ for all $d \in D$ and $\gamma(f) = 0$ otherwise. Is γ a linear functional on V?

6. Let F be a field and let A be a matrix in $\mathcal{M}_{2 \times 2}(F)$ satisfying $A^2 = O$. Show that $tr(A) = 0_F$.

7. Let F be a field and let A be a matrix in $\mathcal{M}_{2 \times 2}(F)$ satisfying $tr(A) = tr(A^2) = 0_F$. Does it follow that $A^2 = O$?

8. Let n be a positive integer and let F be a field of characteristic 0 or of prime characteristic which does not divide n. Let $\delta: \mathcal{M}_{n \times n}(F) \to F$ a linear functional satisfying the following conditions:
   (i) $\delta(I) = n$;
   (ii) $\delta(AB) = \delta(BA)$ for all $A, B \in \mathcal{M}_{n \times n}(F)$.

Show that $\delta = tr$.

9. Let A and B be matrices in $\mathcal{M}_{2\times2}(\mathbb{R})$. Show that $tr(AB) = tr(A) \cdot tr(B)$ if and only if $|A + B| = |A| + |B|$.

10. Let $n > 1$ be an integer. Find a nonsingular matrix $A \in \mathcal{M}_{n\times n}(\mathbb{Q})$ satisfying $tr(A) = 0$.

11. Let n be a positive integer, let F be a field, and let B and C be matrices in $\mathcal{M}_{n\times n}(F)$ satisfying $tr(B) \le tr(C)$. If $A \in \mathcal{M}_{n\times n}(F)$ does it necessarily follow that $tr(AB) \le tr(AC)$?

12. If A is a nonsingular matrix in $\mathcal{M}_{n\times n}(\mathbb{R})$ find a positive integer k such that

$$|A| = \frac{1}{k}\begin{vmatrix} tr(A) & 1 & 0 \\ tr(A^2) & tr(A) & 2 \\ tr(A^3) & tr(A^2) & tr(A) \end{vmatrix}.$$

13. Let $V = \mathbb{R}^3$ and let δ_1, δ_2, and δ_3 be the linear functionals defined by:

$$\delta_1:[a, b, c] \mapsto 2a - b + 3c$$
$$\delta_2:[a, b, c] \mapsto 3a - 5b + c$$
$$\delta_3:[a, b, c] \mapsto 4a - 7b + c.$$

Is $\{\delta_1, \delta_2, \delta_3\}$ a basis for $D(V)$?

14. Let n be a positive integer and let V be the vector space of all polynomial functions from $\mathbb{R}$ to itself having degree at most n. For each $0 \le k \le n$, let $\delta_k:V \to \mathbb{R}$ the function defined by $\delta_k:p \mapsto \int_{-1}^{1} t^k p(t)dt$. Show that $\{\delta_0, \ldots, \delta_n\}$ is a basis for $D(V)$.

15. Let n be a positive integer and let V be the subspace of $\mathbb{R}[X]$ composed of all polynomials of degree at most n. Let $a_0, \ldots, a_n$ be distinct real numbers and, for all $0 \le i \le n$, let δ_i be the linear functional on V defined by $\delta_i:p \mapsto p(a_i)$. Show that $\{\delta_0, \ldots, \delta_n\}$ is a basis for $D(V)$.

16. Find the dual basis for the basis $\{[1, -1, 3], [0, 1, -1], [0, 3, -2]\}$ of $\mathbb{R}^3$ over $\mathbb{R}$.

17. Let F be a field and let V be a vector space of dimension $n > 0$ over F. Let $E = \{\delta_1, \ldots, \delta_n\}$ be a subset of $D(V)$ and assume there exists a vector $0_V \ne v \in V$ satisfying $\delta_i(v) = 0$ for all $1 \le i \le n$. Show that E is linearly dependent.

18. Let V be a vector space over a field F and for each nonempty subset A of V let $E(A) = \{\delta \in D(V) \mid A \subseteq ker(\delta)\}$.
  (i) Show that $E(A)$ is a subspace of $D(V)$ for each $\varnothing \ne A \subset V$;
  (ii) If W' and W'' are subspaces of V, show that $E(W'+W'') = E(W')\cap E(W'')$;
  (iii) If V is finitely generated over F, show that $dim(W)+dim(E(W)) = dim(V)$ for each W;
  (iv) If V is finitely generated over F and $W_1 \ne W_2$ are subspaces of V, show that $E(W_1) \ne E(W_2)$;
  (v) If W is a subspace of V having complement a W' in V, show that $D(V) = E(W) \oplus E(W')$.

Moreover, if $\alpha: V \to Y$ is a linear transformation between vector spaces over F, show that

(vi) $ker(D(\alpha)) = E(im(\alpha))$;

(vii) $im(D(\alpha)) \subseteq E(ker(\alpha))$, where we have equality if both V and Y are finitely generated over F.

19. Show that the matrices

$$\begin{bmatrix} 1 & 2-i & 1+i \\ 4+i & 1+i & 0 \\ 1+i & 1 & 1 \end{bmatrix}$$

and

$$\begin{bmatrix} 1 & 1+i & 2-i \\ 3-i & 1+i & 0 \\ 1 & 27 & 1-i \end{bmatrix}$$

in $\mathcal{M}_{3\times3}(\mathbb{C})$ are not similar.

20. Let n be a positive integer and let V be the subspace of $\mathbb{R}[X]$ composed of all polynomials of degree at most n. What is the dual basis of the basis $\{1, X, X^2, \ldots, X^n\}$ of V?

21. Let V be a vector space over a field F having a maximal subspace W. Show that there indeed exists a linear functional $\delta \in D(V)$ satisfying $W = ker(\delta)$.

22. Let W be a proper subspace of a vector space V over a field F and let $v \in V \backslash W$. Show that there exists a linear functional $\delta \in D(V)$ satisfying $\delta(v) \neq 0_F$ but $\delta(w) = 0_F$ for all $w \in W$.

23. Let $W = \mathbb{R}\{[2,-1,1,-2],[3,-1,0,-2]\} \subseteq \mathbb{R}^4$. Find $\{\delta \in D(V) \mid W \subseteq ker(\delta)\}$.

24. Let V be a vector space finitely generated over a field F. Let W be a subspace of V having a complement Y in V. Show that $D(V) = W' \oplus Y'$, where W' is a subspace of $D(V)$ isomorphic to W and Y' is a subspace of $D(V)$ isomorphic to Y.

CHAPTER XIII

INNER PRODUCT SPACES

Up to now we have considered vector spaces over arbitrary fields. In this chapter we will restrict ourselves to vector spaces over $\mathbb{R}$ or $\mathbb{C}$. Recall that if $z = a + bi \in \mathbb{C}$ then $\bar{z} = a - bi$. In particular, if $a \in \mathbb{R}$ then $\bar{a} = a$. If $A = [a_{ij}] \in \mathcal{M}_{n \times n}(\mathbb{C})$ then we define the *conjugate transpose* $\overline{A}$ of A to be the matrix $[b_{ij}]$, where $b_{ij} = \bar{a}_{ji}$ for all $1 \leq i, j \leq n$.

Let V be a vector space over $\mathbb{R}$ or $\mathbb{C}$. A function $(v, v') \mapsto \langle v, v' \rangle$ from $V \times V$ to the field of scalars is called an *inner product* on V if and only if the following conditions are satisfied:

(1) *(linearity in the first coordinate)* The function $v \mapsto \langle v, v' \rangle$ is a linear functional on V for all $v' \in V$;
(2) *(conjugate symmetry)* $\langle v, v' \rangle = \overline{\langle v', v \rangle}$ for all $v, v' \in V$;
(3) *(positive definiteness)* If $v \in V$ then $\langle v, v \rangle$ is a nonnegative real number, where $\langle v, v \rangle = 0$ if and only if $v = 0_V$.

A vector space over $\mathbb{R}$ or $\mathbb{C}$ on which we have an inner product defined is called an *inner product space*. As we shall see, there are often many ways of defining an inner product on a given vector space, and we must be careful to specify which of these we are talking about.

EXAMPLE

Let n be a positive integer. The function $(v, w) \mapsto vw^T$ is an inner product on $\mathbb{R}^n$, which is usually denoted by $v \cdot w$. Thus, we see that

$$[a_1, \ldots, a_n] \cdot [b_1, \ldots, b_n] = \sum_{i=1}^{n} a_i b_i.$$

This product is called the *dot product* and when we talk about $\mathbb{R}^n$ as an inner product space, we assume that this is the inner product defined on it, unless the contrary is specifically mentioned.

For $\mathbb{C}^n$ the dot product is defined by

$$[a_1, \ldots, a_n] \cdot [b_1, \ldots, b_n] = \sum_{i=1}^{n} a_i \bar{b}_i.$$

EXAMPLE

The construction of the dot product can be generalized. If D is a nonsingular matrix in $\mathcal{M}_{n \times n}(\mathbb{C})$ then for vectors $v = [a_1, \ldots, a_n]$ and $w = [b_1, \ldots, b_n]$ in $\mathbb{C}^n$ we define

$$\langle v, w \rangle = [\bar{b}_1, \ldots, \bar{b}_n] \overline{D} D [a_1, \ldots, a_n]^T$$

and check that this defines an inner product on $\mathbb{C}^n$.

EXAMPLE

Let $a < b$ be real numbers and let V be the vector space over $\mathbb{R}$ consisting of all continuous functions from $[a, b]$ to $\mathbb{R}$. Then we can define an inner product on V by setting

$$\langle f, g \rangle = \int_a^b f(t)g(t)dt.$$

If we take V to be the vector space over $\mathbb{C}$ consisting of all continuous functions from $[a, b]$ to $\mathbb{C}$. Then we can define an inner product on V by setting

$$\langle f, g \rangle = \int_a^b f(t)\overline{g(t)}dt.$$

EXAMPLE

Let $V = \mathcal{M}_{n \times n}(\mathbb{C})$ and define an inner product on V by $\langle A, B \rangle = tr(A\overline{B}) = tr(\overline{A}B)$. It is easy to see that if $A = [a_{ij}]$ and $B = [b_{ij}]$ then

$$\langle A, B \rangle = \sum_{i=1}^n \sum_{j=1}^n a_{ij}\bar{b}_{ij}$$

defines an inner product on V.

EXAMPLE

Let ℓ^2 be the vector space over $\mathbb{C}$ consisting of all infinite sequences $[c_1, c_2, \ldots]$ of complex numbers having the property that $\sum_{i=1}^\infty |c_i|^2 < \infty$. This space is very important in analysis, and it has an inner product defined on it by

$$\langle [c_1, c_2, \ldots], [d_1, d_2, \ldots] \rangle = \sum_{i=0}^\infty c_i\bar{d}_i.$$

Let W be a subspace of an inner product space V. The restriction of this inner product to $W \times W$ defines the structure of an inner product space on W. Therefore we can always assume that any subspace of an inner product space is also an inner product space the product on which is the restriction of the product on the larger space.

(13.1) PROPOSITION. *Let V be an inner product space. If $v, v', v'' \in V$ and if a is a scalar then:*

(1) $\langle v, v' + v'' \rangle = \langle v, v' \rangle + \langle v, v'' \rangle$;
(2) $\langle v, av' \rangle = \bar{a}\langle v, v' \rangle$;
(3) $\langle 0_V, v' \rangle = \langle v, 0_V \rangle = 0$.

PROOF. (1) By definition,

$$\langle v, v' + v'' \rangle = \overline{\langle v' + v'', v \rangle} = \overline{\langle v', v \rangle + \langle v'', v \rangle} = \overline{\langle v, v' \rangle} + \overline{\langle v, v'' \rangle} = \langle v, v' \rangle + \langle v, v'' \rangle.$$

(2) By definition, $\langle v, av' \rangle = \overline{\langle av', v \rangle} = \overline{a\langle v', v \rangle} = \bar{a}\overline{\langle v', v \rangle} = \bar{a}\langle v, v' \rangle$.
(3) By definition, $\langle 0_V, v' \rangle = \langle 0 \cdot 0_V, v' \rangle = 0\langle 0_V, v' \rangle = 0$ and similarly, using (2), we get $\langle v, 0_V \rangle = 0$. $\square$

In particular, by Proposition 13.1 we see that if V is an inner product space over $\mathbb{R}$ then for each $v \in V$ the function $v' \mapsto \langle v, v' \rangle$ is a linear transformation. This is not true, however, if V is an inner product space over $\mathbb{C}$.

(13.2) PROPOSITION. **(Cauchy-Schwartz-Bunyakovsky Inequality)** *If v and v' are vectors in an inner product space V then $|\langle v, v' \rangle|^2 \leq \langle v, v \rangle \langle v', v' \rangle$.*

PROOF. If either v or v' equals 0_V then the result is immediate, so we can assume that this is not the case. Set $a = -\langle v', v \rangle$ and $b = \langle v, v \rangle$. Then $\bar{a} = -\langle v, v' \rangle$ and $\bar{b} = b$. Hence we have:

$$0 \leq \langle av + bv', av + bv' \rangle$$
$$= a\bar{a}\langle v, v \rangle + ab\langle v, v' \rangle + b\bar{a}\langle v, v' \rangle + b^2\langle v', v' \rangle$$
$$= a\bar{a}b - ab\bar{a} - ab\bar{a} + b^2\langle v', v' \rangle$$
$$= b[-a\bar{a} + b\langle v', v' \rangle].$$

Since $v \neq 0_V$, we know that $b > 0$ and so $0 \leq -a\bar{a} + b\langle v', v' \rangle$. But $a\bar{a} = |a|^2 = \langle v, v' \rangle|^2$ and so $|\langle v, v' \rangle|^2 \leq \langle v, v \rangle \langle v', v' \rangle$, as desired. $\square$

The *norm* of a vector v in an inner product space V is defined to be $\|v\| = \sqrt{\langle v, v \rangle}$.

EXAMPLE

Let $V = \mathbb{R}^n$ on which we have the dot product. The norm of a vector $v = [a_1, \ldots, a_n] \in V$ is given by $\|v\| = \sqrt{\sum_{i=1}^n a_i^2}$. This norm is often called the *Euclidean norm* on $\mathbb{R}^n$.

A vector in an inner product space having norm equal to 1 is said to be *normal*. We note immediately that if $0_V \neq v \in V$ then the vector $\frac{1}{\|v\|}v \in V$ is normal.

EXAMPLE

Let V be the vector space over $\mathbb{R}$ consisting of all continuous functions from $[-\pi, \pi]$ to $\mathbb{R}$ on which we define the inner product

$$\langle f, g \rangle = \int_{-\pi}^{\pi} f(t)g(t)dt.$$

and let $f \in V$ be the function $x \mapsto sin(x)$. Then

$$\|f\| = \sqrt{\langle f, f \rangle} = \left[\int_{-\pi}^{\pi} sin^2(x)dx \right]^{1/2} = \sqrt{\pi}$$

and so $g = \frac{1}{\sqrt{\pi}} f$ is a normal vector in V.

(13.3) PROPOSITION. *Let V be an inner product space. If $v, v' \in V$ and if a is a scalar then:*

(1) $\|av\| = |a| \cdot \|v\|$;

(2) $\|v\| \geq 0$, *with equality if and only if $v = 0_V$;*

(3) *(Minkowski Inequality)* $\|v + v'\| \leq \|v\| + \|v'\|$;

(4) *(Parallelogram Identity)* $\|v + v'\|^2 + \|v - v'\|^2 = 2(\|v\|^2 + \|v'\|^2)$.

PROOF. Part (1) follows from the fact that $\|av\| = \sqrt{\langle av, av \rangle} = \sqrt{a\bar{a}\langle v, v \rangle} = |a| \cdot \|v\|$, while (2) is an immediate consequence of the definition.

(3) By the Cauchy-Schwartz-Bunyakovsky inequality we have

$$|\langle v, v' \rangle| = |\langle v', v \rangle| \leq \|v\| \cdot \|v'\|$$

and so

$$\begin{aligned}
\|v + v'\|^2 &= \langle v + v', v + v' \rangle \\
&= \langle v, v \rangle + \langle v, v' \rangle + \langle v', v \rangle + \langle v', v' \rangle \\
&\leq \|v\|^2 + 2\|v\| \cdot \|v'\| + \|v'\|^2 \\
&= (\|v\| + \|v'\|)^2
\end{aligned}$$

and that implies (3).

(4) We know that

$$\|v + v'\|^2 = \langle v + v', v + v' \rangle = \langle v, v \rangle + \langle v', v \rangle + \langle v, v' \rangle + \langle v', v' \rangle$$

and similarly

$$\|v - v'\|^2 = \langle v - v', v - v' \rangle = \langle v, v \rangle - \langle v', v \rangle - \langle v, v' \rangle + \langle v', v' \rangle.$$

We add these two in order to get (4). □

EXAMPLE

Let V be an inner product space and let A be a nonempty set. A function $f \in V^A$ is *bounded* if and only if there exists a real number b_f such that $\|f(x)\| \leq b_f$ for all $x \in A$. If $f, g \in V^A$ are bounded then, by the Minkowski inequality, we have $\|(f+g)(x)\| \leq \|f(x)\| + \|g(x)\| \leq b_f + b_g$ for all $x \in A$ so $f+g$ is bounded. Moreover, if a is a scalar and if $f \in V^A$ is bounded then $\|(af)(x)\| = |a| \cdot \|f(x)\| \leq |a|b_f$ for all $x \in A$ so af is bounded. Therefore the set of all bounded functions in V^A is a subspace of V^A.

It is possible to define the general notion of a *norm* on a vector space over $\mathbb{R}$ or $\mathbb{C}$ as a function $v \mapsto \|v\|$ from V to the field of scalars which satisfies conditions (1) - (3) of Proposition 13.3. The natural question is whether every norm of this form comes from an inner product defined on V. The answer is negative. For example, the function from $\mathbb{R}^2$ to $\mathbb{R}$ defined by $[a, b] \mapsto |a| + |b|$ is a norm in the above sense which cannot possibly come from an inner product on $\mathbb{R}^2$ since condition (4) of Proposition 13.3 is not satisfied. Indeed, condition (4) of Proposition 13.3 is a necessary and sufficient condition for the existence of an inner product in the following sense: let V be a vector space over $\mathbb{R}$ or $\mathbb{C}$ and let $\psi : V \to \mathbb{R}$ be a function satisfying the condition $\psi(v + v')^2 + \psi(v - v')^2 = 2[\psi(v) + \psi(v')]^2$ for all $v, v' \in V$. Define the function $\theta : V \times V \to \mathbb{R}$ by setting $\theta(v, v') = \frac{1}{4}[\psi(v + v')^2 - \psi(v - v')]$. Then:

(1) If the field of scalars is $\mathbb{R}$ then $\langle v, v' \rangle = \theta(v, v')$ is an inner product on V satisfying $\|v\| = \psi(v)$ for all $v \in V$;

(2) If the field of scalars is $\mathbb{C}$ then $\langle v, v' \rangle = \theta(v, v') + i\theta(v, iv')$ is an inner product on V satisfying $\|v\| = \psi(v)$ for all $v \in V$.

If v and v' are vectors in an inner product space V we define the *distance* between v and v' to be $d(v, v') = \|v - v'\|$.

EXAMPLE

If $V = \mathbb{R}^n$ on which the inner product is taken to be the dot product, and if $v = [a_1, \ldots, a_n]$ and $v' = [b_1, \ldots, b_n]$ then

$$d(v, v') = \sqrt{\sum_{i=1}^{n}(a_i - b_i)^2}.$$

This is the usual distance function studied in analytic geometry.

The following proposition shows that this notion of distance satisfies the properties we would expect from the notion of distance.

(13.4) PROPOSITION. *If v, v', v'' are vectors in an inner product space V then:*

(1) $d(v, v') = d(v', v)$;

(2) $d(v, v') \geq 0$, *with equality if and only if $v = w$*;

(3) **(Triangle Inequality)** $d(v, v') + d(v', v'') \geq d(v, v'')$.

PROOF. This is an immediate consequence of Proposition 13.3. □

If V is an inner product space over the field $\mathbb{R}$ and if v, v' are nonzero vectors in V, then by the Cauchy-Schwartz-Bunyakovsky inequality we know that

$$-1 \leq \frac{1}{\|v\| \cdot \|v'\|} \langle v, v' \rangle \leq 1$$

and so there exists a unique real number $0 \leq t \leq \pi$ satisfying

$$cos(t) = (\|v\| \cdot \|v'\|)^{-1} \langle v, v' \rangle.$$

This number t is the *angle* between v and v'.

EXAMPLE

Let V be the vector space of all continuous functions from the interval $[0, 1]$ to $\mathbb{R}$. This is an inner product space over $\mathbb{R}$ with inner product defined, as in a previous example, by $\langle f, g \rangle = \int_0^1 f(t)g(t)dt$. If f and g are the functions in V defined by $f: x \mapsto 5x^2$ and $g: x \mapsto 3x$ then $\|f\| = \sqrt{5}$ and $\|g\| = \sqrt{3}$. Moreover, the angle t between f and g is given by

$$cos(t) = \frac{1}{\sqrt{15}} \int_0^1 (5x^2)(3x)dx = \frac{\sqrt{15}}{4}.$$

Vectors v and v' in an inner product space V are *orthogonal* if and only if $\langle v, v' \rangle = 0$. In this case we write $v \perp v'$. By Proposition 13.3(4) we note that if $v \perp v'$ then $\|v + v'\|^2 = \|v\|^2 + \|v'\|^2$. Similarly, $\|v - v'\|^2 = \|v\|^2 + \|v'\|^2$ which, for the case $V = \mathbb{R}^2$, is just the Pythagonean Theorem. Vectors $\{v_i \mid i \in \Omega\}$ are *mutually orthogonal* if and only if $v_i \perp v_j$ for all $i \neq j \in \Omega$.

EXAMPLE

If $V = \mathbb{R}^n$ and the inner product on V is chosen to be the dot product, then nonzero vectors v and v' in V are orthogonal if and only if the angle between them equals $\frac{\pi}{2}$.

EXAMPLE

Consider $\mathbb{C}^2$ as an inner product space over $\mathbb{C}$, with the product being the dot product. Then $[2 + 3i, -1 + 5i] \perp [1 + i, -i]$.

EXAMPLE

Let V be the vector space of all continuous functions from the closed interval $[-1, 1]$ to $\mathbb{R}$, on which we define an inner product given by

$$\langle f, g \rangle = \int_{-1}^{1} f(t)g(t)dt.$$

For each $i \geq 0$ we define the polynomial function $p_i \in V$ recursively as follows: $p_0(x) = 1$, $p_1(x) = x$, and

$$p_{h+1}(x) = \frac{2h+1}{h+1}xp_h(x) - \frac{h}{h+1}p_{h-1}(x)$$

for all $h \geq 1$. These functions are called the *Legendre polynomials*, and they are mutually orthogonal in V with respect to the given inner product.

Now consider the same space but define on it an inner product given by

$$\langle f, g \rangle = \int_{-1}^{1} \frac{f(t)g(t)}{\sqrt{1 - t^2}} dt.$$

For each $i \geq 0$, define the polynomial function $T_i \in V$ recursively as follows: $T_0(x) = 1$, $T_1(x) = x$, and $T_{h+1}(x) = 2xT_h(x) - T_{h-1}(x)$ for all $h \geq 1$. These functions are called *Chebyshev polynomials*, and they are mutually orthogonal in V with respect to the given inner product.

(13.5) PROPOSITION. *Let V be an inner product space.*

(1) *If $v \in V$ satisfies $v \perp v'$ for all $v' \in V$ then $v = 0_V$;*

(2) *If $\varnothing \neq A \subseteq V$ and if $v \in V$ satisfies $v \perp v'$ for all $v' \in A$ then $v \perp w$ for all $w \in W$, where W is the subspace of V generated by A.*

PROOF. If $v \in V$ satisfies $v \perp v'$ for all $v' \in V$ then, in particular, $v \perp v$ and so $v = 0_V$. This proves (1). Now assume that $\varnothing \neq A \subseteq V$ and that $v \perp v'$ for all $v' \in A$. If $w \in W$ then there exist elements $v_1, \ldots, v_n$ of A and scalars $c_1, \ldots, c_n$ such that $w = \sum_{i=1}^{n} c_i v_i$. Hence

$$\langle v, w \rangle = \left\langle v, \sum_{i=1}^{n} c_i v_i \right\rangle = \sum_{i=1}^{n} \bar{c}_i \langle v, v_i \rangle = 0$$

and so $v \perp w$. $\square$

(13.6) PROPOSITION. *Let V be an inner product space and let $A \neq \varnothing$ be a set of nonzero vectors in V which are mutually orthogonal. Then A is linearly independent.*

PROOF. Suppose $\{v_1, \ldots, v_n\}$ are elements of A and that $c_1, \ldots, c_n$ are scalars satisfying $\sum_{i=1}^{n} c_i v_i = 0_V$. Then for each $1 \leq h \leq n$ we have

$$c_h \langle v_h, v_h \rangle = \sum_{i=1}^{n} c_i \langle v_i, v_h \rangle = \left\langle \sum_{i=1}^{n} c_i v_i, v_h \right\rangle = \langle 0_V, v_h \rangle = 0$$

and so $c_h = 0$. Hence A is linearly independent. $\square$

(13.7) PROPOSITION. (**Gram-Schmidt Theorem**) *An inner product space V finitely generated over its field of scalars has a basis consisting of mutually orthogonal vectors.*

PROOF. We will proceed by induction on the dimension of V. If $dim(V) = 1$ there is nothing to prove. Assume therefore that the result is true for all spaces of dimension k and let V be a space of dimension $k + 1$ having a subspace W of dimension k. By the induction hypothesis there exists a basis $\{v_1, \ldots, v_k\}$ of W consisting of mutually orthogonal vectors. Let $v \in V \setminus W$ and for each $1 \leq i \leq n$ let $c_i = \langle v, v_i \rangle / \langle v_i, v_i \rangle$. Set $v_{k+1} = v - \sum_{i=1}^{k} c_i v_i$. Then $v_{k+1} \notin W$ since $v \notin W$. Moreover, for each $1 \leq j \leq k$ we have

$$
\begin{aligned}
\langle v_{k+1}, v_j \rangle &= \left\langle v - \sum_{i=1}^{k} c_i v_i, v_j \right\rangle \\
&= \langle v, v_j \rangle - \left\langle \sum_{i=1}^{k} c_i v_i, v_j \right\rangle \\
&= \langle v, v_j \rangle - c_j \langle v_j, v_j \rangle \\
&= \langle v, v_j \rangle - \frac{\langle v, v_j \rangle}{\langle v_j, v_j \rangle} \langle v_j, v_j \rangle \\
&= 0
\end{aligned}
$$

and so $v_{k+1} \perp v_j$ for all $1 \leq j \leq k$. Therefore, by Proposition 13.6, the set $\{v_1, \ldots, v_{k+1}\}$ is linearly independent and so is a basis for V. $\square$

Note that the proof of Proposition 13.7 is algorithmic – it gives us a method of actually computing a basis for V consisting of mutually orthogonal vectors. Indeed, given any basis $\{v_1, \ldots, v_n\}$ for such a space we can use this method, called *Gram-Schmidt process*, to construct a basis $\{u_1, \ldots, u_n\}$ consisting of mutually orthogonal normal vectors from it. We can summarize this algorithm as follows:

for $1 \leq k \leq n$ do

$\qquad u_k' = v_k - \sum_{j=1}^{k-1} \langle v_k, u_j \rangle u_j$ and

$\qquad u_k = \|u_k'\|^{-1} u_k'.$

EXAMPLE

Let $V = \mathbb{R}^4$ together with the dot product and consider the vectors $v_1 = [3, 0, 0, 0]$, $v_2 = [0, 1, 2, 1]$, and $v_3 = [0, -1, 3, 2]$ in V. Let W be the subspace of V generated by $\{v_1, v_2, v_3\}$. Then $dim(W) = 3$ and we can construct a basis $\{u_1, u_2, u_3\}$ for W consisting of mutually orthogonal vectors, as follows:

(1) Rv_1 is a one-dimensional subspace of V so choose $u_1 = v_1$;

(2) $v_1 \notin Ru_1$ so choose $u_2 = [0, 1, 2, 1] - \frac{1}{9}([0, 1, 2, 1] \cdot [3, 0, 0, 0])[3, 0, 0, 0] = [0, 1, 2, 1]$

(3) $v_3 \notin R\{u_1, u_2\}$ so choose
$$u_3 = [0, -1, 3, 2] - \frac{1}{9}([0, 1, 2, 1] \cdot [3, 0, 0, 0])[3, 0, 0, 0]$$
$$\qquad - \frac{1}{6}([0, -1, 3, 2] \cdot [0, 1, 2, 1])[0, 1, 2, 1]$$
$$= [0, -\frac{13}{6}, \frac{2}{3}, \frac{5}{6}].$$

and $\{u_1, u_2, u_3\}$ is a basis for W consisting of mutually orthogonal vectors.

EXAMPLE

Let V be the space of all continuous functions from the interval $[-1, 1]$ to $\mathbb{R}$, on which we have the inner product defined by $\langle f, g \rangle = \int_{-1}^{1} f(t)g(t)dt$. For each $i \geq 0$ let $f_i \in V$ be the polynomial function defined by $f_i: x \mapsto x^i$. If $n > 0$ then the set $\{f_0, \ldots, f_n\}$ is linearly independent and so is a basis for a subspace W of V. If we apply the Gram-Schmidt process to this basis we get a basis $\{p_0, \ldots, p_n\}$ of mutually-orthogonal elements of W, where the p_i are the Legendre polynomials we defined above.

There are several ways of modifying above Gram-Schmidt process in order to obtain algorithms having better properties of numerical stability, which can be found in any book on numerical analysis. One of these is the following, which modifies the values of the v_i to create a new basis consisting of mutually orthogonal normal vectors.

for $1 \leq k \leq n$ do

$\qquad v_k = \|v_k\|^{-1}v_k$;
$\qquad$ **for $k + 1 \leq j \leq n$ do** $v_j = v_j - \langle v_j, v_k \rangle v_k$;

There are also alternatives to the Gram-Schmidt process for subspaces of inner product spaces of the form $\mathbb{R}^n$ or $\mathbb{C}^n$. For example, let $\{v_1, \ldots, v_k\}$ be a basis for a subspace W of $\mathbb{R}^n$. Let $A \in \mathcal{M}_{k \times n}(\mathbb{R})$ be the matrix the rows of which are the vectors $v_1, \ldots, v_k$ and consider the extended matrix $[AA^T \ A] \in \mathcal{M}_{k \times (k+n)}(\mathbb{R})$. Using the techniques we learned in Chapter 8, we can transform this matrix using elementary row operations into one of the form $[B \ C]$, where B is an upper-triangular matrix. The rows of C then form a basis for W which are orthogonal with respect to the dot product on $\mathbb{R}^n$. Note that this method is not necessarily faster on a computer, but is often easier for computation by hand.

EXAMPLE

Let W be the subspace of $\mathbb{R}^4$ with basis $\{[0,1,0,1],[-2,3,0,1],[1,1,1,5]\}$ and let

$$A = \begin{bmatrix} 0 & 1 & 0 & 1 \\ -2 & 3 & 0 & 1 \\ 1 & 1 & 1 & 5 \end{bmatrix}.$$

Then

$$AA^T = \begin{bmatrix} 2 & 4 & 6 \\ 4 & 14 & 6 \\ 6 & 6 & 28 \end{bmatrix}.$$

Using elementary row operations, we transform the matrix $[AA^T \; A]$ into the matrix

$$\begin{bmatrix} 2 & 4 & 6 & 0 & 1 & 0 & 1 \\ 0 & 6 & -6 & -2 & 1 & 0 & 1 \\ 0 & 0 & 4 & -1 & -1 & 1 & 1 \end{bmatrix},$$

which gives us the basis $\{[0,1,0,1],[-2,1,0,-1],[-1,-1,1,1]\}$ of W consisting of mutually orthogonal vectors.

For a subspace W of an inner product space V let $W^\perp$ be the set of all vectors $v \in V$ satisfying $\langle v, w \rangle = 0$ for all $w \in W$. By Proposition 13.5 we know that $W^\perp$ is a subspace of V. Since $\langle v, v \rangle \neq 0$ for all $0_V \neq v \in V$, we clearly have $W^\perp \cap W = \{0_V\}$. By Proposition 13.5 we also know that $\{0_V\}^\perp = V$ and $V^\perp = \{0_V\}$. The subspace $W^\perp$ of V is called the *orthogonal complement* of W in V. This name is justified by the following proposition.

(13.8) PROPOSITION. *If V is an inner product space finitely generated over its field of scalars and if W is a subspace of V then $V = W \oplus W^\perp$ and $W = (W^\perp)^\perp$.*

PROOF. Let $\{v_1, \ldots, v_k\}$ be a basis for W, which can be extended to a basis $\{v_1, \ldots, v_n\}$ for V. Use the Gram-Schmidt process to obtain a basis $\{u_1, \ldots, u_n\}$ for V composed of mutually orthogonal vectors. Then $\{u_1, \ldots, u_k\}$ is a basis for W and $u_i \in W^\perp$ for all $k < i \leq n$. This shows that $V = W + W^\perp$ and since we have already seen that $W \cap W^\perp = \{0_V\}$, we must have $V = W \oplus W^\perp$. Moreover, we see that $\{u_{k+1}, \ldots, u_n\}$ is a basis for $W^\perp$ and so, as above, $W = (W^\perp)^\perp$. $\square$

In particular, if V is an inner product space finitely generated over its field of scalars and if W is a subspace of V then we have a canonical projection of V onto W induced from the decomposition $V = W \oplus W^\perp$. This projection is called the *orthogonal projection* of V onto W.

Proposition 13.8 is not necessarily true if the space V is not finitely generated over its field of scalars, as the following example shows.

EXAMPLE

Let $V = \mathbb{R}^{(\mathbb{N})}$ be the vector space over $\mathbb{R}$ consisting of all functions f from the set $\mathbb{N}$ of nonnegative integers to $\mathbb{R}$ satisfying the condition that $f(i) \neq 0$ for only finitely-many values of i. On this space we define an inner product by setting

$$\langle f, g \rangle = \sum_{i=0}^{\infty} f(i)g(i),$$

where this sum is well-defined since only finitely-many of the summands are nonzero. For each $h \geq 0$, let $f_h \in V$ be defined by satisfying $f_h(a_h) = 1$ and $f_h(a_i) = 0$ for all $i \neq h$. It is easy to see that $D = \{f_h \mid h \geq 0\}$ is a basis for V composed of mutually orthogonal vectors. Let W be the subspace of V generated by the set $\{f_0 - f_1, f_1 - f_2, \ldots\}$. Then $W \neq V$ since $f_0 \notin W$. If $0_V \neq g \in W^\perp$ then there exist scalars $a_0, \ldots, a_n$ such that $a_n \neq 0$ and $g = \sum_{i=0}^{n} a_i f_i$. But then $a_n = \langle g, f_n - f_{n+1} \rangle$, which is a contradiction. Therefore we must have $W^\perp = \{0_V\}$ despite the fact that W is a proper subspace of V. In particular, we cannot have $V = W \oplus W^\perp$ and $W \neq (W^\perp)^\perp = V$.

Let V be an inner product space. A nonempty subset A of V is *orthonormal* if and only if every vector in V is normal and these vectors are mutually orthogonal.

EXAMPLE

Let V be the space of all continuous functions from the interval $[-\pi, \pi]$ to $\mathbb{R}$ on which we have the inner product defined by $\langle f, g \rangle = \frac{1}{\pi} \int_{-\pi}^{\pi} f(t)g(t)dt$. Then $\left\{\frac{1}{2}\right\} \cup \left\{sin(nx) \mid n \geq 1\right\} \cup \left\{cos(nx) \mid n \geq 1\right\}$ is an orthonormal subset of V.

We now extend Proposition 13.7

(13.9) PROPOSITION. *Every inner product space V finitely generated over its field of scalars has an orthonormal basis.*

PROOF. By Proposition 13.7 we know that V has a basis $\{v_1, \ldots, v_n\}$ composed of mutually orthogonal vectors. For each $1 \leq i \leq n$ set $w_i = \|v_i\|^{-1}v_i$. Then w_i is normal and for $1 \leq i \neq j \leq n$ we have $\langle w_i, w_j \rangle = \|w_i\|^{-1}\|w_j\|^{-1}\langle v_i, v_j \rangle = 0$. Thus $\{w_1, \ldots, w_n\}$ is an orthonormal basis for V. $\square$

The example after Proposition 13.8 shows that while inner product spaces not finitely generated over their field of scalars may have orthonormal bases, this need not always be the case. However, using the Hausdorff Maximum Principle, it is easy to show that every inner product space V contains a maximal orthonormal subset. Such a subset is known as a *Hilbert subset* of V. If V is finitely generated over its field of scalars then any Hilbert subset is indeed a basis of V, but otherwise that is not necessarily true. By Proposition 13.6 we know that any Hilbert subset is linearly independent.

Needless to say, a given inner product space may have several different orthonormal bases, as the following example shows:

EXAMPLE

If $V = \mathbb{R}^4$ on which the inner product is the dot product then the canonical basis for V is orthonormal. Another orthonormal basis for V is

$$\left\{ [1,0,0,0], [0, \frac{1}{\sqrt{2}}, \frac{1}{\sqrt{2}}, 0], [0, -\frac{1}{\sqrt{2}}, \frac{1}{\sqrt{2}}, 0], [0,0,0,1] \right\}.$$

The advantage of using orthonormal bases is that the given inner product "looks like" the dot product with respect to them, as the following proposition shows.

(13.10) PROPOSITION. *Let V be an inner product space finitely generated over its field of scalars and let $\{v_1, \ldots, v_n\}$ be an orthonormal basis for V. If $v = \sum_{i=1}^n a_i v_i$ and $w = \sum_{i=1}^n b_i v_i$ are vectors in V then:*

(1) $\langle v, w \rangle = \sum_{i=1}^n a_i \bar{b}_i$;
(2) $v = \sum_{i=1}^n \langle v, v_i \rangle v_i$;
(3) **(Parseval Identity)** $\langle v, w \rangle = \sum_{i=1}^n \overline{\langle v_i, v \rangle} \langle v_i, w \rangle$;
(4) **(Bessel Identity)** $\|v\|^2 = \sum_{i=1}^n |\langle v_i, v \rangle|^2$.

PROOF. (1) By definition of the inner product we have:

$$\langle v, w \rangle = \left\langle \sum_{i=1}^n a_i v_i, \sum_{j=1}^n b_j v_j \right\rangle = \sum_{i=1}^n \sum_{j=1}^n a_i \bar{b}_j \langle v_i, v_j \rangle = \sum_{i=1}^n a_i \bar{b}_i,$$

as desired.

(2) For each $1 \le h \le n$ we have $\langle v, v_h \rangle = \langle \sum_{i=1}^n a_i v_i, v_h \rangle = \sum_{i=1}^n a_i \langle v_i, v_h \rangle = a_h \langle v_i, v_h \rangle = v_h$ and so we have (2).

(3) Note that

$$\langle v, w \rangle = \left\langle \sum_{i=1}^n \langle v, v_i \rangle v_i, \sum_{j=1}^n \langle w, v_j \rangle v_j \right\rangle$$

$$= \sum_{i=1}^n \langle v, v_i \rangle \left\langle v_i, \sum_{j=1}^n \langle w, v_j \rangle v_j \right\rangle$$

$$= \sum_{i=1}^n \sum_{j=1}^n \langle v, v_i \rangle \overline{\langle w, v_j \rangle} \langle v_i, v_j \rangle$$

$$= \sum_{i=1}^n \langle v, v_i \rangle \overline{\langle w, v_i \rangle}$$

$$= \sum_{i=1}^n \overline{\langle v_i, v \rangle} \langle v_i, w \rangle$$

which is the desired conclusion.

(4) This follows from (3) when we choose $v = w$. $\square$

The coefficients $\langle v, v_h \rangle$ which appear in Proposition 13.10(2) are called the *Fourier coefficients* of the vector v with respect to the given orthonormal basis of V.

If V is an inner product space then we have seen that for all $v' \in V$ the function from V to its field of scalars given by $v \mapsto \langle v, w \rangle$ is a linear functional on V. When V is finitely generated over its field of scalars, then every linear functional on V is of this form, as the following result shows.

(13.11) PROPOSITION. (**Riesz Representation Theorem**) *Let V be an inner product space finitely generated over its field of scalars. Then for each $\delta \in D(V)$ there exists a unique vector $y \in V$ satisfying $\delta(v) = \langle v, y \rangle$ for all $v \in V$.*

PROOF. Let $\{v_1, \ldots, v_n\}$ be an orthonormal basis for V and let $y = \sum_{i=1}^{n} \overline{\delta(v_i)} v_i$. Then for all $1 \leq h \leq n$ we have

$$\langle v_h, y \rangle = \left\langle v_h, \sum_{i=1}^{n} \overline{\delta(v_i)} v_i, v_i \right\rangle = \sum_{i=1}^{n} \delta(v_i) \langle v_h, v_i \rangle = \delta(v_h)$$

and so $\delta(v) = \langle v, y \rangle$ for all $v \in V$. The vector y is unique since if $\delta(v_h) = \langle v_h, y' \rangle$ for all $1 \leq h \leq n$ then $y' = \sum_{i=1}^{n} \langle y', v_i \rangle v_i = \sum_{i=1}^{n} \overline{\delta(v_i)} v_i = y$. $\square$

We now use Proposition 13.11 for another important construction.

(13.12) PROPOSITION. *Let V and W be inner product spaces finitely generated over the same field of scalars and let $\alpha: V \to W$ be a linear transformation. Then there exists a unique linear transformation $\alpha^*: W \to V$ satisfying $\langle \alpha(v), w \rangle = \langle v, \alpha^*(w) \rangle$ for all $v \in V$ and all $w \in W$.*

PROOF. Let w be a given vector in W. The function $\delta: V \to F$ defined by $\delta: v \mapsto \langle \alpha(v), w \rangle$ can easily seen to be a linear functional in $D(V)$. By Proposition 13.11, there exists a unique vector $y_w \in V$ satisfying $\delta(v) = \langle v, y_w \rangle$ for all $v \in V$. Let $\alpha^*: W \to V$ be the function $w \mapsto y_w$.

We claim that α^* is a linear transformation from W to V. Indeed, if $w_1, w_2 \in W$ then

$$\begin{aligned}
\langle v, \alpha^*(w_1 + w_2) \rangle &= \langle \alpha(v), w_1 + w_2 \rangle \\
&= \langle \alpha(v), w_1 \rangle + \langle \alpha(v), we_2 \rangle \\
&= \langle v, \alpha^*(w_1) \rangle + \langle v, \alpha^*(w_2) \rangle \\
&= \langle v, \alpha^*(w_1) + \alpha^*(w_2) \rangle
\end{aligned}$$

or all $v \in V$, so $\alpha^*(w_1 + w_2) = \alpha^*(w_1) + \alpha^*(w_2)$. If c is a scalar and if $w \in W$ then

$$\langle v, \alpha^*(cw) \rangle = \langle \alpha(v), cw \rangle = \bar{c} \langle \alpha(v), w \rangle = \bar{c} \langle v, \alpha^*(w) \rangle = \langle v, c\alpha^*(w) \rangle$$

or all $v \in V$ and so $\alpha^*(cw) = c\alpha(w)$. Therefore, as claimed, α^* is a linear transformation from W to V, which is uniquely defined since y_w is unique for each $v \in W$. $\square$

Let V and W be inner product spaces over the same field of scalars and let $\alpha: V \to W$ be a linear transformation. A linear transformation $\alpha^*: W \to V$ satisfying $\langle \alpha(v), w \rangle = \langle v, \alpha^*(w) \rangle$ for all $v \in V$ and all $w \in W$ is called an *adjoint* of α. By Proposition 13.12 we know that if both V and W are finitely generated then any $\alpha \in Hom(V, W)$ has a unique adjoint.

EXAMPLE

Sometimes linear transformations between inner product spaces which are not finitely generated over their field of scalars also have adjoints. For example, let V be the space of all polynomial functions in $\mathbb{R}^{\mathbb{R}}$, on which we define an inner product by $\langle f, g \rangle = \int_0^1 f(t)g(t)dt$. Fix $f_0 \in V$ and define the endomorphism α of V by $\alpha: g \mapsto f_0 g$. Then $\langle \alpha(g), h \rangle = \langle g, \alpha(h) \rangle$ for all $g, h \in V$ and so α is its own adjoint. On the other hand, if δ is the differentiation endomorphism $f \mapsto f'$ on V then δ has no adjoint.

(13.13) PROPOSITION. *Let V and W be inner product spaces finitely generated over the same field of scalars having orthonormal bases $B = \{v_1, \ldots, v_n\}$ and $D = \{w_1, \ldots, w_k\}$ respectively. Let $\alpha: V \to W$ be a linear transformation. Then $\Phi_{BD}(\alpha) = [a_{ij}]$, where $a_{ij} = \langle \alpha(v_i), w_j \rangle$, and $\Phi_{DB}(\alpha^*) = [\bar{a}_{ij}]^T$.*

PROOF. For all $1 \leq i \leq n$ let $\alpha(v_i) = \sum_{h=1}^k a_{ih} w_h$. For each $1 \leq j \leq k$ we then have

$$\langle \alpha(v_i), w_j \rangle = \left\langle \sum_{h=1}^k a_{ih} w_h, w_j \right\rangle = a_{ij}.$$

Moreover, $\langle \alpha^*(w_j), v_i \rangle = \overline{\langle v_i, \alpha^*(w_j) \rangle} = \overline{\langle \alpha(v_i), w_j \rangle} = \bar{a}_{ji}$ and we are done $\square$

We now consider some elementary properties of the adjoint.

(13.14) PROPOSITION. *Let V, W and Y be inner product spaces finitely generated over the same field of scalars. Let α and β linear transformations from V to W, let ζ be a linear transformation from W to Y, and let c be a scalar. Then:*

(1) $(\alpha + \beta)^* = \alpha^* + \beta^*$;
(2) $(c\alpha)^* = \bar{c}\alpha^*$;
(3) $(\zeta\alpha)^* = \alpha^*\zeta^*$;
(4) $\alpha^{**} = \alpha^*$.

PROOF. (1) If $v \in V$ and $w \in W$ then

$$\begin{aligned}
\langle v, (\alpha + \beta)^*(w) \rangle &= \langle (\alpha + \beta)(v), w \rangle \\
&= \langle \alpha(v) + \beta(v), w \rangle \\
&= \langle \alpha(v), w \rangle + \langle \beta(v), w \rangle \\
&= \langle v, \alpha^*(w) \rangle + \langle v, \beta^*(w) \rangle \\
&= \langle v, (\alpha^* + \beta^*)(w) \rangle
\end{aligned}$$

and so from the uniqueness of the adjoint we must have $(\alpha + \beta)^* = \alpha^* + \beta^*$.

(2) If $v \in V$ and $w \in W$ then

$$\begin{aligned}
\langle v, (c\alpha)^*(w) \rangle &= \langle ((c\alpha)(v), w \rangle \\
&= \langle \alpha(cv), w \rangle \\
&= \langle c\alpha(v), w \rangle \\
&= c\langle \alpha(v), w \rangle \\
&= c\langle v, \alpha^*(w) \rangle \\
&= \langle v, \bar{c}\alpha^*(w) \rangle \\
&= \langle v, (\bar{c}\alpha^*(w) \rangle
\end{aligned}$$

and so, by uniqueness, we must have $(c\alpha)^* = \bar{c}\alpha^*$.

(3) If $v \in V$ and $y \in Y$ then

$$\begin{aligned}
\langle v, (\zeta\alpha)^*(y) \rangle &= \langle ((\zeta\alpha)(v), y \rangle \\
&= \langle \zeta(\alpha(v)), y \rangle \\
&= \langle \alpha(v), \zeta^*(y) \rangle \\
&= \langle v, \alpha^*(\zeta^*(y)) \rangle \\
&= \langle (v, (\alpha^*\zeta^*)(y) \rangle
\end{aligned}$$

and so, by uniqueness, $(\zeta\alpha)^* = \alpha^*\zeta^*$.

(4) If $v, v' \in V$ then

$$\langle v, \alpha^{**}(v') \rangle = \langle \alpha^*(v), v' \rangle = \overline{\langle v', \alpha^*(v) \rangle} = \overline{\langle \alpha(v'), v \rangle} = \langle v, \alpha(v') \rangle$$

and so, by uniqueness, $\alpha = \alpha^{**}$. $\square$

In particular, if V is an inner product space finitely generated over its field of scalars then we can think of the function $\alpha \mapsto \alpha^*$ and a function from the ring $End(V)$ to itself satisfying

(1) $\sigma_1^* = \sigma_1$,
(2) $(\alpha + \beta)^* = \alpha^* + \beta^*$,
(3) $(\alpha\beta)^* = \beta^*\alpha^*$, and
(4) $\alpha^{**} = \alpha$.

Such a function is called an *involution*. Another ring with involution which we have already studied is the ring of matrices $\mathcal{M}_{n \times n}(F)$ over a field F together with the involution $A \mapsto A^T$. If F is the field of real numbers then the connection between these two rings is given in Proposition 13.13.

(13.15) PROPOSITION. *Let* $\alpha: V \to W$ *be a linear transformation between inner product spaces finitely generated over the same field of scalars.*

(1) *If* α *is a monomorphism then* $\alpha^*\alpha$ *is an automorphism of* V.
(2) *If* α *is an epimorphism then* $\alpha\alpha^*$ *is an automorphism of* W.

PROOF. (1) Since V is finitely generated over its field of scalars, it suffices to show that $\alpha^*\alpha$ is monic. Indeed, if $v \in ker(\alpha^*\alpha)$ then $\langle \alpha(v), \alpha(v) \rangle = \langle v, \alpha^*\alpha(v) \rangle = $ and so $\alpha(v) = 0_V$, proving that $v = 0_V$ since α is a monomorphism.

(2) First we will show that α^* is a monomorphism. Indeed, if $w, w' \in W$ satisfy $\alpha^*(w) = \alpha^*(w')$ then $\langle \alpha(v), w - w' \rangle = \langle v, \alpha^*(w - w') \rangle = 0$ for all $v \in V$ and since α is an epimorphism, this means that $\langle w'', w - w' \rangle = 0$ for all $w'' \in W$ and so $w - w' = 0_W$, i.e. $w = w'$.

Now we will show that $\alpha\alpha^*$ is also a monomorphism. Since W is finitely-generated over its field of scalars, this will suffice to prove (2). Indeed, if $w \in ker(\alpha\alpha^*)$ then $\langle \alpha^*(w), \alpha^*(w) \rangle = \langle \alpha\alpha^*(w), w \rangle = 0$ and so $\alpha^*(w) = 0_V$. This implies that $w = 0_W$ since α^* is monic. $\square$

(13.16) PROPOSITION. *If $\alpha: V \to W$ is an isomorphism between inner product spaces finitely generated over the same field of scalars then $(\alpha^*)^{-1} = (\alpha^{-1})^*$.*

PROOF. Let $\beta = (\alpha^{-1})^*$. Then for all $v, v' \in V$ we have

$$\langle v, v' \rangle = \langle \alpha^{-1}\alpha(v), v' \rangle = \langle \alpha(v), \beta(v') \rangle = \langle v, \alpha^*\beta(v') \rangle.$$

Therefore, by uniqueness, $\alpha^*\beta = \sigma_1$ and so $\beta = (\alpha^*)^{-1}$. $\square$

Let V be an inner product space. An endomorphism α of V is *self-adjoint* or *Hermitian* if and only if $\alpha = \alpha^*$. Self-adjoint endomorphisms of inner product spaces have important applications in mathematical physics, especially in the mathematical modeling of quantum mechanics. As a consequence of Proposition 13.14 and Proposition 13.16 we note that the set of all self-adjoint endomorphisms of V is closed under taking sums and inverses. If V is an inner product space over $\mathbb{R}$ then the set of all self-adjoint endomorphisms of V is in fact a subspace of $End(V)$.

EXAMPLE

Let V be an inner product space finitely generated over its field of scalars. If α is an arbitrary endomorphism of V then $\alpha + \alpha^*$ is self-adjoint.

Let V be an inner product space finitely generated over $\mathbb{C}$ having an orthonormal basis D. If $\alpha \in End(V)$ and $\Phi_{DD}(\alpha) = A = [a_{ij}]$ then we know by Proposition 13.13 that $\Phi_{DD}(\alpha^*) = \overline{A} = [\bar{a}_{ij}]^T$ and so if α is self-adjoint then for all $1 \leq i, j \leq n$ we have $\bar{a}_{ji} = a_{ij}$. In particular, we see that $\bar{a}_{ii} = a_{ii}$ for all $1 \leq i \leq n$ and so the elements of the main diagonal of A are real numbers. A matrix of this form is called a *Hermitian matrix*. The Hermitian matrices with real entries are precisely the symmetric matrices.

It is clear that if A and B are Hermitian matrices, so is $A + B$. However, AB need not be Hermitian.

EXAMPLE

The matrices $\begin{bmatrix} 0 & 1 \\ 1 & 0 \end{bmatrix}$ and $\begin{bmatrix} 1 & 0 \\ 0 & -1 \end{bmatrix}$ in $\mathcal{M}_{2\times 2}(\mathbb{C})$ are Hermitian but the matrix $AB = \begin{bmatrix} 0 & -1 \\ 1 & 0 \end{bmatrix}$ is not.

On the other hand, if A and B are Hermitian then $\frac{1}{2}(AB + BA)$ is Hermitian.

The characterization of the Hermitian matrices among the matrices in $\mathcal{M}_{n \times n}(\mathbb{C})$ as those which satisfy $A = \overline{A}$ is analogous to the characterization of the real numbers among the complex numbers as those which satisfy $z = \overline{z}$. This analogy can be carried further. Thus, if $A = [a_{ij}] \in \mathcal{M}_{n \times n}(\mathbb{C})$ set $B = \frac{1}{2}(A + \overline{A})$ and $C = \frac{-1}{2}i(A - \overline{A})$. Then it is easily verified that both B and C are Hermitian matrices and that $A = B + iC$. This is analogous to the representation of every complex number as $b + ic$ for $b, c \in \mathbb{R}$. We will pursue this analogy further in the next chapter as well.

Problems

1. Is the function $([a, b], [c, d]) \mapsto a(c + d) + b(c + 2d)$ an inner product on $\mathbb{R}^2$?

2. Let c be a positive real number and define a function from $\mathbb{R}^4 \times \mathbb{R}^4$ to $\mathbb{R}$ by

$$(v, w) \mapsto v \begin{bmatrix} 1 & 0 & 0 & 0 \\ 0 & 1 & 0 & 0 \\ 0 & 0 & 1 & 0 \\ 0 & 0 & 0 & -c \end{bmatrix} w^T.$$

Such a function, called a *Minkowski product* on $\mathbb{R}^4$, is very important in theoretical physics (where c is taken to be the speed of light). Is it an inner product on $\mathbb{R}^4$?

3. Let $a < b$ be real numbers and let V be the vector space over $\mathbb{R}$ consisting of all continuous functions from the interval $[a, b]$ to $\mathbb{R}$. Let h_0 be a function in V satisfying the condition that $h_0(t) > 0$ for all $a < t < b$. Show that the function from $V \times V$ to $\mathbb{R}$ defined by

$$(f, g) \mapsto \int_a^b f(t)g(t)h_0(t)dt$$

is an inner product on V.

4. Let c and d be given real numbers. Find a necessary and sufficient condition for the function $([a, b], [a', b']) \mapsto caa' + dbb'$ to be an inner product on $\mathbb{R}^2$.

5. Show that the function

$$([a, b, c], [a', b', c']) \mapsto 2aa' + ac' + ca' + bb' + cc'$$

is an inner product on $\mathbb{R}^3$ and find an orthonormal basis for $\mathbb{R}^3$ relative to this product.

6. Let n be a positive integer and let $A \in \mathcal{M}_{n \times n}(\mathbb{C})$. Show that $tr(A\overline{A}) > 0$.

7. Let V be an inner product space over $\mathbb{R}$ and let $n > 1$ an integer. If $a_1, \ldots a_n$ are given positive real numbers, we can define a function $V^n \times V^n \to \mathbb{R}$ by

$$([v_1, \ldots, v_n], [w_1, \ldots, w_n]) \mapsto \langle\!\langle [v_1, \ldots, v_n], [w_1, \ldots, w_n] \rangle\!\rangle = \sum_{i=1}^{n} a_i \langle v_i, w_i \rangle.$$

Is this an inner product on V^n?

8. Let n be a positive integer and let V be the subspace of $\mathbb{R}[X]$ consisting of all polynomials of degree at most n. Is the function from $V \times V$ to $\mathbb{R}$ defined by $(p, q) \mapsto \langle p, q \rangle = \sum_{i=0}^{n} p(\frac{i}{n})q(\frac{i}{n})$ an inner product on V?

9. Let V be a vector space over $\mathbb{R}$ and let $\phi \colon V \to \mathbb{R}$ a function satisfying

$$\phi(v + w)^2 + \phi(v - w)^2 = 2[\phi(v)^2 + \phi(w)^2]$$

for all $v, w \in V$. Show that there exists an inner product on V which defines a norm satisfying $\|v\| = \phi(v)$ for all $v \in V$.

10. Let n be a positive integer. Is the function from from $\mathbb{C}^n \times \mathbb{C}^n$ to $\mathbb{C}$ defined by

$$([a_1, \ldots, a_n], [b_1, \ldots, b_n]) \mapsto \sum_{i=1}^{n} a_i \bar{b}_{n-i+1}$$

an inner product on $\mathbb{C}^n$?

11. Let W be a subspace of a vector space V over $\mathbb{R}$ and let Y be a complement of W in V. Given inner products $(w, w') \mapsto \langle w, w' \rangle$ on W and $(y, y') \mapsto \langle\!\langle y, y' \rangle\!\rangle$ on Y, is the function from $V \times V$ to V defined by

$$(w + y, w' + y') \mapsto \langle w, w' \rangle + \langle\!\langle y, y' \rangle\!\rangle$$

an inner product?

12. Let V be an inner product space over $\mathbb{R}$ and let v and v' be nonzero vectors in V. Show that $|\langle v, w \rangle|^2 = \langle v, v \rangle \langle w, w \rangle$ if and only if the set $\{v, w\}$ is linearly dependent.

13. Let V be an inner product space and let $v \neq v'$ be vectors in V. Show that there exists a vector $w \in V$ satisfying $\langle v, w \rangle \neq \langle v', w \rangle$.

14. Let n be a positive integer and let $\theta \colon \mathcal{M}_{n \times n}(\mathbb{R}) \to \mathbb{R}^{n^2}$ be the isomorphism of vector spaces over $\mathbb{R}$ defined by

$$\theta \colon [a_{ij}] \mapsto [a_{11}, \ldots, a_{n1}, a_{12}, \ldots, a_{n2}, \ldots, a_{1n}, \ldots, a_{nn}].$$

If $A, B \in \mathcal{M}_{n \times n}(\mathbb{R})$, express $tr(AB)$ in terms of θ and the dot product.

15. Let n be a positive integer. For each $A \in \mathcal{M}_{n \times n}(\mathbb{R})$ we define the real number $\|A\|$ by

$$\|A\| = sup\{\|Av^T\|/\|v^T\| \mid [0, \ldots, 0] \neq v \in \mathbb{R}^n\},$$

where the norm in $\mathbb{R}^n$ is defined by the dot product. Show that the following conditions hold:
   (1) $\|A + B\| \leq \|A\| + \|B\|$ for all $A, B \in \mathcal{M}_{n \times n}(\mathbb{R})$;
   (2) $\|cA\| = c\|A\|$ for all $A \in \mathcal{M}_{n \times n}(\mathbb{R})$ and all $cv \in \mathbb{R}$;
   (3) $\|AB\| \leq \|A\| \cdot \|B\|$ for all $A, B \in \mathcal{M}_{n \times n}(\mathbb{R})$;
   (4) $\|Av^T\| \leq \|A\| \cdot \|v\|$ for all $A \in \mathcal{M}_{n \times n}(\mathbb{R})$ and all $v \in \mathbb{R}^n$.

16. Let V be an inner product space over $\mathbb{R}$ and let $v, w \in V$ be vectors satisfying $\|v + w\| = \|v\| + \|w\|$. Show that $\|av + bw\| = a\|v\| + b\|w\|$ for all $0 \le a, b \in \mathbb{R}$.

17. Let V be the space of all continuous functions from the interval $[0, 1]$ to $\mathbb{R}$ on which we have the inner product defined by $\langle f, g \rangle = \int_0^1 f(t)g(t)dt$. Calculate $\|\cos(t)\|$.

18. Consider the space $\mathbb{C}^2$ together with the dot product and let $G = \{v \in \mathbb{C}^2 \mid \|v\| = 1\}$. Calculate $\{\langle vA, v \rangle \mid v \in G\}$, where $A = \begin{bmatrix} 1 & 0 \\ 0 & 0 \end{bmatrix} \in \mathcal{M}_{2 \times 2}(\mathbb{C})$.

19. Let V be an inner product space over $\mathbb{R}$ and let $\theta: V \times V \times V \to \mathbb{R}$ be the function defined by

$$\theta(v, w, z) \mapsto \|v + w + z\|^2 + \|v + w - z\|^2 - \|v - w - z\|^2 - \|v - w + z\|^2.$$

Show that this function is independent of z.

20. Let V be an inner product space over $\mathbb{R}$ and let $n \ge 2$ be an integer. Let $\theta: V^n \to V$ be the function defined by $\theta: [v_1, \ldots, v_n] \mapsto \frac{1}{n} \sum_{i=1}^n v_i$. Show that $\sum_{i=1}^n \|v_i - \theta(v_1, \ldots, v_n)\| = \sum_{i=1}^n \|v_i\|^2 - n\theta(v_1, \ldots, v_n)$ for all $[v_1, \ldots, v_n] \in V^n$.

21. Let V be a vector space over $\mathbb{R}$ on which we have defined two distinct inner products which, in turn, define distance functions d and d'. Show that $d \ne d'$.

22. Calculate the angle between the vectors $[2, 1, 1]$ and $[3, 1, -2]$ in the space $\mathbb{R}^3$ with the dot product.

23. Given the inner product on $\mathcal{M}_{2 \times 2}(\mathbb{R})$ by

$$\left\langle \begin{bmatrix} a_{11} & a_{12} \\ a_{21} & a_{22} \end{bmatrix}, \begin{bmatrix} b_{11} & b_{12} \\ b_{21} & b_{22} \end{bmatrix} \right\rangle = a_{11}b_{11} + a_{12}b_{12} + a_{21}b_{21} + a_{22}b_{22},$$

find the angle between the matrices $\begin{bmatrix} 1 & 1 \\ 1 & 1 \end{bmatrix}$ and $\begin{bmatrix} 1 & 1 \\ -1 & 1 \end{bmatrix}$.

24. Using the results in this chapter, show that every continuous function f from the unit interval $[0, 1]$ to $\mathbb{R}$ satisfies $\left[\int_0^1 f(t)dt\right]^2 \le \int_0^1 f^2(t)dt$.

25. Let V be the space $\mathbb{R}^4$ together with the dot product. Find a normal vector in V which is orthogonal to each of the vectors $[1, 1, 1, 1]$, $[1, -1, -1, 1]$, and $[2, 1, 1, 3]$.

26. Is it possible to define an inner product on the space $\mathbb{R}^3$ relative to which there exists a nonzero vector orthogonal to each of the vectors $[2, 1, 3]$, $[0, -1, 1]$, and $[2, 0, 4]$?

27. Let V be the space $\mathbb{R}^4$ together with the dot product. Find vectors $v, w, y \in V$ satisfying $v \perp w$, $w \perp y$, but not $v \perp y$.

28. Let $A \in \mathcal{M}_{n \times n}(\mathbb{R})$ be a matrix the rows of which form an orthonormal set of vectors in $\mathbb{R}^n$ (with respect to the dot product). Show that the columns of A also form an orthonormal set.

29. Let $k < n$ be positive integers and let $A \in M_{k \times n}(\mathbb{R})$ be a matrix the rows of which form an orthonormal set of vectors in $\mathbb{R}^n$ (with respect to the dot product). Show that $(A^T A)^2 = A^T A$.

30. Let V be an inner product space over $\mathbb{R}$ and let u and v be elements of V. Show that $\|u\| = \|v\|$ if and only if $(u + v) \perp (u - v)$.

31. Let a, b, c, d be nonzero real numbers satisfying $a^2 + b^2 + c^2 = d^2$ and $ab + ac = bc$. Show that the subset

$$\left\{ [\frac{a}{d}, \frac{b}{d}, \frac{c}{d}], [\frac{b}{d}, -\frac{c}{d}, \frac{a}{d}], [\frac{c}{d}, \frac{a}{d}, -\frac{b}{d}] \right\}$$

of $\mathbb{R}^3$ is orthonormal with respect to the dot product.

32. Let m be a positive integer and let $W = \{f \in \mathbb{C}^{\mathbb{Z}} \mid f(i+m) = f(i)$ for all $i \in \mathbb{Z}\}$, which is a subspace of the vector space $\mathbb{C}^{\mathbb{Z}}$ over $\mathbb{C}$. Define a function from $W \times W$ to $\mathbb{C}$ by $(f, g) \mapsto \langle\!\langle f, g \rangle\!\rangle = \sum_{h=0}^{m-1} f(h)\overline{g(h)}$.

(1) Show that this function is an inner product on W;
(2) For each $0 \leq j < m$ let $f_j \in W$ be the function defined by

$$f_j(h) = \begin{cases} 1, & \text{if } h = j + mi \text{ for } i \in \mathbb{Z} \\ 0, & \text{otherwise.} \end{cases}$$

Show that $\{f_0, \ldots, f_{m-1}\}$ is an orthonormal basis for W with respect to this inner product.

33. Define an inner product on $\mathbb{R}^2$ relative to which the vectors $[-1, 2]$ and $2, 4]$ are orthogonal.

34. A function $f \in \mathbb{R}^{\mathbb{R}}$ has *bounded support* if and only if there exist real numbers $a \leq b$ such that $f(x) = 0$ for all $x \notin [a, b]$.

(1) Show that the set V of all functions having bounded support is a subspace of $\mathbb{R}^{\mathbb{R}}$ and that the function from $V \times V$ to $\mathbb{R}$ given by $(f, g) \mapsto \int_{-\infty}^{\infty} f(t)g(t)dt$ is an inner product on (V).
(2) For each $k \in \mathbb{Z}$ let $f_k \in V$ be the function defined by

$$f_k(x) = \begin{cases} 1, & \text{when } k \leq x \leq k+1 \\ 0, & \text{otherwise.} \end{cases}$$

is an orthonormal subset of V.

35. Let n be a positive integer and let V be the subspace of $\mathbb{R}[X]$ consisting of all polynomials having degree at most n. Define an inner product on V relative to which the set

$$\left\{ 1, X, \frac{X^2}{2!}, \ldots, \frac{X^n}{n!} \right\}$$

is orthonormal.

36. Make use of the Gram-Schmidt process to find an orthonormal basis for $\mathbb{R}^3$ on which we have the dot product, given the initial basis $\{[1, 1, 1], [1, -2, 1], [1, 2, 3]\}$.

37. Let W be the subspace of $\mathbb{R}^4$ with basis $\{[1, 2, 0, 3], [4, 0, 5, 8], [8, 1, 5, 6]\}$. Then W together with the dot product is an inner product space. Beginning with the given basis, use the Gram-Schmidt process to find an orthonormal basis for W.

38. Let V be the space $\mathbb{R}^4$ together with the dot product. Add two vectors to the set $\{[\frac{1}{6}, \frac{1}{6}, \frac{1}{2}, -\frac{5}{6}], [\frac{1}{2}, \frac{1}{2}, \frac{1}{2}, \frac{1}{2}]\}$ to get an orthonormal basis for V.

39. Find an orthonormal basis for the space $\mathbb{R}^2$ on which we have defined the inner product $([a, b], [c, d]) \mapsto ac + \frac{1}{2}(ad + bc) + bd$.

40. Find an orthonormal basis for the subspace W of $\mathbb{R}^4$ (with the dot product) generated by $\{[1, 0, 1, 0], [1, 1, 2, 1], [0, 1, 1, 2]\}$.

41. Let W be the subspace of $\mathbb{R}^4$ generated by

$$\{[2, 1, 0, 0], [0, 0, 1, 2], [4, 2, 2, 4], [1, 1, 1, 1]\}.$$

Given the dot product defined on $\mathbb{R}^4$, find an orthonormal basis for W and an orthonormal basis for $W^\perp$.

42. Let V be the space of all continuous functions from the interval $[-1, 1]$ to $\mathbb{R}$ on which we have an inner product defined by $\langle f, g \rangle = \int_{-1}^1 f(t)g(t)dt$. Let W be the subspace of V consisting of all even functions, i.e. those functions satisfying $f(-t) = f(t)$ for all $0 \le t \le 1$. What is $W^\perp$?

43. Let V be an inner product space having subspaces W and Y. Show that $(W + Y)^\perp = W^\perp \cap Y^\perp$.

44. Let V be the space of all continuous functions from the unit interval $[0, 1]$ to $\mathbb{R}$. Let $C = \{f_1, \ldots, f_n\}$ be a linearly independent subset of V and define the function $K: \mathbb{R} \times \mathbb{R} \to \mathbb{R}$ by $K: (x, y) \mapsto \sum_{j=1}^n f_j(x) \cos^j(x)$. Show that any $h \in V$ for which there exists a $g \in V$ satisfying $\int_0^1 K(x, y)g(y)dy = h(x)$ belongs to the subspace $\mathbb{R}C$ of V.

45. Let F be a set. Find an involution $A \mapsto A^\#$ of the ring $\mathcal{M}_{2 \times 2}(F)$ such that $tr(A)I = A + A^\#$ for all matrices A.

46. Let V be an inner product space having an endomorphism α. Is $\alpha^*\alpha - \sigma_1$ a self-adjoint endomorphism of V?

47. Let V be an inner product space over $\mathbb{C}$ and let α be an endomorphism of V
(1) If $\langle \alpha(v), v \rangle = 0$ for all $v \in V$, show that $\alpha = \sigma_0$.
(2) Use (1) to show that if $\langle \alpha(v), v \rangle \in \mathbb{R}$ for all $v \in V$ then α is self-adjoint.

48. Let V be an inner product space and let α be a self-adjoint endomorphism of V. Show that $ker(\alpha) = ker(\alpha^k)$ for all $k \ge 1$.

49. If A and B are Hermitian matrices different from $\mathbf{O}$, is $AB \ne \mathbf{O}$? Prove or give a counterexample.

50. Let n be a positive integer. Show that a matrix $A \in \mathcal{M}_{n \times n}(\mathbb{C})$ is Hermitian if and only if the matrix $B = iA$ satisfies $B = -\overline{B}$.

51. Let $V = \mathcal{M}_{n \times n}(\mathbb{C})$ and let W be the set of all Hermitian matrices in V Is W a subspace of V over $\mathbb{R}$? Is it a subspace of V as over $\mathbb{C}$?

52. If $A \neq O$ is a Hermitian matrix in $\mathcal{M}_{3 \times 3}(\mathbb{C})$, show that $A^n \neq O$ for all positive integers n.

53. Find complex numbers a and b such that the matrix $\begin{bmatrix} a & 0 & b \\ 0 & 2a & a \\ i & 1 & a \end{bmatrix}$ is Hermitian.

CHAPTER XIV

ENDOMORPHISMS OF INNER PRODUCT SPACES

Let V be an inner product space. If α is a self-adjoint endomorphism of V and $v \in V$ then $\langle \alpha(v), v \rangle = \langle v, \alpha(v) \rangle = \overline{\langle \alpha(v), v \rangle}$ and so $\langle \alpha(v), v \rangle \in \mathbb{R}$. An endomorphism α of V is *positive* if and only if it is self-adjoint and satisfies the condition that $0 < \langle \alpha(v), v \rangle \in \mathbb{R}$ for all $0_V \neq v \in V$. Thus, for example, $\sigma_1 \in End(V)$ is positive. We note immediately that if $\alpha \in End(V)$ is positive then α must be a monomorphism since $\alpha(v) = 0_V$ implies that $\langle \alpha(v), v \rangle = 0$ and so we must have $v = 0_V$.

EXAMPLE

If $\sigma_0 \neq \alpha \in End(V)$ is not positive then it does not follow that $-\alpha$ has to be positive. For example, let $V = \mathbb{R}^2$ on which the inner product is the dot product and let α be the endomorphism of V defined by $\alpha \colon [a, b] \mapsto [a, -b]$. If $v = [1, 0]$ and $v' = [0, 1]$ then $\langle \alpha(v), v \rangle = \langle -\alpha(v'), v' \rangle = 1$, while $\langle \alpha(v'), v' \rangle = \langle -\alpha(v), v \rangle = -1$ so neither α nor $-\alpha$ is positive.

(14.1) PROPOSITION. *Let V be an inner product space on which we have de-fine an endomorphism α. Then α is positive if and only if the function $(v, v') \mapsto \langle\!\langle v, v' \rangle\!\rangle = \langle \alpha(v), v' \rangle$ from $V \times V$ to the field of scalars is an inner product on V.*

PROOF. Let us first assume that α is a positive endomorphism of V. If $v, v', v'' \in V$ then

$$\begin{aligned}
\langle\!\langle v + v', v'' \rangle\!\rangle &= \langle \alpha(v + v'), v'' \rangle \\
&= \langle \alpha(v) + \alpha(v'), v'' \rangle \\
&= \langle \alpha(v), v'' \rangle + \langle \alpha(v'), v'' \rangle \\
&= \langle\!\langle v, v'' \rangle\!\rangle + \langle\!\langle v', v'' \rangle\!\rangle
\end{aligned}$$

and in a similar manner we see that $\langle\!\langle cv, v'' \rangle\!\rangle = c\langle\!\langle v, v'' \rangle\!\rangle$ for all scalars c. Next, we note that

$$\langle\!\langle v, v' \rangle\!\rangle = \langle \alpha(v), v' \rangle = \langle v, \alpha^*(v') \rangle = \langle v, \alpha(v') \rangle = \overline{\langle \alpha(v'), v \rangle} = \overline{\langle\!\langle v', v \rangle\!\rangle}.$$

Finally, we note that $\langle\!\langle v, v \rangle\!\rangle = \langle \alpha(v), v \rangle > 0$ if $v \neq 0_V$ and certainly $\langle\!\langle 0_V, 0_V \rangle\!\rangle = 0$. Thus we see that the function $(v, v') \mapsto \langle\!\langle v, v' \rangle\!\rangle$ is indeed an inner product on V.

Conversely, suppose that the function $(v, v') \mapsto \langle\!\langle v, v' \rangle\!\rangle$ is an inner product on V. Then for all $v, v' \in V$ we have $\langle v, \alpha^*(v') \rangle = \langle \alpha(v), v' \rangle = \langle\!\langle v, v' \rangle\!\rangle = \overline{\langle\!\langle v', v \rangle\!\rangle} =$

$\overline{\langle v, \alpha(v')\rangle} = \langle v, \alpha(v')\rangle$ and so $\alpha^*(v') = \alpha(v')$ for all $v' \in V$, i.e. $\alpha = \alpha^*$. Thus α is self-adjoint. Moreover, from the third condition for being an inner product we see that $\langle \alpha(v), v \rangle = \langle\!\langle v, v \rangle\!\rangle > 0$ for all $0_V \neq v \in V$. $\square$

We now show that if the inner product space V is finitely generated over its field of scalars then all possible inner products definable on V arise in the manner of Proposition 14.1.

(14.2) PROPOSITION. *Let V be an inner product space finitely generated over its field of scalars and let $(v, v') \mapsto \langle\!\langle v, v' \rangle\!\rangle$ be another inner product defined on V. Then there exists a unique positive endomorphism α of V satisfying $\langle\!\langle v, v' \rangle\!\rangle = \langle \alpha(v), v' \rangle$ for all $v, v' \in V$.*

PROOF. Consider a vector $v \in V$. Then the function $v' \mapsto \langle\!\langle v', v \rangle\!\rangle$ is a linear functional on V and so there exists a unique element v'' of V satisfying $\langle\!\langle v', v \rangle\!\rangle = \langle v', v'' \rangle$ for all $v' \in V$. Define the function $\alpha: V \to V$ by $\alpha: v \mapsto v''$. Then

$$\langle \alpha(v), v' \rangle = \overline{\langle v', \alpha(v) \rangle} = \overline{\langle\!\langle v', v \rangle\!\rangle} = \langle\!\langle v, v' \rangle\!\rangle$$

for all $v, v' \in V$. We claim that α is an endomorphism of V. Indeed, if $v_1, v_2 \in V$ then for each $v' \in V$ we have

$$\begin{aligned}
\langle \alpha(v_1 + v_2), v' \rangle &= \langle\!\langle v_1 + v_2, v' \rangle\!\rangle \\
&= \langle\!\langle v_1, v' \rangle\!\rangle + \langle\!\langle v_2, v' \rangle\!\rangle \\
&= \langle \alpha(v_1), v' \rangle + \langle \alpha(v_2), v' \rangle \\
&= \langle \alpha(v_1) + \alpha(v_2), v' \rangle
\end{aligned}$$

and so $\alpha(v_1 + v_2) = \alpha(v_1) + \alpha(v_2)$. Similarly we show that $\alpha(cv_1) = c\alpha(v_1)$ for all $v_1 \in V$ and for all scalars c. Thus we have established that α is indeed an endomorphism of V satisfying the condition that $\langle\!\langle v, v' \rangle\!\rangle = \langle \alpha(v), v' \rangle$ for all $v, v' \in V$ and so this endomorphism must also be positive.

Finally, α must be unique since if $\langle\!\langle v, v' \rangle\!\rangle = \langle \beta(v), v' \rangle$ for all $v, v' \in V$ then $\langle(\alpha - \beta)(v), v' \rangle = \langle \alpha(v) - \beta(v), v' \rangle = 0$ for all $v, v' \in V$. This implies that $(\alpha - \beta)(v)$ is orthogonal to every element of V and so it must be 0_V. Thus $\alpha(v) = \beta(v)$ for all $v \in V$ and so $\alpha = \beta$. $\square$

Next, we give another characterization of positive endomorphisms. Note the analogy between this characterization and the fact that a complex number is real and positive if and only if it is of the form $\bar{z}z$ for some $0 \neq z \in \mathbb{C}$.

(14.3) PROPOSITION. *Let V be an inner product space finitely generated over its field of scalars and let α be an endomorphism of V. Then α is positive if and only if there exists an automorphism β of V satisfying $\alpha = \beta^*\beta$.*

PROOF. Assume there exists an automorphism β of V satisfying $\alpha = \beta^*\beta$. Then, in particular, $\alpha^* = (\beta^*\beta)^* = \beta^*\beta^{**} = \beta^*\beta = \alpha$ and so α is self-adjoint. Moreover, for each $0_V \neq v \in V$ we have $\langle \alpha(v), v \rangle = \langle \beta^*\beta(v), v \rangle = \langle \beta(v), \beta^{**}(v) \rangle = \langle \beta(v), \beta(v) \rangle > 0$. Since the fact that β is an automorphism implies that $\beta(v) \neq 0$ whenever $v \neq 0_V$. Thus α is positive.

Conversely, assume that α is a positive endomorphism of V. Then the function $(v, v') \mapsto \langle\!\langle v, v' \rangle\!\rangle = \langle \alpha(v), v' \rangle$ is an inner product on V. Let $\{v_1, \ldots, v_n\}$ be a basis for V which is orthonormal with respect to the given inner product on V and let $\{w_1, \ldots, w_n\}$ be a basis for V which is orthonormal with respect to this new inner product. By Proposition 4.1 there exists an endomorphism β of V uniquely defined by the condition $\beta(v_i) = w_i$ for all $1 \leq i \leq n$. Clearly β is epic and so it must be an automorphism of V, since V is finitely generated over its field of scalars. Moreover, if $v = \sum_{i=1}^{n} a_i w_i$ and $v' = \sum_{j=1}^{n} b_j w_j$ are vectors in V then

$$\langle \alpha(v), v' \rangle = \langle\!\langle v, v' \rangle\!\rangle$$
$$= \langle\!\langle \sum_{i=1}^{n} a_i w_i, \sum_{j=1}^{n} b_j w_j \rangle\!\rangle$$
$$= \sum_{j=1}^{n} \sum_{i=1}^{n} a_i \bar{b}_j \langle\!\langle w_i, w_j \rangle\!\rangle$$
$$= \sum_{i=1}^{n} a_i \bar{b}_i$$

and

$$\langle \beta^* \beta(v), v' \rangle = \overline{\langle v', \beta^* \beta(v) \rangle}$$
$$= \overline{\langle \beta(v'), \beta(v) \rangle}$$
$$= \langle \beta(v), \beta(v') \rangle$$
$$= \left\langle \beta(\sum_{i=1}^{n} a_i w_i), \beta(\sum_{j=1}^{n} b_j w_j) \right\rangle$$
$$= \sum_{i=1}^{n} \sum_{j=1}^{n} a_i \bar{b}_j \langle v_i, v_j \rangle$$
$$= \sum_{i=1}^{n} a_i \bar{b}_j$$

and thus we see that $\langle \beta^* \beta(v), v' \rangle = \langle \alpha(v), v' \rangle$ for all $v, v' \in V$, which proves that $\alpha = \beta^* \beta$, as desired. $\square$

We now look at the particular case in which the field of scalars is $\mathbb{C}$.

(14.4) PROPOSITION. *Let V be an inner product space finitely generated over $\mathbb{C}$ and let α be an endomorphism of V. If $\langle \alpha(v), v \rangle \in \mathbb{R}$ for all $v \in V$ then α is self-adjoint.*

PROOF. If $v, v' \in V$ then $\langle \alpha(v + v'), v + v' \rangle = \langle \alpha(v), v \rangle + \langle \alpha(v), v' \rangle + \langle \alpha(v'), v \rangle + \langle \alpha(v'), v' \rangle$ and, by hypothesis, each of the numbers $\langle \alpha(v + v'), v + v' \rangle$, $\langle \alpha(v), v \rangle$, and $\langle \alpha(v'), v' \rangle$ are real. From this it follows that $\langle \alpha(v), v' \rangle + \langle \alpha(v'), v \rangle \in \mathbb{R}$. We also have $\langle \alpha(v + iv'), v + iv' \rangle = \langle \alpha(v), v \rangle - i\langle \alpha(v), v' \rangle + i\langle \alpha(v'), v \rangle + \langle \alpha(v'), v' \rangle$ and again, by the same reasoning, $-i\langle \alpha(v), v' \rangle + i\langle \alpha(v'), v \rangle \in \mathbb{R}$. In particular, this number equals its own conjugate. Thus $\langle\alpha(v),v'\rangle+\langle\alpha(v'),v\rangle = \langle v',\alpha(v)\rangle+\langle v,\alpha(v')\rangle$ and $-i\langle\alpha(v),v'\rangle + i\langle\alpha(v'),v\rangle = i\langle v',\alpha(v)\rangle - i\langle v,\alpha(v')\rangle$. Multiply the second of these equations by i and add it to the first to get $2\langle\alpha(v),v'\rangle = 2\langle v,\alpha(v')\rangle$ and so $\langle\alpha(v),v'\rangle = \langle v,\alpha(v')\rangle$. We can therefore conclude that $\alpha = \alpha^*$. $\square$

Let V and W be inner product spaces over the same field of scalars. A linear transformation $\alpha: V \to W$ is an *isometry* if and only if $\langle v,v'\rangle = \langle\alpha(v),\alpha(v')\rangle$ for all $v, v' \in V$. Note that any isometry also preserves distances: if $v, v' \in V$ then $\|v - v'\| = \|\alpha(v - v')\| = \|\alpha(v) - \alpha(v')\|$.

EXAMPLE

Let $V = \mathbb{R}^2$, with inner product being the dot product. Then the only endomorphisms of V which are isometries are represented with respect to a given basis by matrices of the form $\begin{bmatrix} cos(t) & sin(t) \\ sin(t) & -cos(t) \end{bmatrix}$ or $\begin{bmatrix} cos(t) & -sin(t) \\ sin(t) & cos(t) \end{bmatrix}$ for some real number t.

EXAMPLE

Let V be an inner product space over the field $\mathbb{R}$ and let $0_V \neq y \in V$. The vector y defines an endomorphism α_y of V defined by $\alpha_y: v \mapsto -v + 2\langle y,y\rangle^{-1}\langle v,y\rangle y$. This endomorphism is an isometry and satisfies the condition that $\alpha_y^2 = \sigma_1$ and $\alpha_y(y) = Y$.

(14.5) PROPOSITION. *Let V and W be inner product spaces finitely generated over the same field of scalars and having equal dimensions. Then the following conditions on a linear transformation $\alpha: V \to W$ are equivalent:*

(1) α *is an isometry;*
(2) α *is an isomorphism which is an isometry;*
(3) *If $\{v_1,\ldots,v_n\}$ is an orthonormal basis of V then $\{\alpha(v_1),\ldots,\alpha(v_n)\}$ is an orthonormal basis of W.*

PROOF. (1) $\Rightarrow$ (2): If $0_V \neq v \in V$ then $\langle v,v\rangle = \langle\alpha(v),\alpha(v)\rangle$ and so $\alpha(v) \neq 0_W$. Thus we see that $ker(\alpha) = \{0_V\}$ and hence α is an isomorphism, since $dim(V) = dim(W)$.

(2) $\Rightarrow$ (3): If $\{v_1,\ldots,v_n\}$ is an orthonormal basis of V. Since α is an isomorphism it follows that $\{\alpha(v_1),\ldots,\alpha(v_n)\}$ is a basis for W. Moreover, for all $1 \leq i,j \leq n$ we have

$$\langle\alpha(v_i),\alpha(v_j)\rangle = \langle v_i,v_j\rangle = \begin{cases} 1, & \text{when } i = j \\ 0, & \text{otherwise.} \end{cases}$$

Therefore the basis of W is itself orthonormal.

(3) $\Rightarrow$ (1): Let $\{v_1,\ldots,v_n\}$ be an orthonormal basis of V. If $v = \sum_{i=1}^{n} a_i v_i$ and

$v' = \sum_{j=1}^{n} b_j v_j$ are vectors in V then $\langle v, v' \rangle = \sum_{i=1}^{n} a_i \bar{b}_i$. Moreover,

$$\langle \alpha(v), \alpha(v') \rangle = \langle \alpha(\sum_{i=1}^{n} a_i v_i), \alpha(\sum_{j=1}^{n} b_j v_j) \rangle$$

$$= \sum_{i=1}^{n} \sum_{j=1}^{n} a_i \bar{b}_j \langle \alpha(v_i), \alpha(v_j) \rangle$$

$$= \sum_{i=1}^{n} a_i \bar{b}_i$$

$$= \langle v, v' \rangle$$

and this proves (1). $\square$

In particular, if V and W are inner product spaces finitely generated over the same field of scalars and having the same dimension then any isometry $\alpha: V \to W$ is an isomorphism and if $w, w' \in W$ then

$$\langle w, w' \rangle = \langle \alpha \alpha^{-1}(w), \alpha \alpha^{-1}(w') \rangle = \langle \alpha^{-1}(w), \alpha^{-1}(w') \rangle$$

so α^{-1} is also an isometry.

EXAMPLE

Let $W = \{A \in M_{3 \times 3}(\mathbb{R}) \mid A^T = -A\}$. This is a subspace of $M_{3 \times 3}(\mathbb{R})$ having dimension 3. Define an inner product on W as follows: if $A, B \in W$ then $\langle A, B \rangle = \frac{1}{2} tr(AB^T)$. Let $V = \mathbb{R}^3$ on which we have the dot product defined and consider the linear transformation $\alpha: V \to W$ given by

$$\alpha: [a, b, c] \to \begin{bmatrix} 0 & -c & b \\ c & 0 & -a \\ -b & a & 0 \end{bmatrix}.$$

If $A = \begin{bmatrix} 0 & -c & b \\ c & 0 & -a \\ -b & a & 0 \end{bmatrix}$ and $B = \begin{bmatrix} 0 & -f & e \\ f & 0 & -d \\ -e & d & 0 \end{bmatrix}$ then

$$AB^T = \begin{bmatrix} cf + be & -bd & -cd \\ -ae & cf + ad & ce \\ -af & bf & be + ad \end{bmatrix}$$

so $\langle A, B \rangle = [a, b, c] \cdot [d, e, f]$. Thus α is an isometry and hence must be an isomorphism.

EXAMPLE

Proposition 14.5 is not necessarily true for inner product spaces which are not finitely generated. For example, let V be the space of all continuous functions from the interval $[0, 1]$ to $\mathbb{R}$, on which we have the inner product

$$\langle\!\langle f, g \rangle\!\rangle = \int_0^1 f(t)g(t)t^2\, dt$$

and let W be the same space on which we have the inner product

$$\langle f, g \rangle = \int_0^1 f(t)g(t)\, dt.$$

Let $\alpha: V \to W$ be the linear transformation defined by $\alpha: f(x) \mapsto f(x)x$. Then $\langle\!\langle f, g \rangle\!\rangle = \langle \alpha(f), \alpha(g) \rangle$ and so α is an isometry. However, α is not an isomorphism since it is not epic: for example, $x^2 + 1 \notin im(\alpha)$.

Let V be an inner product space. A *unitary automorphism* of V is an automorphism which is an isometry. If V is finitely generated over its field of scalars then by Proposition 14.5 we know that an endomorphism of V is a unitary automorphism if and only if it is an isometry. Moreover, an automorphism of V is unitary in this case if and only if it sends orthonormal bases to orthonormal bases.

(14.6) PROPOSITION. *Let V be an inner product space finitely generated over its field of scalars and let α be an automorphism of V. Then α is unitary if and only if $\alpha^* = \alpha^{-1}$.*

PROOF. If α is a unitary automorphism then for each $v, v' \in V$ we have

$$\langle \alpha(v), v' \rangle = \langle \alpha(v), \alpha\alpha^{-1}(v') \rangle = \langle v, \alpha^{-1}(v') \rangle$$

and so $\alpha^* = \alpha^{-1}$. Conversely, if $\alpha^* = \alpha^{-1}$ and if $v, v' \in V$ then $\langle v, v' \rangle = \langle v, \alpha^*\alpha(v') \rangle = \langle \alpha(v), \alpha(v') \rangle$ and so α is unitary. $\square$

By Proposition 14.6 we know that the set of all unitary automorphisms of an inner product space V is closed under taking products and inverses. Thus it forms a group of automorphisms of V.

Let V be an inner product space of finite dimension n over the field $\mathbb{C}$ and let α be a unitary automorphism of V represented with respect to a given orthonormal basis of V by the matrix $A \in \mathcal{M}_{n \times n}(\mathbb{C})$. Then $A^{-1} = \overline{A}$.

A matrix A in $\mathcal{M}_{n \times n}(\mathbb{C})$ is a *unitary matrix* if and only if $A^{-1} = \overline{A}$. Note that $(A^T)^{-1} = (A^{-1})^T = \overline{A}^T = \overline{A^T}$ and so a matrix A is unitary if and only if A^T is unitary. We also note that if $A = [a_{ij}] \in \mathcal{M}_{n \times n}(\mathbb{C})$ then $A\overline{A} = B$, where $B = [b_{ij}]$ is the matrix defined by $b_{ij} = \sum_{h=1}^n a_{ih}\bar{a}_{jh}$ so that $B = I$ precisely when the set of rows is an orthonormal subset of $\mathbb{C}^n$ (with respect to the dot product) which must perforce be an orthonormal basis of $\mathbb{C}^n$.

If we work instead over the field $\mathbb{R}$ then the condition becomes A satisfies $A^{-1} = A^T$. A matrix in $\mathcal{M}_{n \times n}(\mathbb{R})$ of this sort is called an *orthogonal matrix*. As before, a matrix in $\mathcal{M}_{n \times n}(\mathbb{R})$ is orthogonal if and only if the set of its rows is an orthonormal basis of $\mathbb{R}^n$ with respect to the dot product.

EXAMPLE

A matrix of the form $\begin{bmatrix} cos(t) & -sin(t) \\ sin(t) & cos(t) \end{bmatrix}$ is orthogonal for each real number t.

EXAMPLE

Let n be a positive integer and consider $\mathbb{R}^n$ on which we have the dot product defined. If $v \in \mathbb{R}^n$ is normal then the matrix $A = I - 2v^T v$ is called a *Householder matrix*. Such matrices are always orthogonal. Indeed, if $A = I - 2v^T v$ then clearly A is symmetric and $A^T A = A^2 = (I - 2v^T v)^2 = I - 4v^T v + 4v^T(vv^T)v = I$. Householder matrices have important applications in numerical analysis. Since a Householder matrix is totally determined by the given normal vector v, it is also easy to store in the computer.

(14.7) PROPOSITION. *Let V be an inner product space finitely generated over $\mathbb{R}$ and let α be a unitary automorphism of V, represented with respect to a given orthonormal basis of V by the matrix $A \in \mathcal{M}_{n \times n}(\mathbb{R})$. Then $|A| = \pm 1$.*

PROOF. We know that α^* is represented by the matrix $\overline{A}$ and, since we are working over the real numbers, this equals A^T. By Proposition 14.6 we have $AA^T = I$ and so $|A|^2 = |A||A^T| = |AA^T| = |I| = 1$. Thus $|A| = \pm 1$. $\square$

We now return to a new version of a question which we asked previously: given an endomorphism α of an inner product space V, when does there exist a basis of V composed of eigenvectors of α?

(14.8) PROPOSITION. *Let V be an inner product space and let α be a self-adjoint endomorphism of V. Then:*

(1) *Every eigenvalue of α is real; and*
(2) *Eigenvectors of α associated with distinct eigenvalues are orthogonal.*

PROOF. (1) Let c be an eigenvalue of α. Then there exists a vector $0_V \neq v \in V$ satisfying $\alpha(v) = cv$. In particular,

$$c\langle v, v \rangle = \langle cv, v \rangle = \langle \alpha(v), v \rangle = \langle v, \alpha(v) \rangle = \langle v, cv \rangle = \bar{c}\langle v, v \rangle$$

and so $c = \bar{c}$ since $\langle v, v \rangle \neq 0$. Therefore $c \in \mathbb{R}$.
(2) If $\alpha(v) = cv$ and $\alpha(v') = dv'$ for $c \neq d$ then

$$c\langle v, v' \rangle = \langle \alpha(v), v' \rangle = \langle v, \alpha(v') \rangle = \langle v, dv' \rangle = d\langle v, v' \rangle$$

and so we must have $\langle v, v' \rangle = 0$. $\square$

(14.9) PROPOSITION. *Let V be a nontrivial inner product space finitely generated over its field of scalars. Then every self-adjoint endomorphism of V has an
eigenvector.*

PROOF. Let $\{v_1, \ldots, v_n\}$ be an orthonormal basis for V and let $A = [a_{ij}]$ be
the matrix representing α with respect to this basis. Since α is self-adjoint, we see
that $A = \overline{A}$. Let $W = \mathbb{C}^n$ on which we have defined the dot product. The function
$\beta: w \mapsto wA$ is a self-adjoint endomorphism of W. The degree of the characteristic
polynomial $|XI - A|$ equals $n > 0$ and so by the Fundamental Theorem of Algebra
we know that there exists a complex number c satisfying $|cI - A| = 0$. Therefore
the matrix $cI - A$ is singular and there exists a vector $0_W \neq w \in W$ satisfying
$wA = cw$. Since β is self-adjoint, it follows from Proposition 14.8 that in fact $c \in \mathbb{R}$
and so there exists a vector $0_V \neq v \in V$ satisfying $\alpha(v) = cv$, even if V is defined
over the field $\mathbb{R}$. $\square$

(14.10) PROPOSITION. *Let V be an inner product space finitely generated over
its field of scalars and let α be an endomorphism of V. If W is a subspace of V
stable under α then $W^\perp$ is stable under α^*.*

PROOF. Let $y \in W^\perp$. If $w \in W$ then $\alpha(w) \in W$ and so $\langle w, \alpha^*(y) \rangle = \langle \alpha(w), y \rangle =
0$. Thus $\alpha^*(y) \in W^\perp$, proving that $W^\perp$ is stable under α^*. $\square$

We now answer the question we asked above.

(14.11) PROPOSITION. *Let V be a nontrivial inner product space finitely generated over its field of scalars and let α be a self-adjoint endomorphism of V. Then
there exists an orthonormal basis for V composed of eigenvectors of α.*

PROOF. We will proceed by induction on $n = dim(V)$. If $n = 1$ then by Propo-
sition 14.9 we know that α has an eigenvector $v \in V$. Then $v_1 = \|v\|^{-1}v$ is also an
eigenvector of α and $\{v_1\}$ is the desired basis for V.

Now assume that $n > 1$ and that the result has been established for all spaces
having dimension less than n. Choose v_1 as above and let W be the subspace of V
generated by $\{v_1\}$. Then $V = W \oplus W^\perp$ and by Proposition 14.10 we know that $W^\perp$
is stable under $\alpha^* = \alpha$. Moreover, $W^\perp$ is an inner product space of dimension $n-1$
and α restricts to a self-adjoint endomorphism of $W^\perp$. Therefore, by the induction
hypothesis, $W^\perp$ has an orthonormal basis $\{v_2, \ldots, v_n\}$ composed of eigenvectors
of α. But v_1 is orthogonal to each of the v_i ($2 \leq i \leq n$) since it belongs to W.
Therefore $\{v_1, \ldots, v_n\}$ is the desired basis for V. $\square$

Let V be an inner product space and let α be an endomorphism of V. This
endomorphism is *normal* if and only if $\alpha\alpha^* = \alpha^*\alpha$. As an immediate consequence
of this definition we see that α is normal if and only if α^* is normal. Self-adjoint
endomorphisms of V are surely normal.

(14.12) PROPOSITION. *Let V be an inner product space finitely generated over
its field of scalars and let α be a normal endomorphism of V. Then any eigenvector
of α is also an eigenvector of α^*.*

PROOF. If $v \in V$ then $\|\alpha(v)\|^2 = \langle \alpha(v), \alpha(v) \rangle = \langle v, \alpha^*\alpha(v) \rangle = \langle v, \alpha\alpha^*(v) \rangle =$

$\langle \alpha^*(v), \alpha^*(v) \rangle = \|\alpha^*(v)\|^2$ and so $\|\alpha(v)\| = \|\alpha^*(v)\|$. If c is a scalar then

$$(\alpha - c\sigma_1)^*(\alpha - c\sigma_1) = (\alpha^* - \bar{c}\sigma_1)(\alpha - c\sigma_1)$$
$$= (\alpha - \bar{c}\sigma_1)(\alpha^* - c\sigma_1)$$
$$= (\alpha - c\sigma_1)(\alpha - c\sigma_1)^*$$

and hence $\alpha - c\sigma_1$ is also normal. Thus we see that every vector $v \in V$ satisfies $\|(\alpha - c\sigma_1)(v)\| = \|(\alpha^* - \bar{c}\sigma_1)(v)\|$ and so, in particular, $(\alpha - c\sigma_1)(v) = 0_V$ if and only if $(\alpha^* - \bar{c}\sigma_1)(v) = 0_V$, i.e. v is an eigenvector of α associated with the eigenvalue c if and only if it is an eigenvector of α^* associated with the eigenvalue $\bar{c}$. $\square$

(14.13) PROPOSITION. *Let V be a nontrivial inner product space over $\mathbb{C}$ and let α be a normal endomorphism of V. Then there exists an orthonormal basis for V composed of eigenvectors of α.*

PROOF. Since our field of scalars is $\mathbb{C}$ we know that the characteristic polynomial of α has a root in $\mathbb{C}$ and so that α has an eigenvalue and hence an eigenvector v. Set $v_1 = \|v\|^{-1}v$ and let $W = \mathbb{C}v_1$. Then W is a subspace of V which is stable under α. By Proposition 14.12 we know that W is stable under α^* as well. Therefore $W^\perp$ is stable under $\alpha^{**} = \alpha$. The restriction of α to $W^\perp$ is again a normal endomorphism the adjoint of which is the restriction of α^* to $W^\perp$. We now proceed as in the proof of Proposition 14.11 to get the desired result. $\square$

The next result shows that for a very important class of endomorphisms, the notions of "self-adjoint" and "normal" coincide.

(14.14) PROPOSITION. *Let V be an inner product space finitely generated over its field of scalars. The following conditions on an idempotent endomorphism α of V are equivalent:*

(1) *α is normal;*
(2) *α is self-adjoint;*
(3) *$ker(\alpha) = im(\alpha)^\perp$.*

PROOF. (1) $\Rightarrow$ (2): By (1) we know that $\alpha\alpha^* = \alpha^*\alpha$ and so for all $v \in V$ we have $\|\alpha(v)\|^2 = \langle \alpha(v), \alpha(v) \rangle = \langle v, \alpha\alpha^*(v) \rangle = \langle \alpha^*(v), \alpha^*(v) \rangle = \|\alpha^*(v)\|^2$ whence $\|\alpha(c)\| = \|\alpha^*(v)\|$. In particular, $\alpha(v) = 0_V$ if and only if $\alpha^*(v) = 0_V$ so that $ker(\alpha) = ker(\alpha^*)$. If $v \in V$ and if $v' = v - \alpha(v)$ then $\alpha(v') = \alpha(v) - \alpha^2(v) = \alpha(v) - \alpha(v) = 0_V$ and so $\alpha^*(v') = 0_V$. Therefore $\alpha^*(v) = \alpha^*\alpha(v)$. Indeed, this is true for each $v \in V$ and so $\alpha^* = \alpha^*\alpha$ and $\alpha = \alpha^{**} = (\alpha^*\alpha)^* = \alpha^*\alpha^{**} = \alpha^*\alpha = \alpha^*$, proving (2).

(2) $\Rightarrow$ (3): If $v, v' \in V$ then (2) implies that $\langle \alpha(v), v' \rangle = \langle v, \alpha(v') \rangle$. In particular, if $v \in ker(\alpha)$ then $\langle v, \alpha(v') \rangle = 0$ for all $v' \in V$. That is to say, $v \in im(\alpha)^\perp$. Conversely, suppose that $v \in im(\alpha)^\perp$. Then $\langle \alpha(v), v' \rangle = 0$ for all $v' \in V$. That is to say, $\alpha(v)$ is orthogonal to every element of V, which implies that $\alpha(v) = 0_V$, i.e. $v \in ker(\alpha)$. Thus we have (3).

(3) $\Rightarrow$ (1): Let $v, v' \in V$. Since α is idempotent we have $v - \alpha(v) \in ker(\alpha)$ and $\alpha(v') \in im(\alpha)$. Therefore, by (3),

$$0 = \langle v - \alpha(v), \alpha(v') \rangle = \langle v, \alpha(v') \rangle - \langle \alpha(v), \alpha(v') \rangle = \langle v, \alpha(v') \rangle - \langle v, \alpha^*\alpha(v') \rangle$$

and since this is true for all $v, v' \in V$ it follows that $\alpha = \alpha^*\alpha$. Hence α is self-adjoint and so surely normal, proving (1). $\square$

We now note a result which is a direct consequence of Proposition 5.5 and Proposition 3.12.

(14.15) PROPOSITION. *Let V be an inner product space finitely generated over its field of scalars. Let $W_1, \ldots, W_n$ be subspaces of V and for each $1 \le i \le n$ let α_i be the orthogonal projection of V onto W_i. Then the following conditions are equivalent:*

(1) *$V = \bigoplus_{j=1}^{n} W_j$ and for each $1 \le h \le n$ we have $W_h^\perp = \bigoplus_{j \ne h} W_j$;*
(2) *$\alpha_1 + \cdots + \alpha_n = \sigma_1$ and $\alpha_i\alpha_j = \sigma_0$ for all $1 \le i \ne j \le n$;*
(3) *If D_i is an orthonormal basis for W_i for each $1 \le i \le n$ then $D = \bigcup_{i=1}^{n} D_i$ is an orthonormal basis for V.*

We now return to normal endomorphisms.

(14.16) PROPOSITION. *Let V be an inner product space finitely generated over its field of scalars and let α be a normal endomorphism of V. Then $im(\alpha) \cap ker(\alpha) = \{0_V\}$.*

PROOF. Let $v \in V$ and assume that $\alpha(v) \in im(\alpha) \cap ker(\alpha)$. Then $\alpha^2(v) = 0_V$. Since α is normal, we have $\|\alpha^2(v)\|^2 = \langle \alpha^2(v), \alpha^2(v) \rangle = \langle \alpha(v), \alpha^*\alpha^2(v) \rangle = \langle \alpha(v), \alpha\alpha^*\alpha(v) \rangle = \langle \alpha^*\alpha(v), \alpha^*\alpha(v) \rangle = \|\alpha^*\alpha(v)\|^2$ and so $\|\alpha^*\alpha(v)\| = 0$. This implies that $0 = \langle \alpha^*\alpha(v), v \rangle = \langle \alpha(v), \alpha(v) \rangle$ and so $\alpha(v) = 0_V$. $\square$

In other words, if V is an inner product space finitely generated over its field of scalars and if α be a normal endomorphism of V then $\{im(\alpha), ker(\alpha)\}$ is an independent set of subspaces of V.

(14.17) PROPOSITION. *Let V be an inner product space finitely generated over its field of scalars F. If α is a normal endomorphism of V then so is $p(\alpha)$ for each $p(X) \in F[X]$.*

PROOF. If $p(X) = \sum_{i \ge 0} c_i X^i$ then $p(\alpha) = \sum_{i \ge 0} c_i \alpha^i$ and so we see that $p(\alpha)^* = \sum_{i \ge 0} \bar{c}_i(\alpha^*)^i$. Since $\alpha\alpha^* = \alpha^*\alpha$ it follows that $p(\alpha)p(\alpha)^* = p(\alpha)^*p(\alpha)$. Therefore $p(\alpha)$ is a normal endomorphism of V. $\square$

(14.18) PROPOSITION. *Let V be an inner product space finitely generated over $\mathbb{C}$ and let α be a normal endomorphism of V. Then the minimal polynomial of α has no multiple roots.*

PROOF. Let $f(X) \in \mathbb{C}[X]$ be the minimal polynomial of α. By the Fundamental Theorem of Algebra we know that $f(X) = \prod_{i=1}^{n}(X - c_i)$, where the c_i are complex numbers, not necessarily distinct. In fact, what we have to show is that the c_i are indeed distinct. Assume therefore that $c_i = c_j$ for $i \ne j$, and let us denote this common value by c. Then we can write $f(X) = g(X)(X - c)^2$. Since $f(\alpha) = \sigma_0$ we have $g(\alpha)(\alpha - c\sigma_1)^2 = \sigma_0$ and so for all $v \in V$ we have $g(\alpha)(\alpha - c\sigma_1)^2(v) = 0_V$. From Proposition 14.17 it follows that $\beta = \alpha - c\sigma_1$ is a normal endomorphism. If $v' = g(\alpha)(v)$ then $\beta^2(v') = 0_V$ and so $\beta(v') \in im(\beta) \cap ker(\beta) = \{0_V\}$. Thus we see that $g(\alpha)(\alpha - c\sigma_1)(v) = 0_V$ for all $v \in V$, contradicting the minimality of $f(X)$. $\square$

(14.19) PROPOSITION. (**Spectral Decomposition Theorem**) *Let V be an inner product space finitely generated over $\mathbb{C}$ and let α be a normal endomorphism of V. Then there exist scalars $c_1, \dots, c_n$ in $\mathbb{C}$ and idempotent endomorphisms $\alpha_1, \dots, \alpha_n$ of V such that*

(1) $\alpha = c_1\alpha_1 + \cdots + c_n\alpha_n$;
(2) $\sigma_1 = \alpha_1 + \cdots + \alpha_n$;
(3) $\alpha_i\alpha_j = \sigma_0$ *if* $1 \le i \ne j \le n$.

Moreover, these scalars c_j and idempotent endomorphisms α_j of V are unique. The c_j are precisely the eigenvalues of α and each corresponding α_j comes from the orthogonal projection of V onto the eigenspace W_j associated with c_j.

PROOF. If $dim(V) \le 1$ then the result is trivial, so we can assume that $dim(V) \ge 2$. Let $p(X)$ be the minimal polynomial of α. By Proposition 14.18 we know that $p(X) = \prod_{i=1}^n (X - c_i)$, where the c_i are distinct elements of $\mathbb{C}$. For each $1 \le h \le n$ set $p_h(X) = \prod_{i \ne h}(c_h - c_i)^{-1}(X - c_i)$. Thus we see that

$$p_h(c_j) = \begin{cases} 1, & \text{if } h = j \\ 0, & \text{otherwise.} \end{cases}$$

If $f(X)$ is an arbitrary polynomial of degree less than n then the polynomial $f(X) - \sum_{i=1}^n f(c_i)p_i(X)$ is of degree less than n but has n roots – namely $c_1, \dots, c_n$. Therefore it must be the 0-polynomial. Thus we have

$$f(X) = \sum_{i=1}^n f(c_i)p_i(X).$$

In particular, we have $1 = p_1(X) + \cdots + p_n(X)$ and $X = c_1p_1(X) + \cdots + c_np_n(X)$. For each $1 \le i \le n$ set $\alpha_i = p_i(\alpha)$. Then $\sigma_1 = \alpha_1 + \cdots + \alpha_n$ and $\alpha = c_1\alpha_1 + \cdots + c_n\alpha_n$. Moreover:

(i) We note that $\alpha_i \ne \sigma_0$ since $\alpha_i = p_i(\alpha)$ and $p_i(X)$ is a polynomial of degree less than the degree of the minimal polynomial of α.

(ii) If $i \ne j$ then $\alpha_i\alpha_j = \sigma_0$ since $\alpha_i\alpha_j$ is a multiple of $p(\alpha) = \sigma_0$.

(iii) Each α_i is idempotent since $\alpha_i = \alpha_i\sigma_1 = \alpha_i(\alpha_1 + \cdots + \alpha_n) = \alpha_i^2$.

Since the minimal polynomial and the characteristic polynomial of α have the same roots, we see that the c_i are precisely the eigenvalues of α. Let W_i be the eigenspace associated with c_i. In order to prove that $W_i = im(\alpha_i)$ we have to show that $v \in im(\alpha_i)$ if and only if $\alpha(v) = c_iv$. Indeed, if $\alpha(v) = c_iv$ then

$$c_i\left[\sum_{j=1}^n \alpha_j(v)\right] = c_iv = \alpha(v) = \sum_{j=1}^n (c_j\alpha_j)(v)$$

and this implies that $\sum_{j=1}^n [(c_i - c_j)\alpha_j](v) = 0_V$. Therefore, for all $j \ne i$ we have $\alpha_j(v) = 0_V$ and from this we conclude that $v = \alpha_i(v) \in im(\alpha_i)$. Conversely, if $v \in im(\alpha_i)$ then $v = \alpha_i(v)$ and we have already seen that this implies that $\alpha(v) = c_iv$.

Finally, we see that α_i comes from the orthogonal projection of V onto W_i since α_i is normal by Proposition 14.17 and Proposition 14.16. $\square$

Note that if $\alpha = c_1\alpha_1 + \cdots + c_n\alpha_n$ as in Proposition 14.19 then for each positive integer k we have $\alpha^k = (c_1\alpha_1 + \cdots + c_n\alpha_n)^k = c_1^k\alpha_1 + \cdots + c_n^k\alpha_n$ and, more generally, if $p(X) \in F[X]$ then $p(\alpha) = p(c_1)\alpha_1 + \cdots + p(c_n)\alpha_n$.

(14.20) PROPOSITION. *Let V be a finitely-generated inner product space. A self-adjoint endomorphism α of V is positive if and only if each of its eigenvalues is positive.*

PROOF. If α is positive and if c is an eigenvalue of α with associated eigenvector v then $0 < \langle \alpha(v), v \rangle = \langle cv, v \rangle = c\langle v, v \rangle$. Since $\langle v, v \rangle > 0$, this implies that $c > 0$. Conversely, assume that each of the eigenvalues of α is positive. By Proposition 14.19 we can write $\alpha = c_1\alpha_1 + \cdots + c_n\alpha_n$, where the c_i are the eigenvalues of α and the α_i are idempotent endomorphisms of V satisfying $\sigma_1 = \alpha_1 + \cdots + \alpha_n$ and $\alpha_i\alpha_j = \sigma_0$ for $i \neq j$. If $0_V \neq v \in V$ then

$$\langle \alpha(v), v \rangle = \sum_{i=1}^{n}\sum_{j=1}^{n} c_i\langle \alpha_i(v), \alpha_j(v) \rangle = \sum_{i=1}^{n} c_i\|\alpha_i(v)\|^2 > 0$$

and so α is positive. $\square$

As a consequence of Proposition 14.20 and the remarks before it we see that if α is a positive endomorphism of a finitely-generated inner product space V there exists a positive endomorphism $\sqrt{\alpha}$ of V satisfying $(\sqrt{\alpha})^2 = \alpha$. Indeed,

$$\sqrt{\alpha} = (\sqrt{c_1})\alpha_1 + \cdots + (\sqrt{c_n})\alpha_n,$$

where the c_i are the eigenvalues of α and the α_i are given in Proposition 14.19.

Problems

1. Let V be the space $\mathbb{R}^2$ on which we have defined the dot product. Is the endomorphism α of V given by $\alpha: [a, b] \mapsto [a + b, a + b]$ positive?

2. Let V be an inner product space and let α be a positive endomorphism of V. Must α^2 also be positive?

3. Let V be the space $\mathbb{R}^4$ on which we have the dot product defined and let α be the endomorphism of V represented with respect to the canonical basis by the matrix

$$\begin{bmatrix} 1 & 0 & -1/4 & -1/4 \\ 0 & 1 & -1/4 & -1/4 \\ -1/4 & -1/4 & 1 & 0 \\ -1/4 & -1/4 & 0 & 1 \end{bmatrix}.$$

Is α positive?

4. Let n be a positive integer and let V be the space $\mathbb{R}^n$ on which we have defined the dot product. Let α be a positive endomorphism of V represented with respect to the canonical basis by the matrix $A \in \mathcal{M}_{n \times n}(\mathbb{R})$. Show that $|A| > 0$.

5. Let n be a positive integer. Is the endomorphism of $\mathbb{C}^n$ defined by $v \mapsto iv$ normal?

6. Let α and β be normal endomorphisms of an inner product space V. Is $\beta\alpha$ necessarily normal?

7. Let α be the endomorphism of $\mathbb{C}^3$ represented with respect to the canonical

basis by the matrix $\begin{bmatrix} 6 & -2 & 3 \\ 3 & 6 & -2 \\ -2 & 3 & 6 \end{bmatrix}$. Show that α is normal and find an orthonor-

mal basis of $\mathbb{C}^3$ (with respect to the dot product) composed of eigenvectors of α.

8. Let n be an integer greater than 1 and let V be the subspace of $\mathbb{C}[X]$ composed of all polynomials of degree less than or equal to n. Let $0 \neq c \in \mathbb{C}$ and let α be the endomorphism of V defined by $\alpha: p(X) \mapsto p(X + c)$. Is it possible to define an inner product on V so that α is normal?

9. Let n be a positive integer. A matrix $A \in M_{n \times n}(\mathbb{C})$ is *normal* if and only if $\overline{A}A = A\overline{A}$. Show that any normal upper-triangular matrix is diagonal.

10. Let V be an inner product space over $\mathbb{C}$. If $\alpha, \beta \in End(V)$ we write $\alpha > \beta$ if and only if the endomorphism $\alpha - \beta$ of V is positive. Find endomorphisms α and β of $\mathbb{C}^2$ satisfying $\alpha > \beta$ but not $\alpha^2 > \beta^2$.

11. Let V be an inner product space finitely generated over $\mathbb{R}$ and let S be the set of all endomorphisms of V which preserve distances. Is S a subring of the ring $End(V)$?

12. Let α be a nilpotent endomorphism of an inner product space V having index of nilpotency k. Show that α^* is also nilpotent and has index of nilpotency k.

13. Let V be the space $\mathbb{R}^3$ on which we have defined the dot product and let W be the space of all polynomial functions of degree less than 3 from $\mathbb{R}$ to itself, on which we have defined the inner product $\langle f, g \rangle = \int_0^1 f(t)g(t)dt$. Let α be the linear transformation from V to W defined by

$$\alpha: [a, b, c] \mapsto 1 + \frac{1}{2}b + \frac{1}{6}c + (b - c)X + cX^2.$$

Is this transformation an isometry?

14. Let V be an inner product space and let α be an endomorphism of V satisfying the condition that $\alpha^*\alpha = \sigma_0$. Show that $\alpha = \sigma_0$.

15. Let n be a positive integer. If $A \in M_{n \times n}(\mathbb{R})$, show that there exists a vector $v \in \mathbb{R}^n$ satisfying $vAv^T = tr(A)$ and $v \cdot v = n$.

16. Let V be the space $\mathbb{R}^3$ on which we have defined the dot product. Is the automorphism of V defined by $\alpha: [a, b, c] \mapsto [-c, -b, -a]$ unitary?

17. Is the matrix $\begin{bmatrix} 0 & 0 & 0 & i \\ 0 & 0 & 1 & 0 \\ 0 & 1 & 0 & 0 \\ i & 0 & 0 & 0 \end{bmatrix} \in M_{4 \times 4}(\mathbb{C})$ unitary?

18. Find a real number a such that the matrix

$$a \begin{bmatrix} -9 + 8i & -10 - 4i & -16 - 18i \\ -2 - 24i & 1 + 12i & -10 - 4i \\ 4 - 10i & -2 - 24i & -9 + 8i \end{bmatrix} \in M_{3 \times 3}(\mathbb{C})$$

unitary?

19. Given a real number a, show that the matrix

$$\begin{bmatrix} -sin^2(a) + icos^2(a) & (1+i)sin(a)cos(a) \\ (1+i)sin(a)cos(a) & -cos^2(a) + isin^2(a) \end{bmatrix} \in M_{2\times2}(\mathbb{C})$$

is unitary.

20. Is the matrix $\begin{bmatrix} 1/2 & 1/2 & 1/2 & 1/2 \\ 1/2 & 1/2 & -1/2 & -1/2 \\ 1/2 & -1/2 & 1/2 & -1/2 \\ 1/2 & -1/2 & -1/2 & 1/2 \end{bmatrix} \in M_{4\times4}(\mathbb{R})$ orthogonal?

21. Let n be a positive integer and let $A \in M_{n\times n}(\mathbb{R})$ be a matrix satisfying the condition that in every row and every column of A there is precisely one nonzero entry, and this entry equals 1. Show that A is orthogonal.

22. Let n be a positive integer. Show that the set of all orthogonal matrices in $M_{n\times n}(\mathbb{R})$ is closed under multiplication and taking inverses.

23. Let α be the endomorphism of $\mathbb{R}^3$ represented with respect to the canonical basis by the matrix $\begin{bmatrix} 14/3 & 2/3 & 14/3 \\ 2/3 & -1/3 & -16/3 \\ 14/3 & -16/3 & 5/3 \end{bmatrix}$. Show that α is self-adjoint and find an orthonormal basis of $\mathbb{R}^3$ (with respect to the dot product) composed of eigenvectors of α.

24. If v is an arbitrary vector in $\mathbb{R}^2$, show that there exists an orthogonal matrix $A \in M_{2\times2}(\mathbb{R})$ satisfying $vA \in \mathbb{R}[1,0]$.

25. Let a and b be real numbers, at least one of which is nonzero. Show that the matrix

$$\frac{1}{a^2 + ab + b^2} \begin{bmatrix} ab & a(a+b) & b(a+b) \\ a(a+b) & -b(a+b) & ab \\ b(a+b) & ab & -a(a+b) \end{bmatrix}$$

is orthogonal.

26. Let n be a positive integer and let $A \in M_{n\times n}(\mathbb{R})$. Show that A is orthogonal if and only if the set of rows of A is an orthonormal subset of $\mathbb{R}^n$ (with the dot product).

27. Find a real number a such that the matrix $\begin{bmatrix} 2a & -2a & a \\ -2a & -a & 2a \\ a & 2a & 2a \end{bmatrix}$ is orthogonal.

28. If $A = \begin{bmatrix} 1 & 2 & 0 & 0 \\ 2 & 1 & 0 & 0 \\ 0 & 0 & 1 & 2 \\ 0 & 0 & 2 & 1 \end{bmatrix} \in M_{4\times4}(\mathbb{R})$ find an orthogonal matrix $P \in M_{4\times4}(\mathbb{R})$ such that the matrix $P^T AP$ is diagonal.

29. Let n be a positive integer. For each permutation π of the set $\{1, \ldots, n\}$ let H_π be the matrix $[a_{ij}] \in \mathcal{M}_{n \times n}(\mathbb{R})$ defined by

$$a_{ij} = \begin{cases} 1, & \text{if } j = \pi(i) \\ 0, & \text{otherwise.} \end{cases}$$

Show that this matrix is orthogonal.

30. Find a Householder matrix $A \in \mathcal{M}_{3 \times 3}(\mathbb{R})$ satisfying $A \begin{bmatrix} 1 \\ 1 \\ -1 \end{bmatrix} = \begin{bmatrix} -1 \\ -1 \\ 1 \end{bmatrix}$.

31. For

$$A = \begin{bmatrix} \frac{1}{2} & -\frac{1}{2} & \sqrt{2} \\ \sqrt{2} & \sqrt{2} & 0 \\ -\frac{1}{2} & \frac{1}{2} & \sqrt{2} \end{bmatrix} \in \mathcal{M}_{3 \times 3}(\mathbb{R})$$

find infinitely-many pairs (P, Q) of orthogonal matrices such that the matrix PAQ is diagonal.

32. Let $A = \begin{bmatrix} 2 & 1 & 1 \\ 1 & 2 & 1 \\ 0 & 0 & 1 \end{bmatrix} \in \mathcal{M}_{3 \times 3}(\mathbb{R})$, which has distinct eigenvalues $c_1 = 1$ and $c_2 = 3$. Find matrices $P_1, P_2 \in \mathcal{M}_{3 \times 3}(\mathbb{R})$ satisfying the following conditions:

(1) $P_1^2 = P_1$ and $P_2^2 = P_2$;
(2) $P_1 P_2 = O = P_2 P_1$;
(3) $P_1 + P_2 = I$;
(4) $c_1 P_1 + c_2 P_2 = A$;
(5) $AP_1 = P_1 A$ and $AP_2 = P_2 A$.

CHAPTER XV

THE MOORE-PENROSE PSEUDOINVERSE

Let V and W be inner product spaces finitely generated over the same field of scalars and let $\alpha: V \to W$. We know that there exists a linear transformation $\alpha^{-1}: W \to V$ such that $\alpha^{-1}\alpha$ is the identity automorphism on V and $\alpha\alpha^{-1}$ is the identity automorphism on W if and only if $dim(V) = dim(W)$ and α is an isomorphism. If this is not the case, we would like to weaken the notion of "inverse" so that some similar condition can exist. There are many possibilities of doing this, and we will bring only one of them here. Given a linear transformation $\alpha: V \to W$ as above, we will say that a linear transformation $\beta: W \to V$ is the *Moore-Penrose pseudoinverse* of α if and only if the following conditions are satisfied:

(1) $\alpha\beta\alpha = \alpha$;
(2) $\beta\alpha\beta = \beta$;
(3) $\beta\alpha: V \to V$ is self-adjoint;
(4) $\alpha\beta: W \to W$ is self-adjoint.

We will denote this linear transformation β by α^+. Of course, in order to justify this terminology, we must show that β exists and is unique. We will begin with uniqueness.

(15.1) PROPOSITION. *Let V and W be inner product spaces finitely generated over the same field of scalars and let $\alpha: V \to W$ be a linear transformation. If α^+ exists, it is unique.*

PROOF. If β and β' are linear transformations from W to V satisfying the conditions of the Moore-Penrose pseudoinverse then $\beta = \beta\alpha\beta = (\beta\alpha)^*\beta = \alpha^*\beta^*\beta = (\alpha\beta'\alpha)^*\beta^*\beta = (\beta'\alpha)^*\alpha^*\beta^*\beta = \beta'\alpha\alpha^*\beta^*\beta = \beta'\alpha(\beta\alpha)^*\beta = \beta'\alpha\beta\alpha\beta = \beta'\alpha\beta = \beta'\alpha\beta'\alpha\beta = \beta'(\alpha\beta')^*\alpha\beta = \beta'\beta'^*\alpha^*\alpha\beta = \beta'\beta'^*\alpha^*(\alpha\beta)^* = \beta'\beta'^*(\alpha\beta\alpha)^* = \beta'\beta'^*\alpha^* = \beta'(\alpha\beta')^* = \beta'\alpha\beta' = \beta'$. $\square$

(15.2) PROPOSITION. *Let V and W be inner product spaces finitely generated over the same field of scalars and let $\alpha: V \to W$ be a linear transformation.*

(1) *If α is a monomorphism then α^+ exists and equals $(\alpha^*\alpha)^{-1}\alpha^*$. Moreover, in this case $\alpha^+\alpha$ is the identity automorphism on V.*
(2) *If α is an epimorphism then α^+ exists and equals $\alpha^*(\alpha\alpha^*)^{-1}$. Moreover, in this case $\alpha\alpha^+$ is the identity automorphism on W.*

PROOF. (1) By Proposition 13.15 we know that if α is a monomorphism then $\alpha^*\alpha$ is an automorphism of V and so the linear transformation $(\alpha^*\alpha)^{-1}$ exists. Set $\beta = (\alpha^*\alpha)^{-1}\alpha^*$. Then $\beta\alpha$ is the identity endomorphism on V and so $\beta\alpha$ is surely self-adjoint and satisfies $\alpha\beta\alpha = \alpha$ and $\beta\alpha\beta = \beta$. Finally, $(\alpha\beta)^* = [\alpha(\alpha^*\alpha)^{-1}\alpha^*]^* =$

$\alpha[(\alpha^*\alpha)^{-1}]^*\alpha^* = \alpha[(\alpha^*\alpha)^*]^{-1}\alpha^* = \alpha(\alpha^*\alpha)^{-1}\alpha^* = \alpha\beta$ and so $\alpha\beta$ is also self-adjoint. Thus $\beta = \alpha^+$.

(2) By Proposition 13.15 we know that if α is an epimorphism then $\alpha\alpha^*$ is an automorphism of W and so the linear transformation $(\alpha\alpha^*)^{-1}$ exists and, as in the proof of (1), we see that $\alpha^+ = \alpha^*(\alpha\alpha^*)^{-1}$. $\square$

EXAMPLE

Let $\alpha: \mathbb{R}^2 \rightarrow \mathbb{R}^3$ be the linear transformation

$$\alpha: [a, b] \mapsto [a + 2b, -a + 3b, 2a + 4b].$$

This is a monomorphism and so by Proposition 16.2 we know that if α is represented with respect to the canonical basis by the matrix $A \in \mathcal{M}_{3\times 2}(\mathbb{R})$ then α^+ is represented by the matrix $(A^T A)^{-1} A^T \in \mathcal{M}_{2\times 3}(\mathbb{R})$. This matrix equals $\frac{1}{25}\begin{bmatrix} 3 & -10 & 6 \\ 1 & 5 & 2 \end{bmatrix}$ and so α^+ is given by $[a, b, c] \mapsto \frac{1}{25}[3a - 10b + 6c, a + 5b + 2c]$.

We now show that Moore-Penrose pseudoinverses always indeed exist.

(15.3) PROPOSITION. *If V and W are inner product spaces finitely generated over the same field of scalars then every linear transformation $\alpha: V \rightarrow W$ has a Moore-Penrose pseudoinverse.*

PROOF. Let $W' = im(\alpha)$ and write $\alpha = \mu\beta$, where $\beta: V \rightarrow W'$ is an epimorphism and $\mu: W' \rightarrow W$ is the inclusion monomorphism. By Proposition 15.2 we know that β^+ and μ^+ exist and satisfy the conditions that $\mu^+\mu$ and $\beta\beta^+$ are the identity endomorphisms on W'. Therefore $(\mu\beta)\beta^+\mu^+(\mu\beta) = \mu\beta$ and $(\beta^+\mu^+)(\mu\beta)(\beta^+\mu^+) = \beta^+\mu^+$. Moreover, we see that the linear transformations $(\beta^+\mu^+)(\mu\beta) = \beta^+\beta$ and $(\mu\beta)(\beta^+\mu^+) = \mu\mu^+$ are self-adjoint. Therefore α has a Moore-Penrose pseudoinverse, and $\alpha^+ = \beta^+\mu^+$. $\square$

If $\alpha: F^k \rightarrow F^n$ is a linear transformation, where $F = \mathbb{R}$ or $F = \mathbb{C}$ and where k and n are positive integers, and if A is the matrix representing α with respect to the canonical bases, then we will denote the matrix representing α^+ with respect to these same bases by A^+.

EXAMPLE

If $\alpha: \mathbb{R}^2 \rightarrow \mathbb{R}^2$ is the linear transformation represented with respect to the canonical basis by the matrix $A = \begin{bmatrix} 2 & 6 \\ 1 & 3 \end{bmatrix}$ then α is not invertible but it has a Moore-Penrose pseudoinverse, represented by the matrix $A^+ = \frac{1}{50}\begin{bmatrix} 2 & 1 \\ 6 & 3 \end{bmatrix}$.

If A and B are invertible matrices of the same size then we know that AB is invertible and, indeed, $(AB)^{-1} = B^{-1}A^{-1}$. No such relation exists between Moore-Penrose pseudoinverses, as the following example shows.

EXAMPLE

If $\alpha, \beta\colon \mathbb{R}^2 \to \mathbb{R}^2$ are the linear transformations represented with respect to the canonical basis by the matrices $A = \begin{bmatrix} 2 & 6 \\ 1 & 3 \end{bmatrix}$ and $B = \begin{bmatrix} 1 & 2 \\ 2 & 4 \end{bmatrix}$ then $AB = \begin{bmatrix} 14 & 28 \\ 7 & 14 \end{bmatrix}$ and we have

$$A^+ = \frac{1}{50} \begin{bmatrix} 2 & 1 \\ 6 & 3 \end{bmatrix},$$

$$B^+ = \frac{1}{25} \begin{bmatrix} 1 & 2 \\ 2 & 4 \end{bmatrix},$$

$$(AB)^+ = \frac{1}{175} \begin{bmatrix} 2 & 1 \\ 4 & 2 \end{bmatrix},$$

and

$$B^+ A^+ = \frac{7}{1250} \begin{bmatrix} 2 & 1 \\ 4 & 2 \end{bmatrix}.$$

Let $A \in \mathcal{M}_{k \times n}(F)$, where $F = \mathbb{R}$ or $F = \mathbb{C}$. Let us see what the matrix A^+ tells us about a system of linear equations of the form $AX = w^T$. First of all, we note that this system is consistent, i.e. that it has at least one solution, if and only if $AA^+w^T = w^T$. Indeed, if the system has a solution $v \in F^n$ then $(AA^+)w^T = (AA^+)(Av^T) = (AA^+A)v = Av^T = w^T$. Conversely, if $(AA^+)w^T = w^T$ then the vector $v = w(A^+)^T$ is a solution to the system. Next, we note that if $y \in F^n$ then $A(I - A^+A)y^T = [0, \ldots, 0]^T$ and so we see that

$$w(A^+)^T + y(I - A^+A)^T = w(A^+)^T + y[I - A^T(A^+)^T]$$

is also a solution of $AX = w^T$. Conversely, every solution to this system is of the form $w(A^+)^T + y$, where $Ay^T = [0, \ldots, 0]^T$ and so $(I - A^+A)y^T = y^T$.

(15.4) PROPOSITION. *Let F be the field $\mathbb{R}$ or $\mathbb{C}$, let $A = [a_{ij}] \in \mathcal{M}_{k \times n}(F)$ and let $w \in F^k$. If the system of linear equations $AX = w^T$ is consistent then the set of solutions to this system has a unique solution having minimal norm, namely the vector $w(A^+)^T$.*

PROOF. If $u \in F^n$ is a solution to the given system of linear equations then we have seen that u is of the form $w(A^+)^T + y[I - A^T(A^+)^T]$. Since

$$\langle w(A^+)^T, y[I - A^T(A^+)^T] \rangle = \langle w(A^+AA^+)^T, y[I - A^T(A^+)^T] \rangle$$
$$= \langle w(A^+)^T, y[I - A^T(A^+)^T](A^+A)^T \rangle$$
$$= \langle w(A^+)^T, y[A(A^+)^T - A^T(A^+)^T A^T(A^+)^T] \rangle$$
$$= \langle w(A^+)^T, [0, \ldots, 0] \rangle$$
$$= 0$$

We have

$$
\begin{aligned}
\|u\|^2 &= \langle u, u \rangle \\
&= \langle w(A^+)^T + y[I - A^T(A^+)^T], w(A^+)^T + y[I - A^T(A^+)^T] \rangle \\
&= \langle w(A^+)^T, w(A^+)^T \rangle + \langle y[I - A^T(A^+)^T], y[I - A^T(A^+)^T] \rangle \\
&= \|w(A^+)^T\|^2 + \|y[I - A^T(A^+)^T]\|^2
\end{aligned}
$$

and so $\|w(A^+)^T\|^2 \le \|u\|^2$. $\square$

EXAMPLE

Let $w = [1, 2] \in \mathbb{R}^2$ and let $A = \begin{bmatrix} 2 & -1 & 1 \\ 1 & 1 & 1 \end{bmatrix}$. Then

$$
A^+ = \frac{1}{14} \begin{bmatrix} 4 & 2 \\ -5 & 8 \\ 1 & 4 \end{bmatrix}
$$

and a solution of minimal norm to the system of linear equations $AX = w^T$ is $w(A^+)^T = \frac{1}{14}[8, 11, 9]$.

To emphasize the utility of Proposition 15.4 we consider the notion of the *least squares method*, which is very important in applied mathematics and statistics. This method was first introduced by C. F. Gauss and A. M. Legendre at the beginning of the 19th century. We assume that we have a set of real numerical values $c_1, \ldots, c_n$ corresponding to real numerical values of some parameter $t_1, \ldots, t_n$. Our theory tells us that the set of points $\{(t_i, c_i) \mid 1 \le i \le n\}$ lies on some line in $\mathbb{R}^2$. However, since the values c_i were obtained experimentally or by some other external means, it is highly unlikely that this turns out to be the case, since one can expect that there were various errors in collecting and evaluating the data. However, we would like to find the equation of the line which fits most closely to this data we have on hand, where by "fits most closely" we mean that the sum of the squares of the distances from these points to the closest corresponding points on the line should be as small as possible. In other words, we want a solution having minimal norm to the system of linear equations $X_1 + t_i X_2 = c_i$ ($1 \le i \le n$). As we have seen, this solution is precisely

$$
[c_1, \ldots, c_n] \left(\begin{bmatrix} 1 & t_1 \\ 1 & t_2 \\ \vdots & \vdots \\ 1 & t_n \end{bmatrix}^+ \right)^T.
$$

EXAMPLE

We want to find the line in the plane which fits the points

$$\{(1,3),(2,7),(3,8),(4,11)\}$$

most closely. If

$$A = \begin{bmatrix} 1 & 1 \\ 1 & 2 \\ 1 & 3 \\ 1 & 4 \end{bmatrix}$$

then $A^+ = \frac{1}{20}\begin{bmatrix} 20 & 10 & 0 & -10 \\ -6 & -2 & 2 & 6 \end{bmatrix}$ and $[3,7,8,11](A^+)^T = [1,\frac{5}{2}]$. Therefore the desired line is $\{(t, 1 + \frac{5}{2}t) \mid t \in \mathbb{R}\}$.

We can also use the same technique to find other sorts of curves most closely. Thus, for example, to find the parabola most closely fitting a given set of points $\{(t_i, c_i) \mid 1 \leq i \leq n\}$ we want to find the solution having minimal norm of the system of equations $X_1 + t_i X_2 + t_i^2 X_3 = c_i$ $(1 \leq i \leq n)$. This solution is $w(A^+)^T$, where

$$A = \begin{bmatrix} 1 & t_1 & t_1^2 \\ 1 & t_2 & t_2^2 \\ \vdots & \vdots & \vdots \\ 1 & t_n & t_n^2 \end{bmatrix}.$$

EXAMPLE

We want to find the parabola in the plane which fits the points

$$\{(1,3),(2,7),(3,8),(4,11)\}$$

most closely. If

$$A = \begin{bmatrix} 1 & 1 & 1 \\ 1 & 2 & 4 \\ 1 & 3 & 9 \\ 1 & 4 & 16 \end{bmatrix}$$

then

$$A^+ = \frac{1}{20}\begin{bmatrix} 45 & -15 & -25 & 15 \\ -31 & 23 & 27 & -19 \\ 5 & -5 & -5 & 5 \end{bmatrix}$$

and $[3,7,8,11](A^+)^T = \frac{1}{4}[-1,15,-1]$ so that the parabola is

$$\{(t, -\frac{1}{4} + \frac{15}{4}t - \frac{1}{4}t^2) \mid t \in \mathbb{R}\}.$$

Problems

1. Calculate $\begin{bmatrix} 1 & 1 \\ -1 & 1 \\ 2 & 3 \end{bmatrix}^{+} \in \mathcal{M}_{2\times 3}(\mathbb{R})$.

2. Calculate $[5 \quad 0 \quad 0]^{+} \in \mathcal{M}_{3\times 1}(\mathbb{Q})$.

3. Calculate $\begin{bmatrix} 7/3 & 8 \\ -10/3 & 10 \\ 34/3 & -4 \end{bmatrix}^{+} \in \mathcal{M}_{2\times 3}(\mathbb{R})$.

4. Calculate $\begin{bmatrix} 2 & 2 & 0 \\ 1 & 2 & 1 \\ 1 & 2 & 1 \end{bmatrix}^{+} \in \mathcal{M}_{3\times 3}(\mathbb{R})$.

5. Let V and W be inner product spaces finitely generated over the same field of scalars and let $\alpha: V \rightarrow W$ be a linear transformation. For any scalar $c \neq 0$, show that $(c\alpha)^{+} = \frac{1}{c}\alpha^{+}$.

6. Let n be a positive integer and let $A \in \mathcal{M}_{n\times n}(\mathbb{R})$ be a diagonal matrix. Calculate A^{+}.

7. Let V and W be inner product spaces finitely generated over the same field of scalars and let $\alpha: V \rightarrow W$ be a linear transformation. Show that $(\alpha^{+})^{*} = (\alpha^{*})^{+}$.

8. Let V be the space $\mathbb{R}^{2}$ on which we have defined the dot product. Let $\alpha: V \rightarrow \mathbb{R}$ be the linear transformation defined by $\alpha: [a, b] \mapsto a$ and let $\beta: \mathbb{R} \rightarrow V$ be the linear transformation defined by $\beta: a \mapsto [a, a]$. Show that $(\alpha\beta)^{+} \neq \beta^{+}\alpha^{+}$.

9. Let n be a positive integer and let $A \in \mathcal{M}_{n\times n}(\mathbb{R})$ be the matrix all of the entries of which are equal to 1. Show that $A^{+} = n^{-2}A$.

10. Let k and n be positive integers, and let $A \in \mathcal{M}_{k\times n}(\mathbb{R})$ be a matrix which can be written in the form $\begin{bmatrix} C & O \\ O & O \end{bmatrix}$, where C is an invertible diagonal matrix of size $t \times t$. Show that $A^{+} = \begin{bmatrix} C^{-1} & O \\ O & O \end{bmatrix}$.

11. Let $V = \mathbb{R}^{n}$ on which the dot product is defined and let $O \neq A = [a_1, \dots, a_n] \in \mathcal{M}_{1\times n}(\mathbb{R})$. Calculate A^{+}.

12. Let V and W be inner product spaces finitely-generated over the same field of scalars and let $\alpha: V \rightarrow W$ be a linear transformation. Show that $\alpha^{++} = \alpha$.

CHAPTER XVI

BILINEAR TRANSFORMATIONS AND FORMS

Let V, W, and Y be vector spaces over a field F. A *bilinear transformation* from $V \times W$ to Y is a function $f: V \times W \to Y$ which satisfies the following conditions:

(1) For each given vector $w \in W$, the function from V to Y given by $v \mapsto f(v, w)$ is a linear transformation;

(2) For each given vector $v \in V$, the function from W to Y given by $w \mapsto f(v, w)$ is a linear transformation.

EXAMPLE

Let F be a field and let k, n, and t be positive integers. Let $V = \mathcal{M}_{k \times n}(F)$, let $W = \mathcal{M}_{n \times t}(F)$, and let $Y = \mathcal{M}_{k \times t}(F)$. Then there exists a bilinear transformation $V \times W \to Y$ defined by $(A, B) \mapsto AB$. More generally, every matrix $C \in \mathcal{M}_{n \times n}(F)$ defines a bilinear transformation $V \times W \to Y$ by $(A, B) \mapsto ACB$.

A bilinear transformation f from $V \times W$ to F is called a *bilinear form*. In the case $V = W$ we say that f is a *bilinear form on V*. We will denote the set of all bilinear forms from $V \times W$ to F by $Bil(V \times W)$. This is surely a subset of $F^{V \times W}$. It is easy to verify that if $f, g \in Bil(V \times W)$ and if $c \in F$ then $f + g \in Bil(V \times W)$ and $cf \in Bil(V \times W)$ so that we see that $Bil(V \times W)$ is in fact a subspace of $F^{V \times W}$.

EXAMPLE

If V is an inner product space over $\mathbb{R}$ then the function from $V \times V$ to $\mathbb{R}$ given by $(v, v') \mapsto \langle v, v' \rangle$ is a bilinear form on V. This would not the case if the field of scalars were $\mathbb{C}$.

EXAMPLE

Let V be a vector space over a field F having dual space $D(V)$. Then there exists a bilinear form $D(V) \times V \to F$ given by $(\delta, v) \mapsto \delta(v)$. More generally, for each vector space Y over F then there exists a bilinear transformation $Hom(V, Y) \times V \to Y$ defined by $(\alpha, v) \mapsto \alpha(v)$.

EXAMPLE

Let V and W be vector spaces over a field F. If $\alpha \in D(V)$ and $\beta \in D(W)$ then we have a bilinear form on $V \times W$ defined by $(v, w) \mapsto \alpha(v)\beta(w)$.

EXAMPLE

Let F be a field and let k, n be positive integers. Every matrix $A \in \mathcal{M}_{k \times n}(F)$ defines a bilinear form on $F^k \times F^n$ by $(v, w) \mapsto vAw^T$.

If $f: V \times W \to F$ is a bilinear form then we also have a bilinear form $g: W \times V \to F$ defined by $g: (w, v) \mapsto f(v, w)$. This form is called the *opposite form* of f.

We will now see that in the case that V and W are vector spaces finitely generated over the same field F then every bilinear form from $V \times W$ to F is given as in the last of the above examples. Indeed, let us choose a basis $B = \{v_1, \ldots, v_k\}$ for V over F and a basis $D = \{w_1, \ldots, w_n\}$ for W over F. Every form $f \in Bil(V \times W)$ defines a matrix $T_{BD}(f) \in \mathcal{M}_{k \times n}(F)$ given by $T_{BD}(f) = [f(v_i, w_j)]$. Thus the bases B and D define a function $T_{BD}: Bil(V \times W) \to \mathcal{M}_{k \times n}(F)$.

(16.1) PROPOSITION. *Let V and W be vector spaces finitely generated over the same field F and having bases $B = \{v_1, \ldots, v_k\}$ and $D = \{w_1, \ldots, w_n\}$ over F respectively. Then the function $T_{BD}: Bil(V \times W) \to \mathcal{M}_{k \times n}(F)$ is an isomorphism. Moreover, for all $v = \sum_{i=1}^{k} a_i v_i \in V$ and all $w = \sum_{j=1}^{n} b_j w_j \in W$ we have*

$$f(v, w) = [a_1, \quad \ldots, \quad a_k] T_{BD}(f) \begin{bmatrix} b_1 \\ \vdots \\ b_n \end{bmatrix}.$$

PROOF. As an immediate consequence of the definition we see that T_{BD} is a linear transformation. If $v = \sum_{i=1}^{k} a_i v_i \in V$ and all $w = \sum_{j=1}^{n} b_j w_j \in W$ then

$$f(v, w) = f(\sum_{i=1}^{k} a_i v_i, \sum_{j=1}^{n} b_j w_j)$$

$$= \sum_{i=1}^{k} \sum_{j=1}^{n} a_i b_j f(v_i, w_j)$$

$$= \sum_{i=1}^{k} \sum_{j=1}^{n} a_i f(v_i, w_j) b_j$$

$$= [a_1, \quad \ldots, \quad a_k] T_{BD}(f) \begin{bmatrix} b_1 \\ \vdots \\ b_n \end{bmatrix}$$

as desired. It is also therefore clear that if $f, g \in Bil(V \times W)$ satisfy $T_{BD}(f) = T_{BD}(g)$ then $f(v, w) = g(v, w)$ for all $v \in V$ and all $w \in W$ so $f = g$. Thus the function T_{BD} is monic. If $C = [c_{ij}] \in M_{k \times n}(F)$ then we can define a function $f \in F^{V \times W}$ by

$$f : \left(\sum_{i=1}^{k} a_i v_i, \sum_{j=1}^{n} b_j w_j \right) \mapsto [a_1, \quad \ldots, \quad a_k] C \begin{bmatrix} b_1 \\ \vdots \\ b_n \end{bmatrix}.$$

This function is well-defined and it is easy to verify that in fact $f \in Bil(V \times W)$ and that $T_{BD}(f) = C$. Thus the function T_{BD} is epic as well and so it is an isomorphism. $\square$

EXAMPLE

Let F be a field and let $V = F^2$. Let f be the bilinear form on V defined by

$$f : ([a, b], [c, d]) \mapsto (a + b)(c + d).$$

If B is the canonical basis of V then $T_{BB}(f) = \begin{bmatrix} 1 & 1 \\ 1 & 1 \end{bmatrix}$. If $D = \{[1, -1], [1, 1]\}$ then $T_{DD}(f) = \begin{bmatrix} 0 & 0 \\ 0 & 4 \end{bmatrix}$. Indeed, if $w_1 = [1, -1]$ and $w_2 = [1, 1]$ then $[a, b] = \frac{1}{2}(a - b)w_1 + \frac{1}{2}(a + b)w_2$ for all $a, b \in F$ and so

$$[\tfrac{1}{2}(a - b), \quad \tfrac{1}{2}(a + b)] \begin{bmatrix} 0 & 0 \\ 0 & 4 \end{bmatrix} \begin{bmatrix} \frac{1}{2}(c - d) \\ \frac{1}{2}(c + d) \end{bmatrix} = (a + b)(c + d).$$

(16.2) PROPOSITION. *Let V and W be vector spaces finitely generated over the same field F. Let $B = \{v_1, \ldots, v_k\}$ and $C = \{x_1, \ldots, x_k\}$ be bases for V and let $D = \{w_1, \ldots, w_n\}$ and $E = \{y_1, \ldots, y_n\}$ be bases for W. Let P be the change of basis matrix from B to C and let Q be the change of basis matrix from D to E. Then for any bilinear form $f \in Bil(V \times W)$ we have $T_{CE}(f) = PT_{BD}(f)Q^T$.*

PROOF. Recall that the matrices $P = [p_{ir}]$ and $Q = [q_{js}]$ are defined by $x_i = \sum_{r=1}^{k} p_{ir} v_r$ for all $1 \le i \le k$ and $y_j = \sum_{s=1}^{n} q_{js} w_s$ for all $1 \le j \le n$. Therefore, by definition, we have

$$f(x_i, y_j) = f\left(\sum_{r=1}^{k} p_{ir} v_r, \sum_{s=1}^{n} q_{js} w_s \right) = \sum_{r=1}^{k} \sum_{s=1}^{n} p_{ir} f(v_r, w_s) q_{js},$$

and this is precisely the (i, j) entry of $PT_{BD}(f)Q^T$. $\square$

In particular, we see that if f is a bilinear form on a vector space V finitely generated over a field F and if B and C are bases for V over F then $T_{CC}(f) = PT_{BB}P^T$, where P is a suitable nonsingular matrix square matrix over F. In general, if F is a field and n is a positive integer then matrices A and B in $M_{n \times n}(F)$ are *congruent* if there exists a nonsingular matrix $P \in M_{n \times n}(F)$ satisfying $B =$

PAP^T In this case we write $A \simeq B$. Congruence is clearly an equivalence relation between matrices in $\mathcal{M}_{n \times n}(F)$.

We note that congruent matrices have the same rank, and so we see that the rank of a matrix of the form $T_{BB}(f)$ depends solely on the bilinear form f and not on the basis B which was chosen. This rank will be called the *rank* of the bilinear form f.

EXAMPLE

The matrix $A = \begin{bmatrix} 1 & -6 & -3 \\ -6 & 40 & 19 \\ -3 & 19 & 10 \end{bmatrix} \in \mathcal{M}_{3 \times 3}(\mathbb{R})$ is congruent to I since $I =$

PAP^T, where $P = \begin{bmatrix} 1 & 0 & 0 \\ 3 & 1/2 & 0 \\ \sqrt{3} & -1/2\sqrt{3} & 2/\sqrt{3} \end{bmatrix}$.

(16.3) PROPOSITION. *Let V and W be vector spaces finitely generated over the same field F and let f be a nonzero bilinear form from $V \times W$ to F. Then there exist bases $\{v_1, \ldots, v_k\}$ for V over F and $\{w_1, \ldots, w_n\}$ for W over F and there exists an integer t satisfying $1 \le t \le min\{k, n\}$ such that*

$$f(v_i, w_j) = \begin{cases} 1_F, & \text{if } i = j \le t \\ 0_F, & \text{otherwise.} \end{cases}$$

PROOF. Since the function f is nonzero we know that there exists vectors $v_1 \in V$ and $y_1 \in W$ satisfying $f(v_1, y_1) \ne 0_F$. Thus, if $w_1 = f(v_1, y_1)^{-1}y_1$, we have $f(v_1, w_1) = 1_F$. Let $V_1 = Fv_1$ and $W_1 = Fw_1$. It is easy to see that the set $W_2 = \{w \in W \mid f(v_1, w) = 0_F\}$ is in fact a subspace of W satisfying $W_1 \cap W_2 = \{0_W\}$. We claim that in fact $W = W_1 \oplus W_2$. Indeed, if $w \in W$ and if $c = f(v_1, w)$ then $f(v_1, w - cw_1) = f(v_1, w) - cf(v_1, w_1) = c - c = 0_F$ and so $w - cw_1 \in W_2$. Thus we can write $w = cw_1 + (w - cw_1) \in W_1 + W_2$, which establishes the claim. In a similar manner, we note that $V_2 = \{v \in V \mid f(v, w_1) = 0_F\}$ is a subspace of V satisfying $V = V_1 \oplus V_2$. Moreover, $f(v, w) = 0_F$ whenever $v \in V_1$ and $w \in W_2$ or $v \in V_2$ and $w \in W_1$.

Replacing f by its oppose form if necessary, we can assume that $dim(V) \le dim(W)$. We now proceed by induction on $k = dim(V)$. Suppose, initially, that $k = 1$, i.e. that $V = V_1$. If $\{w_2, \ldots, w_n\}$ is a basis for W_2 then we have already noted that $f(v, w_j) = 0_F$ for all $2 \le j \le n$ and so $\{v_1\}$ and $\{w_1, \ldots, w_n\}$ are the bases we seek in order to prove the proposition. Assume, therefore, that $k > 1$ (which implies that $n > 1$) and that the result has been established for all vector spaces V of dimension less than k. Consider the restriction of f to $V_2 \times W_2$. By the induction hypothesis there exist bases $\{v_2, \ldots, v_k\}$ for V_2 over F and $\{w_2, \ldots, w_n\}$ for W_2 over F and there exists an integer $t \le k$ satisfying

$$f(v_i, w_j) = \begin{cases} 1_F, & \text{if } i = j \le t \\ 0_F, & \text{otherwise.} \end{cases}$$

Then $\{v_1, \ldots, v_k\}$ and $\{w_1, \ldots, w_n\}$ are the bases we seek for V and W. $\square$

If V is a vector space over a field F then a bilinear form f on V is *symmetric* if and only if it equals its opposite form. That is to say, f is symmetric if and only if $f(v, v') = f(v', v)$ for all $v, v' \in V$.

(16.4) PROPOSITION. *Let V be a vector space finitely generated over a field F and let $D = \{v_1, \ldots, v_n\}$ be a basis for V over F. Then a bilinear form f on V is symmetric if and only if the matrix $T_{DD}(f)$ is symmetric.*

PROOF. If f is symmetric and if $T_{DD}(f) = [a_{ij}]$ then for all $1 \le i, j \le n$ we have $a_{ij} = f(v_i, v_j) = f(v_j, v_i) = a_{ji}$ and so the matrix $T_{DD}(f)$ is symmetric. Conversely, if $T_{DD}(f)$ is symmetric then for each $1 \le i, j \le n$ we have $f(v_i, v_j) = a_{ij} = a_{ji} = f(v_j, v_i)$ and so the form f is symmetric. $\square$

EXAMPLE

If D is the canonical basis for $\mathbb{R}^3$ over $\mathbb{R}$ and if $A = \begin{bmatrix} 1 & -5 & 3 \\ -5 & 1 & 1 \\ 3 & 1 & 4 \end{bmatrix}$ then $A = T_{DD}(f)$, where f is the bilinear form on $\mathbb{R}^3$ which sends $([a_1, a_2, a_3], [b_1, b_2, b_3])$ to

$$a_1 b_1 - 5a_2 b_1 + 3a_3 b_1 - 5a_1 b_2 + a_2 b_2 + a_3 b_2 + 3a_1 b_3 + a_2 b_3 + 4a_3 b_3.$$

This form is symmetric.

(16.5) PROPOSITION. *Let F be a field of characteristic different from 2 and let V be a vector space finitely generated over F on which we have a symmetric bilinear form f. Then there exists a basis $D = \{v_1, \ldots, v_n\}$ for V over F such that $T_{DD}(f)$ is a diagonal matrix.*

PROOF. If f is the 0-function then the result is trivially true, so that we can assume that this is not the case. We now proceed by induction on $n = dim(V)$. If $n = 1$ the result is immediate — since any 1×1 matrix is diagonal — and so we can assume that $n > 1$ and that the result has already been established for all vector spaces having dimension less than n.

Our first claim is that there exists a vector $v_1 \in V$ satisfying $f(v_1, v_1) \ne 0_F$. Indeed, assume that this is not the case. Then for all $v, v' \in V$ we have $0_F = f(v+v', v+v') = f(v, v) + 2f(v, v') + f(v', v') = 2f(v, v')$ and since the characteristic of F differs from 2, we conclude that $f(v, v') = 0_F$ for all $v, v' \in V$. But this happens only when f is the 0-function, a case which we have already ruled out. Thus there must exist a vector $v_1 \in V$ satisfying $f(v_1, v_1) \ne 0_F$.

Let $V_1 = Fv_1$ and let $V_2 = \{v' \in V \mid f(v_1, v') = 0_F\}$. This is a subspace of V satisfying $V_1 \cap V_2 = \{0_V\}$ and, indeed, as in the proof of Proposition 16.3 we can check that in fact $V = V_1 \oplus V_2$. In particular, $dim(V_2) = n - 1$ and so by the induction hypothesis we note that if f' is the restriction of f to V_2 then there exists a basis $C = \{v_2, \ldots, v_n\}$ for V_2 such that $T_{CC}(f')$ is a diagonal matrix. Since $f(v_1, v_i) = 0_F = f(v_i, v_1)$ for all $2 \le i \le n$, it follows that $D = \{v_1, \ldots, v_n\}$ is a basis for V over F satisfying the condition that $T_{DD}(f)$ is a diagonal matrix. $\square$

COROLLARY. *Every symmetric matrix over a field of characteristic other than 2 is congruent to a diagonal matrix.*

(16.6) PROPOSITION. *Let V be a vector space finitely generated over $\mathbb{C}$ and let f be a symmetric bilinear form on V having rank r. Then there exists a basis $D = \{v_1, \ldots, v_n\}$ for V over $\mathbb{C}$ satisfying the following conditions:*

(1) *The matrix $T_{DD}(f)$ is symmetric;*

(2) $f(v_i, v_i) = \begin{cases} 1_F, & \text{if } 1 \le i \le r \\ 0_F, & \text{otherwise.} \end{cases}$

PROOF. By Proposition 16.5 we see that there exists a basis $D = \{v_1, \ldots, v_n\}$ of V satisfying the condition that $T_{DD}(f)$ is a diagonal matrix. This matrix has rank r and so by renumbering the elements of the basis if necessary we can assume that $f(v_i, v_i) = 0$ if and only if $1 \le i \le r$. Since our field of scalars is $\mathbb{C}$, we know that for each $1 \le i \le r$ there exists a scalar c_i satisfying $c_i^2 = f(v_i, v_i)^{-1}$. If we replace v_i by $c_i v_i$ for all $1 \le i \le r$ we will obtain a basis having the desired properties. $\square$

Let f be a bilinear form defined on a vector space V over a field F. The function $q: V \to F$ defined by $q: v \mapsto f(v, v)$ is called the *quadratic form* defined by f. If the characteristic of F is not 2 then the bilinear form f is itself totally determined by this quadratic form, since $f(v, v') = \frac{1}{2}[q(v + v') - q(v) - q(v')]$ for all $v, v' \in V$. The systematic study of quadratic forms in Gauss' work on number theory at the beginning of the 19th century. The classification of quadratic forms also plays an important part in classical analytic geometry.

EXAMPLE

Let n be a positive integer and let f be the symmetric bilinear form defined on $\mathbb{R}^n$ by $f: (v, v') \mapsto v \cdot v'$. Then the quadratic form defined by f is the function $v \mapsto \|v\|^2$.

By Proposition 16.6 we note that if V is a vector space finitely generated over $\mathbb{C}$ and if f is a symmetric bilinear form of rank r defined on V then we can choose a basis $\{v_1, \ldots, v_n\}$ for V over $\mathbb{C}$ such that the quadratic form $q: V \to \mathbb{C}$ defined by f is of the form $q: \sum_{i=1}^n a_i v_i \mapsto \sum_{i=1}^r a_i^2$.

Let V be a vector space finitely generated over $\mathbb{R}$ and let f be a symmetric bilinear form defined on V which defines a quadratic form q. The bilinear form f is *nonnegative* if and only if $q(v) \ge 0$ for all $v \in V$ and it is *positive* if and only if $q(v) > 0$ for all $0_V \ne v \in V$.

EXAMPLE

Let n be a positive integer. The symmetric bilinear form $f: (v, v') \mapsto v \cdot v'$ defined on $\mathbb{R}^n$ is positive. If $1 \le k < n$ then the symmetric bilinear form defined on $\mathbb{R}^n$ by $([a_1, \ldots, a_n], [b_1, \ldots, b_n]) \mapsto \sum_{i=1}^k a_i b_i$ is nonnegative but not positive.

EXAMPLE

The symmetric bilinear form f defined on the vector space $\mathbb{R}^2$ defined by $f: ([a, b], [c, d]) \mapsto (a - b)(c - d)$ is nonnegative but not positive.

EXAMPLE

Let n be a positive integer and let $A = [a_{ij}] \in \mathcal{M}_{n \times n}(\mathbb{R})$ be the symmetric matrix defined by

$$a_{ij} = \begin{cases} min\{i, j\}, & \text{if } i \neq j \\ i, & \text{if } i = j. \end{cases}$$

This matrix is of the form BB^T, where $B = [b_{ij}]$ is the lower triangular matrix defined by

$$b_{ij} = \begin{cases} -1, & \text{if } j < i \\ 1 & \text{if } i = j. \end{cases}$$

The symmetric bilinear form f defined on $\mathbb{R}^n$ by $f: (v, w) \mapsto vAw^T = (vB) \cdot (wB)$ is positive.

(16.7) PROPOSITION. *Let V be a vector space finitely generated over the field $\mathbb{R}$ and let D be a basis for this space. If f is a symmetric bilinear form defined on V then:*

(1) *f is positive if and only if all of the eigenvalues of $T_{DD}(f)$ are positive.*

(2) *f is nonnegative if and only if all of the eigenvalues of $T_{DD}(f)$ are nonnegative.*

PROOF. (1) Let $A = T_{DD}(f)$. This matrix is real and symmetric and so $A = \overline{A}$. In particular, it represents an endomorphism α of $\mathbb{R}^n$. The change of basis from the canonical basis to the basis D is represented by a unitary matrix P satisfying $P^T = P^{-1}$. Thus we see that $PA\overline{P} = PAP^{-1}$ is a diagonal matrix $D = [d_{ij}]$. Since A is similar to D, both matrices have the same eigenvalues, namely those nonzero elements of the diagonal of D. But A and D are also congruent and so D represents f with respect to a suitable basis $\{v_1, \dots, v_n\}$ of V. If $v = \sum_{i=1}^n a_i v_i \in V$ then

$$q(v) = f(v, v) = [a_1 \quad \dots, a_n] D \begin{bmatrix} a_1 \\ \vdots \\ a_n \end{bmatrix} = \sum_{i=1}^n d_{ii} a_i^2$$

and so it is clear that if all of the d_{ii} are positive then also $q(v) > 0$ for all $v \in V$ and if f is a positive form then $d_{ii} = f(v_i, v_i) > 0$ for all i.

(2) The proof of this is similar. $\square$

As in the case of the inner product, a general bilinear form defines a general notion of orthogonality. If V is a vector space over a field F and if f is a bilinear form on V then vectors $v, w \in V$ are *f-orthogonal* if and only if $f(v, w) = 0_F$. In this case we write $v \perp_f w$. If W is a subspace of V then the *f-orthogonal complement* of W is $W^{\perp_f} = \{y \in V \mid w \perp_f y$ for all $w \in W\}$. The situation, however, may be radically different than in the case of inner products.

EXAMPLE

Let $F = \mathbb{Z}/(2)$ and let $V = F^4$. Define the bilinear form f on V by

$$f: ([a_1, \ldots, a_4], [b_1, \ldots, b_4]) \mapsto \sum_{i=1}^{4} a_i b_i.$$

Then $W = \{[0,0,0,0], [1,1,1,1], [1,0,1,0], [0,1,0,1]\}$ is a subspace of V and $W^{\perp_f} = W$.

However, we do note the following result.

(16.8) PROPOSITION. *Let V be a vector space finitely generated over a field F and let f be a bilinear form on V satisfying the condition that for each $0_V \neq v \in V$ there exists a vector $v' \in V$ satisfying $f(v, v') \neq 0_F$. If W is a subspace of V then:*

(1) *If $\delta \in D(W)$ there exists a vector $v \in V$ satisfying $\delta(w) = f(w, v)$ for all $w \in W$;*

(2) *$dim(W) + dim(W^{\perp_f}) = dim(V)$.*

PROOF. (1) Each vector $v \in V$ defines a linear functional $\delta_v \in D(V)$ by $\delta_v: v' \mapsto f(v', v)$. Moreover, the function $\phi: V \to D(V)$ defined by $\phi: v \mapsto \delta_v$ is a linear transformation which, by the given condition on f, is in fact a monomorphism. Since $dim(V) = dim(D(V))$ by Proposition 12.2, this means that it is in fact an isomorphism. Now let $\delta \in D(W)$ and let W' be a complement of W in V. Then the function from $V = W \oplus W'$ to F given by $w + w' \mapsto \delta(w)$ belongs to $D(V)$ and so equals δ_v for some $v \in V$. In particular, we then have $\delta(w) = \delta_v(w) = f(w, v)$ for all $w \in W$.

(2) The function $\psi: V \to D(W)$ which assigns to each $v \in V$ the restriction of δ_v to W is a linear transformation which, by (1), is in fact an epimorphism. The kernel of ψ is $\{v \in V \mid f(w, v) = 0_F$ for all $w \in W\} = W^{\perp_f}$. and so, by Proposition 1.9, we have $dim(W) + dim(W^{\perp_f}) = dim(V)$. $\square$

In particular, we notice that, under the conditions of Proposition 16.8, a necessary and sufficient condition for $V = W \oplus W^{\perp_f}$ is that $W \cap W^{\perp_f} = \{0_V\}$ or, equivalently, that $V = W + W^{\perp_f}$.

We now return to general bilinear transformations. Let V and W be vector spaces over the same field F and let $G = F^{(V \times W)}$ be the set of all functions from $V \times W$ to F which assign a nonzero value to only finitely-many members of $V \times W$. Then G is a subspace of $F^{V \times W}$ having a basis $\{g_{v,w} \mid v \in V; w \in W\}$, where $g_{v,w}$ is the function satisfying the condition that $g_{v,w}(v', w') = 1_F$ if $(v', w') = (v, w)$ and $g_{v,w}(v', w') = 0_F$ otherwise. Let H be the subspace of G generated by the set of all functions of the form $g_{v+v',w} - g_{v,w} - g_{v',w}, g_{v,w+w'} - g_{v,w} - g_{v,w'}, g_{av,w} - ag_{v,w}$ or $g_{v,aw} - ag_{v,w}$, where $v, v' \in V$, $w, w' \in W$, and $a \in F$. We choose a complement for H in G, which we will denote by $V \otimes W$. By Proposition 5.6 we know that this subspace is unique up to isomorphism. If τ is the projection of G onto $V \otimes W$ which comes from the direct-sum decomposition $G = H \oplus (V \otimes W)$ then for each $v \in V$ and $w \in W$ we will denote $\tau(g_{v,w})$ by $v \otimes w$. Then $\{v \otimes w \mid v \in V; w \in W\}$ is a generating set for $V \otimes W$. Moreover, if $v, v' \in V$ and $w \in W$ then $[v+v'] \otimes w - v \otimes w - v' \otimes w = \tau(g_{v+v',w} - g_{v,w} - g_{v',w}) = 0_G$ and so $[v+v'] \otimes w = v \otimes w + v' \otimes w$. A similar argument show that for all $v \in$ and all $w, w' \in W$ we have $v \otimes [w + w'] = v \otimes w + v \otimes w'$ and for all $v \in V$, all $w \in W$, and all $a \in F$ we have $av \otimes w = a[v \otimes w] = v \otimes aw$. The space $V \otimes W$ is called the *tensor product* of V and W over F. As we have seen, the function $t: V \times W \to V \otimes W$ defined by $t: (v, w) \mapsto v \otimes w$ is a bilinear transformation. This transformation has special significance in light of the following proposition, which allows us to move from bilinear transformations to linear transformations.

(16.9) PROPOSITION. *Let V, W, and Y be vector spaces over the same field F. For every bilinear transformation $f: V \times W \to Y$ there exists a unique linear transformation $\alpha: V \otimes W \to Y$ satisfying $f = \alpha t$.*

PROOF. Let $G = G^{(V \times W)}$. Given a bilinear transformation f there exists a linear transformation $\beta: G \to Y$ defined on elements of a basis defined by $\beta(g_{v,w}) = f(v, w)$. Since f is a bilinear transformation, the subspace H of G defined above is contained in $ker(\beta)$ and so we can define a linear transformation $\alpha: V \otimes W \to Y$ by

$$\alpha: \sum_{i=1}^{n} a_i[v_i \otimes w_i] \mapsto \sum_{i=1}^{n} a_i f(v_i, w_i).$$

This function is indeed well-defined since if $\sum_{i=1}^{n} a_i[v_i \otimes w_i] = \sum_{i=1}^{n} b_i[v_i \otimes w_i]$ then $\sum_{i=1}^{n}(a_i - b_i)g_{v_i, w_i} \in H$ and so $\beta(\sum_{i=1}^{n}(a_i - b_i)g_{v_i, w_i}) = 0_Y$. This implies that $\alpha(\sum_{i=1}^{n} a_i[v_i \otimes w_i]) = \alpha(\sum_{i=1}^{n} b_i[v_i \otimes w_i])$. Clearly α is a linear transformation which satisfies $f = \alpha t$.

We are left to show uniqueness. Suppose that $\alpha': V \otimes W \to Y$ is a linear transformation satisfying $f = \alpha' t$. For each $v \in V$ and $w \in W$ we then have $\alpha'(v \otimes w) = \alpha' t(v, w) = f(v, w) = \alpha t(v, w) = \alpha(v \otimes w)$. Since $\{v \otimes w \mid v \in V; w \in W\}$ is a generating set for $V \otimes W$, this proves that $\alpha' = \alpha$. □

EXAMPLE

Let V and W be vector spaces over the same field F. Let $\delta: V \to F$ and $\delta': W \to F$ be linear functionals. Then there exists a bilinear form in $Bil(V \times W)$ defined by $(v, w) \mapsto \delta(v)\delta'(w)$. By Proposition 16.9 there exists a linear functional $\delta \otimes \delta' \in D(V \otimes W)$ satisfying

$$\delta \otimes \delta': \sum_{i=1}^{n} c_i(v_i \otimes w_i) \mapsto \sum_{i=1}^{n} c_i \delta(v_i)\delta'(w_i).$$

EXAMPLE (Continued)

More generally, if V and W are vector spaces over the same field F, if α is an endomorphism of V, and if β is an endomorphism of W then the function from $V \times W$ to $V \otimes W$ defined by $(v, w) \mapsto \alpha(v) \otimes \beta(w)$ is a bilinear transformation and so defines an endomorphism $\alpha \otimes \beta$ of $V \otimes W$ satisfying

$$\alpha \otimes \beta: \sum_{i=1}^{n} c_i(v_i \otimes w_i) \mapsto \sum_{i=1}^{n} c_i[\alpha(v_i) \otimes \beta(w_i)].$$

(16.10) PROPOSITION. *If V and W are vector spaces finitely generated over the same field F then $V \otimes W$ is also finitely generated over F and $dim(V \otimes W) = dim(V)dim(W)$.*

PROOF. Let $\{v_1, \ldots, v_k\}$ be a basis for V over F and let $\{w_1, \ldots, w_n\}$ be a basis for W over F. If $v = \sum_{i=1}^{k} a_i v_i \in V$ and $w = \sum_{j=1}^{n} b_j w_j \in W$ then $v \otimes w = \sum_{i=1}^{k} \sum_{j=1}^{n} a_i b_j (v_i \otimes w_j)$ and so $\{v_i \otimes w_j \mid 1 \leq i \leq k; 1 \leq j \leq n\}$ is a generating set for $V \otimes W$ over F. Thus $dim(V \otimes W) \leq kn = dim(V)dim(W)$. We must prove the reverse inequality, and to do so it suffices to find a vector space Y of dimension kn and an epimorphism from $V \otimes W$ to Y.

If $D(V)$ is the dual space of V then by Proposition 12.2 we note that the dimension of $D(V)$ equals k and so by Proposition 6.1 we know that the vector space $Hom(D(V), W)$ has dimension kn over F. For each $v \in V$ and each $w \in W$ there exists a linear transformation $\theta_{v,w}: D(V) \rightarrow W$ defined by $\theta_{v,w}: \delta \mapsto \delta(v)w$ and it is easy to verify that the function $f: V \times W \rightarrow Hom(D(V), W)$ defined by $f: (v, w) \mapsto \theta_{v,w}$ is a bilinear transformation. Therefore there exists a linear transformation $\alpha: V \otimes W \rightarrow Hom(D(V), W)$ satisfying $\alpha t = f$, where $t: V \times W \rightarrow V \otimes W$ is the bilinear transformation in the definition of the tensor product. We will be done when we show that α is an epimorphism. Indeed, by Proposition 12.2 we know that there exists a basis $\{\delta_1, \ldots, \delta_n\}$ of $D(V)$ composed of linear functionals satisfying the condition

$$\delta_i(v_h) = \begin{cases} 1_F, & \text{when } i = h \\ 0_F, & \text{otherwise.} \end{cases}$$

Therefore we have a basis $\{\theta_{ij} \mid 1 \leq i \leq k; 1 \leq j \leq n\}$ for $Hom(D(V), W)$ over F, where θ_{ij} satisfies the condition

$$\theta_{ij}(\delta_h) = \begin{cases} w_j, & \text{when } i = h \\ 0_W, & \text{otherwise.} \end{cases}$$

But then $\alpha(v_i \otimes w_j) = f(v_i, w_j) = \theta_{ij}$. Thus the image of α contains a basis for $Hom(D(V), W)$ and so must equal $Hom(D(V), W)$, as desired. $\square$

EXAMPLE

Let n be a positive integer, let F be a field, and let V be a vector space of finite dimension k over F. Pick a basis $\{v_1, \ldots, v_k\}$ for V over F. Then the vector space $\mathcal{M}_{n \times n}(V)$ has dimension $n^2 k$ over F. Let $f: \mathcal{M}_{n \times n}(F) \times V \to \mathcal{M}_{n \times n}(V)$ be the bilinear transformation given by $([a_{ij}], v) \mapsto [a_{ij} v]$. This bilinear transformation in turn defines a linear transformation $\alpha: \mathcal{M}_{n \times n}(F) \otimes V \to \mathcal{M}_{n \times n}(V)$ which is clearly an epimorphism. But by Proposition 16.10 we know that

$$dim(\mathcal{M}_{n \times n}(F) \otimes V) = n^2 k$$

and so α must be an isomorphism. Thus $\mathcal{M}_{n \times n}(F) \otimes V \cong \mathcal{M}_{n \times n}(V)$.

EXAMPLE

Let F be a field, let k, n, s, and t be positive integers, and let

$$f: \mathcal{M}_{k \times n}(F) \times \mathcal{M}_{s \times t}(F) \to \mathcal{M}_{ks \times nt}(F)$$

be the function defined by

$$f: ([a_{ij}], B) \mapsto \begin{bmatrix} a_{11}B & \ldots & a_{1n}B \\ & \ldots & \\ a_{k1}B & \ldots & a_{kn}B \end{bmatrix},$$

which is a bilinear transformation of vector spaces over F and so defines a linear transformation $\alpha: \mathcal{M}_{k \times n}(F) \otimes \mathcal{M}_{s \times t}(F) \to \mathcal{M}_{ks \times nt}(F)$. Let D be a matrix in $\mathcal{M}_{ks \times nt}(F)$, which we will write as $\begin{bmatrix} D_{11} & \ldots & D_{1n} \\ & \ldots & \\ D_{k1} & \ldots & D_{kn} \end{bmatrix}$, where each D_{ij} is a block of size $s \times t$. Then $D = \sum_{i=1}^{k} \sum_{j=1}^{n} G(i, j)$, where $G(i, j)$ is the matrix in $\mathcal{M}_{ks \times nt}(F)$ which, when written in such a block form, has D_{ij} in the (i, j)-position and $\mathbf{O}$ elsewhere. In other words, $G(i, j) = \alpha(E_{ij} \otimes D_{ij})$, where $E_{ij} \in \mathcal{M}_{k \times n}(F)$ is the matrix having 1_F as its (i, j) entry and 0_F elsewhere. From this it follows that α is an epimorphism and so by Proposition 16.10 we see that

$$dim(\mathcal{M}_{k \times n}(F) \otimes \mathcal{M}_{s \times t}(F)) = knst = \mathcal{M}_{ks \times nt}(F)$$

and so α is an isomorphism of vector spaces over F. Thus we can write $A \otimes B$ instead of $f(A, B)$. This matrix is called the *Kronecker product* of the matrices A and B. Kronecker products play an important part in advanced matrix theory.

Next, we present an important characterization of the tensor product of two finitely-generated vector spaces.

(16.11) PROPOSITION. *If V and W are vector spaces finitely generated over a field F then $V \otimes W \cong D(Bil(V \times W))$.*

PROOF. Let $dim(V) = k$ and $dim(W) = n$. For each $v \in V$ and each $w \in W$ define a function $\psi_{v,w}: Bil(V \times W) \to F$ by setting $\psi_{v,w}: f \mapsto f(v,w)$. One easily verifies that this is in fact a linear functional and so belongs to $D(Bil(V \times W))$. Moreover, the function $\Psi: V \times W \to D(Bil(V \times W))$ defined by $\Psi: (v,w) \mapsto \psi_{v,w}$ is a monic bilinear transformation and so defines a monomorphism from $V \otimes W$ to $D(Bil(V \times W))$. But by Proposition 16.10 we know that $dim(V \otimes W) = kn$ and by Propositions 12.2 and 16.1 we know that

$$dim(D(Bil(V \times W))) = dim(Bil(V \times W)) = dim(\mathcal{M}_{k \times n}(F)) = kn$$

and so this monomorphism is in fact an isomorphism. □

Finally, we show that tensor products are, up to isomorphism, commutative and associative.

(16.12) PROPOSITION. *If V, W, and Y are vector spaces over a field F then:*
(1) *There exists an isomorphism $V \otimes W \to W \otimes V$ satisfying $v \otimes w \mapsto w \otimes v$ for all $v \in V$ and $w \in W$.*
(2) *There exists an isomorphism $(V \otimes W) \otimes Y \to V \otimes (W \otimes Y)$ satisfying $(v \otimes w) \otimes y \mapsto v \otimes (w \otimes y)$ for all $v \in V$, $w \in W$, and $y \in Y$.*

PROOF. (1) Let $f: V \times W \to W \otimes V$ be the function defined by $f: (v,w) \mapsto w \otimes v$. This is surely a bilinear transformation and so, by Proposition 16.9, there exists a unique linear transformation $\alpha: V \otimes W \to W \otimes V$ satisfying $\alpha: v \otimes w \mapsto w \otimes v$ for all $v \in V$ and $w \in W$. By a similar argument, there exists a unique linear transformation $\alpha': W \otimes V \to V \otimes W$ satisfying $\alpha': w \otimes v \mapsto v \otimes w$. Then $\alpha'\alpha$ is the identity function on $V \otimes W$ and $\alpha\alpha'$ is the identity function on $W \otimes V$, proving that α is an isomorphism with $\alpha' = \alpha^{-1}$.
(2) For a fixed $y \in Y$ we have a bilinear transformation $g_y: V \times W \to V \otimes (W \otimes Y)$ defined by $g_y: (v,w) \mapsto v \otimes (w \otimes y)$ for all $v \in V$ and $w \in W$ and so, by Proposition 16.9, there exists a unique linear transformation $\beta_y: V \otimes W \to V \otimes (W \otimes Y)$ satisfying $\beta_y: v \times w \mapsto v \otimes (w \otimes y)$ for all $v \in V$ and $w \in W$. Similarly, the function $g: (V \otimes W) \times Y \to V \otimes (W \otimes Y)$ defined by $g: (x,y) \mapsto g_y(x)$ for each $x \in V \otimes W$ and each $y \in Y$ is a bilinear transformation and so defines a unique linear transformation $\beta: (V \otimes W) \otimes Y \to V \otimes (W \otimes Y)$ satisfying $\beta: (v \otimes w) \otimes y \mapsto v \otimes (w \otimes y)$ for all $v \in V$, $w \in W$, and $y \in Y$. In the same manner, we can also define a linear transformation $\beta': V \otimes (W \otimes Y) \to (V \otimes W) \otimes Y$ satisfying $\beta': v \otimes (w \otimes y) \mapsto (v \otimes w) \otimes y$ for all $v \in V$, $w \in W$, and $y \in Y$. Clearly $\beta'\beta$ is the identity function on $(V \otimes W) \otimes Y$ and $\beta\beta'$ is the identity function on $V \otimes (W \otimes Y)$ and so β is an isomorphism with $\beta' = \beta^{-1}$. □

Problems

1. Let V be the vector space over $\mathbb{R}$ consisting of all functions in $\mathbb{R}^{\mathbb{R}}$ differentiable infinitely-many times. Let $\mu: V \times V \to V$ be the function defined by $\mu: (f,g) \mapsto (fg)'$. Is μ a bilinear transformation?

2. Let D be the canonical basis for $\mathbb{R}^2$ over $\mathbb{R}$. Find a bilinear form f on V satisfying $T_{DD}(f) = \begin{bmatrix} 2 & 2 \\ 4 & -1 \end{bmatrix}$.

3. Let $F = \mathbb{Z}/(2)$ and let A be a finite nonempty set. Define a function $u: F^A \times F^A \to F$ as follows: if $f, g \in F^A$ and if the number of elements $a \in A$ such that $f(a) = g(a)$ is even then $u(f, g) = 1$. Otherwise, $u(f, g) = 0$. Is u a bilinear form on F^A?

4. Let D be the canonical basis for $\mathbb{R}^3$ over $\mathbb{R}$. Find $T_{DD}(f)$, where f is the bilinear form on $\mathbb{R}^3$ defined by

$$f([a, b, c], [d, e, f]) = ad + 2bf + cf + 2ce - ae + be - bd.$$

5. Let $\mathbb{R}^3 \times \mathbb{R}^3 \to \mathbb{R}$ be the bilinear form defined by $f: (v, w) \mapsto vAw^T$, where $A = \begin{bmatrix} 0 & 1 & 0 \\ 1 & 0 & 2 \\ 0 & 1 & 1 \end{bmatrix}$. Find the matrix representing f with respect to the basis $\{[1, 0, 0], [0, 1, 1], [1, 0, 1]\}$ of $\mathbb{R}^3$.

6. Let V be a vector space finitely generated over a field F. Characterize the bilinear forms of rank 1 on V.

7. Let V and W be vector spaces over the same field F and let $\alpha: V \to W$ be a given linear transformation. For each bilinear form g on W define the function $g_\alpha: V \times V \to F$ by $g_\alpha: (v, v') \mapsto g(\alpha(v), \alpha(v'))$. Show that g_α is a bilinear form on V and that the function $g \mapsto g_\alpha$ is a linear transformation from $Bil(W \times W)$ to $Bil(V \times V)$.

8. Find a bilinear form on the vector space $\mathbb{R}^3$ over $\mathbb{R}$ which defines the quadratic form $[a, bc] \mapsto a^2 - 2ab + 4ac - 2bc + 2c^2$ from $\mathbb{R}^3$ to $\mathbb{R}$.

9. Let f be the bilinear form on $\mathbb{R}^4$ defined by

$$f: (v, w) \mapsto v \begin{bmatrix} 1 & 0 & 0 & 0 \\ 0 & 1 & 0 & 0 \\ 0 & 0 & 1 & 0 \\ 0 & 0 & 0 & -1 \end{bmatrix} w^T.$$

Find a basis $\{v_1, v_2, v_3, v_4\}$ of $\mathbb{R}^4$ satisfying $f(v_i, v_i) = 0_F$ for all $1 \le i \le 4$.

10. Are the matrices $\begin{bmatrix} 0 & 0 & 0 \\ 1 & 1 & 0 \\ 1 & 1 & 1 \end{bmatrix}$ and $\begin{bmatrix} 1 & -1/5 & 7/5 \\ 0 & -3/25 & 21/25 \\ 0 & -4/25 & 28/25 \end{bmatrix}$ in $\mathcal{M}_{3\times3}(\mathbb{R})$ congruent?

11. Find an upper triangular matrix in $\mathcal{M}_{3\times3}(\mathbb{R})$ congruent to $\begin{bmatrix} 1 & 0 & -2 \\ -1 & 1 & 0 \\ 0 & -2 & 4 \end{bmatrix}$.

12. Let F be a field. Show that every upper triangular matrix in $\mathcal{M}_{3\times3}(F)$ is both congruent and similar to a lower triangular matrix.

13. Let n be a positive integer and let $A \in \mathcal{M}_{3\times3}(\mathbb{C})$ be a symmetric nonsingular matrix. Show that A and A^{-1} are congruent.

14. Find diagonal matrices with entries in $\mathbb{R}$ congruent to each of the following matrices:

(i) $\begin{bmatrix} 1 & 2 & 3 & 2 \\ 2 & 3 & 5 & 8 \\ 3 & 5 & 8 & 10 \\ 2 & 8 & 10 & -8 \end{bmatrix}$;

(ii) $\begin{bmatrix} 2 & 1 & 3 \\ 1 & 0 & 1 \\ 3 & 1 & 3 \end{bmatrix}$;

(iii) $\begin{bmatrix} 1 & 1 & 1 & 1 \\ 1 & 1 & 1 & 1 \\ 1 & 1 & 1 & 1 \\ 1 & 1 & 1 & 1 \end{bmatrix}$.

15. Show that the matrix $\begin{bmatrix} 1 & i & 1+i \\ i & 0 & 2-i \\ 1+i & 2-i & 10+2i \end{bmatrix} \in \mathcal{M}_{3\times3}(\mathbb{C})$ is congruent to I.

16. Let n be a positive integer and let α be a positive endomorphism of $\mathbb{R}^n$ represented by a matrix A with respect to the canonical basis. If A' is a matrix congruent to A, does it also represent a positive endomorphism of $\mathbb{R}^n$ with respect to this same basis?

17. Let F be a field of characteristic other than 2 and let V be a vector space finitely generated over it. If $0 \neq f: V \times V \to F$ is a symmetric bilinear form, show that there exists a vector $y \in V$ satisfying $f(y,y) \neq 0_F$.

18. Let n be a positive integer, let F be a field, and let $A \in \mathcal{M}_{n\times n}(F)$. Show that there exists a symmetric matrix $B \in \mathcal{M}_{n\times n}(F)$ such that $vAv^T = vBv^T$ for all $v \in F^n$.

19. Is the symmetric bilinear form f on $\mathbb{R}^3$ defined by

$$f: ([a,b,c],[a',b',c']) \mapsto [a,b,c] \begin{bmatrix} 2 & -1 & 0 \\ -1 & 2 & -1 \\ 0 & -1 & 2 \end{bmatrix} \begin{bmatrix} a' \\ b' \\ c' \end{bmatrix}$$

positive?

20. Let f be the symmetric bilinear form on $\mathbb{R}^3$ defined by the matrix

$$\begin{bmatrix} -3 & 1 & 0 \\ 1 & -6 & 1 \\ 0 & 1 & 7 \end{bmatrix}.$$

Find the quadratic form defined by f.

21. Find the symmetric bilinear form on $\mathbb{R}^3$ which defines the quadratic form $[a,b,c] \mapsto 2ab + 4ac + 6bc$.

22. Let V be a vector space over F. Show that $V \cong F \otimes V$.

23. Let F be a subfield of a field K and consider K as a vector space over F. If V is a vector space over F, show that $K \otimes V$ is a vector space over K.

24. Let V be a vector space over a field F. Is the function from $V \otimes V$ to itself defined by $\sum[v_i \otimes y_i] \mapsto \sum[y_i \otimes v_i]$ an endomorphism of $V \otimes V$?

25. Let $F = \mathbb{Z}/(2)$, let $V = F^4$, and let $W = F\{[0,1,0,0],[0,1,0,1]\}$. If f is the bilinear form on V defined by $f\colon ([a_1, a_2, a_3, a_4], [b_1, b_2, b_3, b_4]) \mapsto \sum_{i=1}^{4} a_i b_i$, find $W^{\perp_f}$.

26. Let F be a field of characteristic other than 2 and let V be a vector space over F. Show that the following conditions are equivalent for a bilinear form f on V:

(1) $f(v, v) \neq 0_F$ for all $0_V \neq v \in V$;
(2) If W is a nontrivial subspace of V then for each $0_V \neq w \in W$ there exists an element $w' \in W$ satisfying $f(w, w') \neq 0_F$.

27. Let F be a field of characteristic 0 and let f be the bilinear form defined on $V = F^3$ by $f\colon ([a, b, c], [d, e, f]) \mapsto ad + be - cf$. Let W be a subspace of V satisfying the condition that $f(w, w') = 0_F$ for all $w, w' \in W$. Is W necessarily trivial?

28. Let V be a vector space of finite dimension n over a field F and let Y be the subspace of $V \otimes V$ generated by the set of all vectors of the form $v \otimes v' - v' \otimes v$ for $v, v' \in V$. Calculate $dim(Y)$.

29. Let V be a vector space of finite dimension over a field F and let v and v' be nonzero elements of V satisfying $v \otimes v' = 0_{V \otimes V}$. Show that $v' \in Fv$.

CHAPTER XVII

ALGEBRAS OVER A FIELD

Up to now we have not considered the possibility of multiplying two vectors to obtain another vector, though we have noted that this is possible in certain cases. For example, we can multiply elements of the vector space $\mathcal{M}_{n \times n}(F)$ over a field F. A vector space V over a field F is an *algebra* over F if and only if there exists a bilinear transformation $(v, w) \mapsto vw$ from $V \times V$ to V satisfying the following additional conditions:

(1) $u(v + w) = uv + uw$;
(2) $(u + v)w = uw + vw$; and
(3) $a(vw) = v(aw) = (av)w$

for all $a \in F$ and all $u, v, w \in V$. In case the additional condition

(4) $u(vw) = (uv)w$

is satisfied for all $u, v, w \in V$ the algebra V is *associative*. A subspace of V which is closed under this multiplication is called a *subalgebra* of V.

EXAMPLE

If F is a subfield of a field K then K is an associative algebra over F. Thus, in particular, $\mathbb{R}$ is an associative algebra over $\mathbb{Q}$ and $\mathbb{C}$ is an associative algebra over $\mathbb{R}$.

Hence we see that the vector space $\mathbb{R}^2$ is an algebra over $\mathbb{R}$ with multiplication defined by $[a, b][c, d] = [ac - bd, ad + bc]$. (This, of course, is just the multiplication in $\mathbb{C}$, which is isomorphic to $\mathbb{R}^2$ as a vector space over $\mathbb{R}$.) We can also define another product on $\mathbb{R}^2$ which would turn it into an associative algebra over $\mathbb{R}$, namely $[a, b][c, d] = [ac, bd]$. Thus we see that the addition in an algebra over a field does not necessarily determine the nature of the multiplication.

EXAMPLE

If F is a field and n is a positive integer then $\mathcal{M}_{n \times n}(F)$ is an associative algebra over F. More generally, if V is a vector space over F then $End(V)$ is a vector space over F with scalar multiplication defined by $a\alpha: v \mapsto a\alpha(v)$ for all $a \in F$ and $\alpha \in End(V)$. Moreover, $End(V)$ is an associative algebra over F.

EXAMPLE

If F is a field then $F[X]$ is an associative algebra over F. This example can be generalized as well. Let H be a nonempty set on which we have defined an associative operation $*$. Thus, for example, we can take H to be the set of all positive integers under addition or under multiplication. Let V be a vector space over F having a basis $\{v_h \mid h \in H\}$ and define multiplication of elements of V as follows: if $v = \sum_{g \in H} a_g v_g$ and $v' = \sum_{h \in H} b_h v_h$ are vectors in V, then vv' is the vector $\sum_{g \in H} \sum_{h \in H} a_g b_h v_{g*h}$. (This sum is well-defined since only finitely-many of the coefficients a_g and b_h are nonzero.) Then V can easily be checked to be an associative algebra over F, which is often denoted by $F[H]$.

If $H = \{X^i \mid i \geq 0\}$ under the operation of multiplication, this construction yields the algebra $F[X]$.

EXAMPLE

Let $\{v_1, v_2, v_3, v_4\}$ be the canonical basis for $\mathbb{R}^4$ over $\mathbb{R}$ and define an operation $*$ on $\mathbb{R}^4$ by

$$\left(\sum_{i=1}^4 a_i v_i\right) * \left(\sum_{j=1}^4 b_j v_j\right) = \sum_{i=1}^4 \sum_{j=1}^4 a_i b_j (v_i * v_j)$$

where the elements $v_i * v_j$ of $\mathbb{R}^4$ are given by:

$$v_1 * v_1 = -v_2 * v_2 = -v_3 * v_3 = -v_4 * v_4 = v_1$$
$$v_1 * v_2 = v_2 * v_1 = v_3 * v_4 = -v_4 * v_3 = v_2$$
$$v_1 * v_3 = -v_2 * v_4 = v_3 * v_1 = v_4 * v_2 = v_3$$
$$v_1 * v_4 = v_2 * v_3 = -v_3 * v_2 = v_4 * v_1 = v_4$$

This operation defines on $\mathbb{R}^4$ the structure of an associative algebra over R. In fact, $(\mathbb{R}^4, +, *)$ satisfies all of the conditions for being a field *except* for the commutativity of multiplication. This algebra is called the *algebra of real quaternions*. It was introduced in 1844 by the Irish mathematician and physicist William R. Hamilton as a generalization of the field of complex numbers.

The algebra of real quaternions can also be represented in terms of matrices. Consider $\mathcal{M}_{2 \times 2}(\mathbb{C})$ as a vector space over $\mathbb{R}$. Then the subspace of $\mathcal{M}_{2 \times 2}(\mathbb{C})$ generated over $\mathbb{R}$ by the set

$$\left\{ \begin{bmatrix} 1 & 0 \\ 0 & 1 \end{bmatrix}, \begin{bmatrix} -i & 0 \\ 0 & i \end{bmatrix}, \begin{bmatrix} 0 & 1 \\ 1 & 0 \end{bmatrix}, \begin{bmatrix} 0 & -i \\ -i & 0 \end{bmatrix} \right\}$$

is closed under taking products of matrices and every nonzero matrix in it is invertible. Indeed, its multiplication table is the same as given above, so this is just the algebra of real quaternions.

(17.1) PROPOSITION. *If V is an associative algebra over a field F having a* *neutral element with respect to multiplication then V is isomorphic to a subalgebra of $End(V)$.*

PROOF. If $v \in V$ define the endomorphism α_v of V by $\alpha_v: w \mapsto wv$. For each $v, v' \in V$ and all $c \in F$ we have $\alpha_{v+v'} = \alpha_v + \alpha_{v'}$, $\alpha_{vv'} = \alpha_v \alpha_{v'}$ and $c\alpha_v = \alpha_{cv}$. Thus $\{\alpha_v \mid v \in V\}$ is a subalgebra of $End(V)$ and we have a linear transformation from V to this algebra defined by $v \mapsto \alpha_v$. Surely this function is epic; it is also monic since $\alpha_v = \alpha_{v'}$ for $v, v' \in V$ implies that $v = \alpha_v(1) = \alpha_{v'}(1) = v'$. This we have the desired isomorphism. □

COROLLARY. *An associative algebra finitely generated over a field F is isomorphic to a subalgebra of $\mathcal{M}_{n \times n}(F)$ for some positive integer n.*

We now turn to consider algebras which are not necessarily associative and begin by considering a very important example of such an algebra, having important applications in analytic geometry.

EXAMPLE

Define multiplication $\times$ between vectors in $\mathbb{R}^3$ as follows: if $v = [a_1, a_2, a_3]$ and $w = [b_1, b_2, b_3]$ then

$$v \times w = [a_2 b_3 - a_3 b_2, a_3 b_1 - a_1 b_3, a_1 b_2 - a_2 b_1].$$

Thus, for example, if $\{v_1, v_2, v_3\}$ is the canonical basis of $\mathbb{R}^3$ then

$$v_1 \times v_2 = v_3;$$
$$v_2 \times v_3 = v_1;$$
$$v_3 \times v_1 = v_2.$$

It is easy to check that this product, known as the *cross product*, induces the structure of an algebra on $\mathbb{R}^3$ and satisfies the condition

(*) $\qquad\qquad v \times w = -(w \times v)$ for all $v, w \in \mathbb{R}^3$.

From this condition we see that $v \times v = 0_V$ for all $v \in \mathbb{R}^3$. Moreover, the cross product is not associative since $(v_1 \times v_2) \times v_2 = v_3 \times v_2 = -v_1$ while $v_1 \times (v_2 \times v_2) = v_1 \times 0_V = 0_V$.

Let us look at this example in more detail. As a direct consequence of the definition we see that if $u, v, w \in \mathbb{R}^3$ then

$$u \times (v \times w) = (u \cdot w)v - (u \cdot v)w$$

and

$$(u \times v) \times w = (u \cdot w)v - (v \cdot w)u.$$

The cross product is related to the dot product on $\mathbb{R}^3$ by the relations

**) $\qquad\qquad v \cdot (v \times w) = 0$

and

$$(***) \qquad\qquad \|v \times w\|^2 = \|v\|^2\|w\|^2 - (v \cdot w)^2$$

for all $v, w \in \mathbb{R}^3$. Equality (**) says, in particular, that in the inner product space $\mathbb{R}^3$ (with the dot product) the vectors $v \times w$ and v are orthogonal and, by the same argument, the vectors $v \times w$ and w are orthogonal. Thus we see if $\{v, w\}$ is a linearly-independent subset of $\mathbb{R}^3$ then $\{v, w, v \times w\}$ is a basis for $\mathbb{R}^3$. In particular, if $u, v, w \in \mathbb{R}^3$ then

$$u \times (v \times w) + v \times (w \times u) + w \times (u \times v) = [0, 0, 0].$$

(17.2) PROPOSITION. *If $v = [a_1, a_2, a_3]$ and $w = [b_1, b_2, b_3]$ are nonzero vectors in $\mathbb{R}^3$ satisfying $v \times w = [0, 0, 0]$ then $\mathbb{R}v = \mathbb{R}w$ and $v \cdot w = 0$.*

PROOF. Since these vectors are nonzero we can assume, without loss of generality, that $b_1 \neq 0$. Since $v \times w = [0, 0, 0]$ we have $a_2 b_3 - a_3 b_2 = a_3 b_1 - a_1 b_3 = a_1 b_2 - a_2 b_1 = 0$ and so $a_i = (a_1 b_1^{-1}) b_i$ for $i = 1, 2, 3$. Thus v is a scalar multiple of w and so $\mathbb{R}v = \mathbb{R}w$. Moreover, by (***) we must have $v \cdot w = 0$. $\square$

Recall that if $v, w \in \mathbb{R}^3$ then the angle θ between v and w satisfies the condition $v \cdot w = (\|v\| \cdot \|w\|)cos(\theta)$.

(17.3) PROPOSITION. *If $v, w \in \mathbb{R}^3$ and if θ is the angle between these vectors then $\|v \times w\| = (\|v\| \cdot \|w\|)|sin(\theta)|$.*

PROOF. By (***) we see that

$$\begin{aligned}
\|v \times w\|^2 &= \|v\|^2\|w\|^2 - (v \cdot w)^2 \\
&= \|v\|^2\|w\|^2[1 - cos^2(\theta)] \\
&= \|v\|^2\|w\|^2 sin^2(\theta)
\end{aligned}$$

and from here we get the desired result. $\square$

(17.4) PROPOSITION. *If $u, v, w \in \mathbb{R}^3$ then:*

(1) $u \cdot (v \times w) = v \cdot (w \times u) = w \cdot (u \times v)$;
(2) $u \cdot (w \times v) = -w \cdot (u \times v)$;
(3) $u \cdot (v \times w) = 0$ *if and only if two of these vectors are equal or the set $\{u, v, w\}$ is linearly dependent.*

PROOF. If $u = [a_1, a_2, a_3]$, $v = [b_1, b_2, b_3]$, and $w = [c_1, c_2, c_3]$ are vectors in $\mathbb{R}^3$ then

$$u \cdot (v \times w) = a_1(b_2 c_3 - b_3 c_2) - a_2(b_3 c_1 - b_1 c_3) + a_3(b_1 c_2 - b_2 c_1)$$

$$= \begin{vmatrix} a_1 & a_2 & a_3 \\ b_1 & b_2 & b_3 \\ c_1 & c_2 & c_3 \end{vmatrix}$$

and from this (1), (2), and (3) follow immediately. $\square$

Note that if n is a positive integer then it is always possible to define a product on $\mathbb{R}^n$ which would turn this space into an associative algebra over $\mathbb{R}$ – simply set $[a_1, \ldots, a_n][b_1, \ldots, b_n] = [a_1 b_1, \ldots, a_n b_n]$. However, it is not clear that it is possible to define a product similar to the cross product on $\mathbb{R}^n$ for other values of n. In other words, it is not clear that there exist positive integers n such that we can on $\mathbb{R}^n$ it is possible to define a nontrivial vector product $\times$ turning $\mathbb{R}^n$ into an algebra satisfying (*), (**), and (***)? We have seen one way of doing it for $n = 3$. There is another possible product $\times'$ on $\mathbb{R}^3$, defined by $v \times' w = -(v \times w)$ for all $v, w \in \mathbb{R}^3$. If $n = 7$ we can define such a product as follows: write an element of $\mathbb{R}^7$ in the form $[v, a, v']$, where $a \in \mathbb{R}$ and $v, v' \in \mathbb{R}^3$. The dot product on $\mathbb{R}^7$ is given by

$$[v, a, v'] \cdot [w, b, w'] = v \cdot w + ab + v' \cdot w'$$

and we can define a product $\times$ on $\mathbb{R}^7$ by setting $[v, a, v'] \times [w, b, w']$ equal to

$$[aw' - bv' + (v \times w) - (v' \times w'), -v \cdot w' + v' \cdot w, bv - aw - (v \times w') - (v' \times w)].$$

And that is all! It can be shown that there is no possibility of defining the structure of an algebra satisfying the desired conditions on $\mathbb{R}^n$ for any other positive integer $n \geq 3$.

To end things, we mention briefly the theoretical context of the cross product. Let V be an algebra over a field F, the product in which is denoted by $*$. This algebra is a *Lie algebra* over F if and only if

(1) $v * w = -(w * v)$ and
(2) *(Jacobi identity)* $u * (v * w) + v * (w * u) + w * (u * v) = 0_V$

for all $u, v, w \in V$.

EXAMPLE

As we have already seen, the cross product turns $\mathbb{R}^3$ into a Lie algebra over $\mathbb{R}$.

If V is an associative algebra over a field F in which multiplication is not commutative, we can define a new product $*$ on V by setting $a * b = ab - ba$. This product induces the structure of a Lie algebra over F.

EXAMPLE

If n is a positive integer and F is a field then $S = \{A \in \mathcal{M}_{n \times n}(F) \mid tr(A) = 0_F\}$ is a Lie algebra over F with the above multiplication.

Lie algebras are very important structures in mathematics, which arise naturally in physics and especially in quantum mechanics. Therefore you will encounter them frequently in more advanced courses. They were first introduced in 1876 by the Norwegian mathematician Sophus Lie in connection with his work on infinitesimal transformations.

Problems

1. Let F be a field and define an operation $*$ on the vector space F^4 as follows: if $[a_1, a_2, a_3, a_4]$ and $[b_1, b_2, b_3, b_4]$ are vectors in F^4 then $[a_1, a_2, a_3, a_4] * [b_1, b_2, b_3, b_4] = [c_1, c_2, c_3, c_4]$ where

$$c_1 = a_1 b_1 + a_2 b_2 + a_3 b_3 + a_4 b_4,$$
$$c_2 = a_1 b_2 + a_2 b_1 + a_3 b_4 + a_4 b_3,$$
$$c_3 = a_1 b_3 + a_2 b_4 + a_3 b_1 + a_4 b_2, and$$
$$c_4 = a_1 b_4 + a_2 b_3 + a_3 b_2 + a_4 b_1.$$

Does this operation turn F^4 into an algebra over F?

2. Let F be a field and let A be a nonempty set. Let V be the vector space over F with basis $\{v_B \mid B \subseteq A\}$ and define a product on V by setting

$$\left(\sum a_B v_B \right) \left(\sum d_C v_C \right) = \sum a_B d_C v_{B \cup C},$$

where the sums range over all subsets of A. Is V an algebra over F?

3. Let F be a field and let n be a positive integer. Is the set of all scalar matrices a subalgebra of $\mathcal{M}_{n \times n}(F)$ over F?

4. Let F be a field and let n be a positive integer. Is $\mathcal{UT}_{n \times n}(F)$ a subalgebra of $\mathcal{M}_{n \times n}(F)$ over F?

5. Let V be a vector space over a field F and let α be a fixed endomorphism of V. Let $S = \{\beta \in End(V) \mid \beta \alpha = \alpha \beta\}$. Is S a subalgebra of $End(V)$ over F?

6. Let S be the subset of $\mathcal{M}_{4 \times 4}(\mathbb{Q})$ composed of all matrices of the form $\begin{bmatrix} a & -b & -c & -d \\ b & a & -d & c \\ c & d & a & -b \\ d & -c & b & a \end{bmatrix}$. Show that S is a subalgebra of $\mathcal{M}_{4 \times 4}(\mathbb{Q})$. Is it a field?

7. Let F be a field and let n be a positive integer. Let S be the set of all matrices A in $\mathcal{M}_{n \times n}(F)$ of the form

$$\begin{bmatrix} a_1 & a_2 & \cdots & a_n \\ a_n & a_1 & \cdots & a_{n-1} \\ & & \cdots & \\ a_2 & a_3 & \cdots & a_1 \end{bmatrix},$$

where the a_i are elements of F. Show that S is a subalgebra of $\mathcal{M}_{n \times n}(F)$ which is stable under the function $A \mapsto A^T$.

8. Let V be a vector space over a field F and let W be a subspace of V. Let S be the set of all endomorphisms of V relative to which W is stable. Is S a subalgebra of $End(V)$ over F?

9. Let V and W be algebras, not necessarily associative, over a field F. Define addition and multiplication on $V \times W$ componentwise: $(v, w) + (v', w') = (v + v', w + w')$ and $(v, w)(v', w') = (vv', ww')$ for all $v, v' \in V$ and $w, w' \in W$. Do these operations turn $V \times W$ into an algebra over F?

10. Let F be a field and let V be a vector space finitely generated over F. Assume, moreover, that we have a product defined on V which turns V into an integral domain which is also an F-algebra. Show that V is in fact a field.

11. For $v, w \in \mathbb{R}^3$, simplify the expression $(v + w) \times (v - w)$.

12. For $u, v, w \in \mathbb{R}^3$, simplify the expression $(u + v + w) \times (v + w)$.

13. Let u, v, w be vectors in $\mathbb{R}^3$ and let x, y, z be the vectors defined by

$$x = a_1 u + b_1 v + c_1 w,$$
$$y = a_2 u + b_2 v + c_2 w,$$
$$z = a_3 u + b_3 v + c_3 w.$$

Show that $x \cdot (y \times z) = [u \cdot (v \times w)] \begin{vmatrix} a_1 & b_1 & c_1 \\ a_2 & b_2 & c_2 \\ a_3 & b_3 & c_3 \end{vmatrix}$.

14. Find a vector in $\mathbb{R}^3$ which is orthogonal to the vectors $[-2, 1, 0]$ and $[2, 1, -3]$.

15. If u, v, w are vectors in $\mathbb{R}^3$, is it necessarily always true that $u \cdot (v \times w) = (u \times v) \cdot w$?

16. Let F be a field and let $\{V_i \mid i \in \Omega\}$ be a family of algebras over F, where the multiplication in V_i is denoted by $*_i$. Set $V = \prod_{i \in \Omega} V_i$ and define an operation $*$ on V by setting $(f * g)(i) = f(i) *_i g(i)$ for all $i \in \Omega$. Does this operation turn V into an algebra over F?

17. Let V be an algebra over $\mathbb{R}$ and let W be a subspace of V satisfying $w^2 \in W$ for all $w \in W$. Show that W is a subalgebra of V over $\mathbb{R}$.

18. Let $V = \mathbb{R}^{\mathbb{R}}$, which is an algebra over $\mathbb{R}$ under the usual addition and multiplication of functions. Let W be the set of all functions in V having continuous derivatives everywhere on $\mathbb{R}$. Is W a subalgebra of V over $\mathbb{R}$?

19. For $u, v, w, y \in \mathbb{R}^3$, show that

$$(u \times v) \cdot (w \times y) = \begin{vmatrix} u \cdot w & u \cdot y \\ v \cdot w & v \cdot y \end{vmatrix}.$$

20. For $u, v, w \in \mathbb{R}^3$, simplify $[u \times (v + w)] + [v \times (w + u)] + [w \times (u + v)]$.

21. Let V be an associative algebra over a field F and let W and Y be subalgebras of V over F satisfying the condition that $wy = yw$ for all $w \in W$ and $y \in Y$. Show that the set of all finite sums of the form $\sum_{i=1}^{n} w_i y_i$, where the w_i belong to W and the y_i belong to Y, is itself a subalgebra of V. If $dim(W) = k < \infty$ and

$dim(Y) = n < \infty$, what is the dimension of this subalgebra as a vector space over F?

22. Let F be a field of characteristic other than 2 in which a and b are nonzero elements. Let V be a vector space of dimension 4 over F having a basis $\{v_1, v_2, v_3, v_4\}$. Define multiplication on V by

$$\left(\sum_{i=1}^{4} c_i v_i\right)\left(\sum_{i=1}^{4} d_j v_j\right) = \sum_{i=1}^{4}\sum_{j=1}^{4} c_i d_j v_i v_j,$$

where the vectors $v_i v_j$ are defined as follows:

$$v_i v_j = \begin{cases} v_1, & \text{if } i = 1 \text{ or } j = 1 \\ a v_1, & \text{if } i = j = 2 \\ b v_2, & \text{if } i = j = 3 \\ -v_4, & \text{if } i = 3 \text{ and } j = 2. \end{cases}$$

Complete the definition of $v_i v_j$ for the remaining values of i and j so that V becomes an associative algebra over F.

23. Let V be an associative algebra over a field F and let $L = \{\alpha \in End(V) \mid \alpha(vw) = \alpha(v)w = v\alpha(w) \text{ for all } v, w \in V\}$. Define an operation $*$ on L by setting $\alpha * \beta = \alpha\beta - \beta\alpha$. Does this product, together with the usual addition of endomorphisms, turn L into a Lie algebra over F?

24. For $u, v, w, y \in \mathbb{R}^3$, show that $(w \times y) \times (u \times v) = [(w \times y) \cdot v]y - [(y \times u) \cdot v]w$.

25. Let n be a positive integer. Let V be a vector space over a field F and let $f: V \times V \to V$ be a bilinear form. For matrices $A = [v_{ij}]$ and $B = [w_{ij}]$ in $\mathcal{M}_{n \times n}(V)$, define $A * B$ to be the matrix $[y_{ij}] \in \mathcal{M}_{n \times n}(V)$ given by $y_{ij} = \sum_{k=1}^{n} f(v_{ik}, w_{kj})$ for all $1 \leq i, j \leq n$. Does this product, together with the usual matrix addition, turn $\mathcal{M}_{n \times n}(V)$ into an algebra over F? If so, is this algebra associative?

26. Let n be a positive integer and let $S = \{A \in \mathcal{M}_{n \times n}(\mathbb{C}) \mid A + \overline{A} = O\}$. Show that S is a Lie algebra over $\mathbb{C}$ with respect to the product $*$ defined by $A * B = AB - BA$.

27. Let V be an algebra (not necessarily associative) over a field F. A function $\delta \in V^V$ is a *derivation* on V if and only if the following conditions are satisfied:

(i) $\delta(v + v') = \delta(v) + \delta(v')$,
(ii) $\delta(vv') = \delta(v)v' + v\delta(v')$,
(iii) $\delta(av') = a\delta(v')$

for all $v, v' \in V$ and all $a \in F$. Show that the set $Der(V)$ of all derivations of V is a subspace of V^V which is a Lie algebra under the vector multiplication $\delta * \delta' = \delta\delta' - \delta'\delta$.

28. Let F be a field of characteristic other than 2 and let V be an algebra over F. Let f be a bilinear form on V satisfying the condition $f(uv, uv) = f(u, u)f(v, v)$ for all $u, v \in V$. Show that $f(uw, vw) = f(u, v)f(w, w)$ for all $u, v, w \in V$.

29. Let $\mathbb{N}$ be the set of all nonnegative integers and let V be a vector space over a field F. Define the operation $*$ on the space $V^{\mathbb{N}}$ by setting $(f * g)(2k + 1) = g(k)$ and $(f * g)(2k) = f(k)$ for all $k \in \mathbb{N}$. Does this operation define the structure of an algebra on $V^{\mathbb{N}}$.

30. Let V be a vector space over a field F of dimension greater than 1 and let S be a subalgebra of $End(V)$ over F satisfying the condition that for each $\alpha \in S$ there exists a positive integer n (which depends on α) such that $\alpha^n = \sigma_0$. Show that there exists a proper nontrivial subspace W of V which is stable under each element of S.

INDEX

Kluwer Texts in the Mathematical Sciences

1. A.A. Harms and D.R. Wyman: *Mathematics and Physics of Neutron Radiography.* 1986   ISBN 90-277-2191-2
2. H.A. Mavromatis: *Exercises in Quantum Mechanics.* A Collection of Illustrative Problems and Their Solutions. 1987   ISBN 90-277-2288-9
3. V.I. Kukulin, V.M. Krasnopol'sky and J. Horácek: *Theory of Resonances.* Principles and Applications. 1989   ISBN 90-277-2364-8
4. M. Anderson and Todd Feil: *Lattice-Ordered Groups.* An Introduction. 1988   ISBN 90-277-2643-4
5. J. Avery: *Hyperspherical Harmonics.* Applications in Quantum Theory. 1989   ISBN 0-7923-0165-X
6. H.A. Mavromatis: *Exercises in Quantum Mechanics.* A Collection of Illustrative Problems and Their Solutions. Second Revised Edition. 1992   ISBN 0-7923-1557-X
7. G. Micula and P. Pavel: *Differential and Integral Equations through Practical Problems and Exercises.* 1992   ISBN 0-7923-1890-0
8. W.S. Anglin: *The Queen of Mathematics.* An Introduction to Number Theory. 1995   ISBN 0-7923-3287-3
9. Y.G. Borisovich, N.M. Bliznyakov, T.N. Fomenko and Y.A. Izrailevich: *Introduction to Differential and Algebraic Topology.* 1995   ISBN 0-7923-3499-X
10. J. Schmeelk, D. Takači and A. Takači: *Elementary Analysis through Examples and Exercises.* 1995   ISBN 0-7923-3597-X
11. J.S. Golan: *Foundations of Linear Algebra.* 1995   ISBN 0-7923-3614-3

KLUWER ACADEMIC PUBLISHERS – DORDRECHT / BOSTON / LONDON